THIS BOOK IS ITEM NU
1.00000000000001
FROM THE LIBABELRARY
BRUC
OF
ESTATE
R

Paradigm Shift Knuckle Sandwich

& other examples of P.N.T.
(Perverse Number Theory)

- tENTATIVELY, a cONVENIENCE

"The problem here is that I really do want to give a respectful reading of [the many bks quoted herein] in wch the reader will be able to hopefully follow some sort of elementary thread running thru all this math - but I want to do it in such a way that my own mindset is revealed as being equally valid by virtue of its elastic humor & creativity. I'd hate for the authors of all the bks I'm quoting to dislike this text, I'd much rather have them laugh along w/ it & excuse my mathematical inadequacies when they rear their ugly heads."

- quote from page 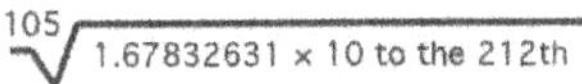

a list of the parts of this book
organized in the order in which the parts appear

PART ONE

Part Two

Part Three

Part One

Paradigm Shift Knuckle Sandwich

& other examples of P.N.T. (Perverse Number Theory)

- *tENTATIVELY, a cONVENIENCE*

****

INTRODUCTION

When I was a child, I was a very good mathematician. I was also good at drawing. I remember thinking, when I was 9 yrs old, something to the effect that a mathematician's life was probably too socially isolated & not much fun & I decided that, therefore, it might be better for me to be an artist.

Since I was bad at sports, when I was around 10, the school gym teacher offered to teach them to me & sd that I cd teach him math in return. I didn't take him up on this.

In the 6th grade, at age 11, I invented some sort of formula for solving a math problem. I have no idea what it was anymore. I taught it to my class. One of the 'good' students, a fellow named Larry Hewitt, sd that it didn't work for a particular number. Since I was a trouble-maker & not particularly respected as a result, Larry's word was apparently taken on the matter & no-one refuted him. It seemed that my system was flawed. I double-checked it & I was correct.

By the time I was a teenager, I had little interest in math anymore. Nonetheless, at 1st, I was still good enuf at it to pass classes w/ high grades regardless of whether I pd attn or not. One girl that I was attracted to kidded me by saying that she wdn't like me if I continued to do so good on the tests.

By my senior yr, students wd sit next to me in math classes during exams so I cd tell them the correct answers. They passed. I was so bored I just wrote punchlines as answers on the tests. I failed. The teacher never realized what was happening.

In my late teens my abilities in math had deteriorated to little more than an 'idiot savant' parlor trick. My friend Alfred Harris wd give me numbers to multiply in my head & then I'd race his use of a calculator. I always won. The technique was simple enuf: I'd fine-tune from rounded-off multiplications that I already knew the answers to.

From ages 20 to 23, my creative output was primarily concentrated on experimental writing. One of these pieces was all numbers. There wasn't much to it, as I recall; I mainly typed out a counting that was to restart when a mistake was made. *SO*, in the detail from page 1 reproed below, I start w/ "1/4" & progress from left to right thru to the beginning of "733/4" - the "3/4" of wch begins page 2.

¼½3/411¼1½13/422¼2½23/433¼3½33/444¼4½43/455¼5½53/466¼6½63/477
¼7½73/488¼8½83/499¼9½93/41010¼10½103/41111¼11½113/41212¼12½12
3/41313¼13½133/41414¼14½143/41515¼15½153/41616¼16½163/41717¼1
7½173/41818¼18½183/41919¼19½193/42020¼20½203/42121¼21½213/422
22¼22½223/42323¼23½233/42424¼24½243/42525¼25½253/42626¼26½263
/42727¼27½273/42828¼28½283/42929¼29½293/43030¼30½303/43131¼31
½313/43232¼32½323/43333¼33½333/43434¼34½343/43535¼35½353/4363
6¼36½363/43737¼37½373/43838¼38½383/43939¼39½393/44040¼40½403/
44141¼41½413/44242¼42½423/44343¼43½433/44444¼44½443/44545¼45½
453/44646¼46½463/44747¼47½473/44848¼48½483/44949¼49½493/45050
¼50½503/45151¼51½513/45252¼52½523/45353¼53½533/45454¼54½543/4
5555¼55½553/45656¼56½563/45757¼57½573/45858¼58½583/45959¼59½5
93/46060¼60½603/46161¼61½613/46262¼62½623/46363¼63½633/46464¼
64½643/46565¼65½653/46666¼66½663/46767¼67½673/46868¼68½683/46
969¼69½693/47070¼70½703/47171¼71½713/47272¼72½723/47373¼73½73

However, deviations from this process appear w/ each new page. This number writing will be presented in greater detail later in this bk. For now, I just drop it in as an example of where my math mind was at.

At 22, I created a score for a sound piece called "decomposition". It had 13 basic pages - 5 of wch were to have overlays on them. The top sheets had sections cut out to reveal parts of the underneath that were to modify the tops.

Since any of the 5 overlays cd be on any of the 5 bottoms requiring them, there was a specific number of possible combinations. Having forgotten what little algebra I might've learned in high school, I decided to calculate the permutations in the simplest (& only) way I cd conceive of.

This resulted in a 25 page document in wch I concluded that 3,125 (5^5) variations existed. The 1st page of this is reproduced below. Naturally, this problem cd've been solved w/ much smaller effort!

I felt a little less foolish or stupid about pursuing such a tedious process unnecessarily when I read many yrs later that the great mathematician Alan Turing went thru a similarly primitive process to solve a problem when he was young.

Since tENT asked me to write this marginalia for PSKS let me admit to two things left off the side of the ball! 1. I AM NOT a mathematician — there will be quite a bit here that I am unable to comment on. 2. I AM VERY likely to digress into areas peripheral to the concerns of this book. Did I say peripheral? Good! — lets start there — at the margins where I am writing where I am socially located. where things begin to disappear

& again where different things come together

11111 11112 11113 11114 11115
11121 11122 11123 11124 11125
11131 11132 11133 11134 11135
11141 11142 11143 11144 11145
11151 11152 11153 11154 11155
11211 11212 11213 11214 11215
11221 11222 11223 11224 11225
11231 11232 11233 11234 11235
11241 11242 11243 11244 11245
11251 11252 11253 11254 11255
11311 11312 11313 11314 11315
11321 11322 11323 11324 11325
11331 11332 11333 11334 11335
11341 11342 11343 11344 11345
11351 11352 11353 11354 11355
11411 11412 11413 11414 11415
11421 11422 11423 11424 11425
11431 11432 11433 11434 11435
11441 11442 11443 11444 11445
11451 11452 11453 11454 11455
11511 11512 11513 11514 11515
11521 11522 11523 11524 11525
11531 11532 11533 11534 11535
11541 11542 11543 11544 11545
11551 11552 11553 11554 11555

Like tENTS ideas & mine of perversity & marginality

In a way P#T is marginal math right? outside the orthodox) but also toward the excluded, the exceptional, the impossible the beyond & the outer rim.

There is a certain concept of human beauty as the perception of averages — whatever falls too far from the center in any direction is relegated to the outside — is excluded — from its compass. Normativity is defined by quantities of sameness — what is proper, acceptable,

Perhaps the story about Turing is the one recounted at the bottom of page 25 of the 2000 edition of Andrew Hodges' biography: Alan Turing: The Enigma. Turing is sd to've "found the infinite series for the 'inverse tangent function'" w/o using even elementary calculus. This, when he was 15 yrs old! Of course, such a feat was no doubt more precocious than anything I ever did w/ math! Still, I felt a little kinship.

I basically no longer thought much about math at all except thru counting. Counting was to play a particularly important part in my 1st bk (culled from the experimental writings mentioned earlier but not including the number writing piece):

t he book
t he referent 4 wch consists of
t he non-materialized punch-outs from a transparent
letter/whatever stencil

The role of counting in that bk is analyzed somewhat in depth in the 1st section of my 10th bk: footnotes - esp in the "Dos & Don'ts of Dating" subsection.

By the time of my collaborative (M)Usic projects of the late 1980s & early 1990s, counting also played an important role in the producing & playing of scores such as that for my "78 (Diminishing Puceno)" (wch certainly didn't win me any popularity contests w/ some of my band-mates not naturally inclined toward such rigorousness). The **bold 1s** indicate accents in the rhythmic progression:

1 2 3, **1** 2, **1** 2, **1** 2 3, **1** 2
1 2 3, **1** 2, **1** 2, **1** 2 3, **1**
1 2 3, **1** 2, **1** 2, **1** 2 3
1 2 3, **1** 2, **1** 2, **1** 2
1 2 3, **1** 2, **1** 2, **1**
1 2 3, **1** 2, **1** 2
1 2 3, **1** 2, **1**
1 2 3, **1** 2
1 2 3, **1**
1 2 3
1 2
1

An important exception to this counting-only use of math were references to the idea of "imaginary numbers". Another exception is the idea of "Folk Math" probably conceived of by Kirby Malone in 1979.

In 1982, my lover Hannah Aviva & I made an unsplit 8mm film entitled "A Double Negative As Not A Positive". More about this later.

In early 1980, I published my 1st cassette compilation of my audio work & called it "Usic - 1" as a take-off of "Music - 1" records. By 1983, this turned into "Usic &/or Usicality - √-1" for a 9 hr radio show of my sounds broadcast on WJHU by my friend Ron Cummings. In 1995, I released a record album entitled "Usic - √-1 = A Plethora of No-Longer Neglected Audio/Conceptual Obstacle Courses".

All of these were somewhat informed by the notion that if something can be imagined then it can be *applied*. Such were the seeds of what I now call "P.N.T." (Perverse Number Theory).

Then, probably in early 2003 when I was 49 yrs old, I was browsing in a used bk store for something new to read when I saw a copy of Simon Singh's Fermat's Enigma.

I felt like it was time for me to break my reading habits a bit by getting away from fiction, poetry, music & film theory, political history, etc. Reading a pop math history bk seemed the ticket! Maybe I still had a little of the relevant propensity left. I thoroughly enjoyed reading it & even broke my preference for not writing in bks by jotting notes on the front page.

In fairly quick succession, I then read Hodges' Turing bio, Douglas R. Hofstadter's Gödel, Escher, Bach: an Eternal Golden Braid, Sylvia Nasar's biography about John Forbes Nash, Jr.: a Beautiful Mind, Paul Hoffman's The Man Who Loved Only Numbers - The Story of Paul Erdös and the Search for Mathematical Truth, Eli Maor's e: The Story of a Number, Petr Beckmann's A History of π (PI), Charles Seife's ZERO - The Biography of a Dangerous Idea, David Foster Wallace's Everything and More - A Compact History of ∞, & Barry Mazur's Imagining Numbers (particularly the square root of minus fifteen).

Reading each bk generated notes re whatever resonated w/ me the most & from these inspirations grew my desire to create *something* to represent my own, hopefully somewhat unique, perspective. Texts, rubber stamps, a tattoo, a haircut, a movie, &, now, this bk (wch tries to quasi-document it all), were to follow.

BEFORE I GO ANY FURTHER: I shd explain that I am NOT even an amateur entry-level mathematician. As such, this bk will probably be little more than a target for derision from mathematicians - in the unlikely event that any read it at all. So who exactly is this bk aimed at? Who is its highly hypothetical readership? Other people 'like me'. Do you exist? & what exactly do I mean by 'like me'?

Math, esp "pure math" or number theory, fascinates me as a potential tool for doing what I call "undermining 'reality' maintenance

valuable, π can
(or human)
be defined
by equations
& curves
with varying
degrees of
insidious purpose
(Graphological
analysis will
marginalize this author
as an anti social
individual of limited
intellect.)
Perhaps P#T can
begin to cast
doubt upon
the alliance of
such oppressive
ideologies & the
numbers & the
systems which
support them.

traps". Given that the most abstruse mathematical theory, such as imaginary numbers or Non-Euclidean Geometry, can be used to explain & control practical physical phenomena, I like to think that P.N.T. might lead to an even deeper fluidity in our relation to what usually seem to be restrictive physical 'laws'.

Tired of not being able to fly? Of 'needing' to eat? Of not being able to be 'invisible'? Of yr mortality? Of yr perceptual input range? Can profound mathematics open new pathways out of these restrictions? My intention isn't to evoke 'spiritual' or 'occult' 'means'. While the writings of Carlos Casteneda & his ilk may be driven by similar frustrations w/ 'limits', their path is not the path for me.

Perhaps I can identify more w/ the protagonist in Darren Aronofsky's 1997 film "π" whose mathematical understanding becomes so profound that Kabbalists hotly pursue him for his knowledge. I wonder how many mathematicians delve into the field w/ the ambition to understand 'reality' to the point of no longer being controlled by it. Romantic? Foolish? Perhaps. Perhaps not.

As a child & a young man I imagined that I might be able to be invisible if I cd just vibrate in such a way that light no longer reflected off me but went thru me instead. Ridiculous? Perhaps. But there's the hope that, as I wrote earlier, "if something can be imagined then it can be *applied.*"

Don't misunderstand, my purpose here isn't to claim that these simple ideas I'm putting forth can be used to produce any fantastic changes in anyone's life. I'm just hoping that *someone* will write notes in their copy of this bk & find some inspiration to at least *imagine* something that they might not've imagined otherwise. &, maybe, eventually, *someone* will create an application somewhere along the line of this catalytic chain that'll liberate us a little further from our inabilities.

****

FOUNDATION

In many respects, by writing this bk, I feel as one of my favorite writers, Alfred Jarry, is sd to've felt. According to Roger Shattuck's Introduction to the marvelous Evergreen Press' Selected Works, "At twenty-five Jarry suggested he was writing over everyone's head, including his own". If that's what I'm doing here, I may as well get started on being as thorough as I can stand to be:

Since I've already mentioned Alfred Jarry, I may as well start w/ 'Pataphysics. When I was 15 I heard a band called "The Soft Machine" on a rock radio stn - probably their song "Hope for Happiness" from their 1st LP. I loved it! I had no idea that they'd taken their name from a William S. Burroughs novel. I bought the record in 1968.

When I graduated from high school in 1971, I realized that I'd have to be more investigative on my own as far as looking for new things to read was concerned - now that there weren't teachers around to expose me to bks. One of the 1st things I did was walk to a library 3.32 miles away from home & look up Jarry because I'd recently discovered Marcel Duchamp & Duchamp had mentioned Jarry as a favorite writer. I probably just found Jarry's "Ubu Roi".

cover design: Byron Goto/Henry Epstein

In 1972, I acquired The Soft Machine's "Volume Two". On its "Rivmic Melodies" side, there're 2 Robert Wyatt songs: "Pataphysical Introduction-Pt.1" & "Pt. 2". In the 1st, Wyatt introduces The Soft Machine as "The Official Orchestra of the College of Pataphysics". I doubt that I knew what that meant at the time.

&, yet, by 1994, I was a proud recipient of a certificate from that very same venerable & exalted institution:

In the penultimate "paragraphos" of his essay "Marginalia," Edgar Allen Poe writes the following: "It may be as well to observe, however, that just as the goodness of your true pun is in the direct ratio of its intolerability, so is nonsense the essential sense of the Marginal Note." Abiding Poe's suggestion & following tENT's lead in designating the spirit of Alfred Jarry as one of the significant

COLLEGIVM PATAPHYSICVM

NOUS

Vice-Curateur-Fondateur du Collège de Pataphysique

Président par Interim Perpétuel du Conseil Suprême
de l'Ordre de la Grande Gidouille

A tous ceux qui les présentes verront Salut et Jubilation…

AIllent pataphysiquement consideré que la Pataphysique ce suffit à elle même & n'a point même besoin d'un Collège, ni de qui ou quoi que ce soit Que le suprême Hommage dudit Collège au Monsieur tENTATIVELY, a cONVENIENCE équarrisseux de 1e classe, est digne, juste, équitable, salutair & stomachique, Vus les Statuts dudit Collège titre II, article 6, & titre IV, article 10 (83), [illegible] 5 (a)

Ouis nos Féaux & Loyaux Provéditeurs,
& après Opination du Modérateurs Amovible provisoire
du Corps des Satrapes.

De par l'Autorité & le Magistère du Docteur FAUSTROLL

Curateur Inamovible du Collège de Pataphysique

avons élucidé & élucidons, divulgue et divulguons ubique

CAPITALEMENT

que le dit Monsieur est équarisseux tENTATIVELY, a cONVENIENCE aux Instances des Transcendants Satrapes du Collège de Pataphysique révéré ad perpetuam in memoriam comme Satrrape dudit Collège.

avec les Sublimes Privilèges que de droit

& si quelqu'un contrevient à cette élucidation, divulgation, qu'il encoure la colère du Docteur Faustroll.

De nos Pataphysique Demeures.
Berlin, le 15 Janvier 1994, fête du retardement éternel

Sur l'Ordre de la Magnificence, le Provéditeur Générale

Vice Curateur du Collège de Pataphysique

AIllent pataphysiquement consideré que la Patáphysique ce suffit à elle même & n'a point même besoin d'un Collège, ni de qui ou quoi que ce soit Que le suprème Hommage dudit Collège au Monsieur tENTATIVELY, a cONVENIENCE équarrisseux de 1e classe, est digne, juste, équitable, salutair & stomachique, Vus les Statuts dudit Collège titre II, article 6, & titre IV, article 10 (§3), [illegible] 5 (a)

This, of course, isn't nearly as important as the many diplomas I have from my very own **NUCLEAR BRAIN PHYSICS SURGERY SCHOOL**, but, nonetheless, to those in the No(-No Class) it's still preposterously gigantictic.

one side of my 1st Nuclear Brain Physics Surgery's cool diploma - 1978

To demonstrate the supreme importance of NBPS, I provide, FOR THE 1ST TIME, a list of its graduates:

1st graduating class:

id entity (myself) (Baltimore)
Sumu Pretzler (Doug Retzler) (Baltimore)
Miss Stepp (Laurie Stepp) (Baltimore)
Lynda Cohen (Baltimore)
Richard Ellsberry (Baltimore)
Dick Baublitz (Baltimore)

muses of this volume, I will work my way through part one of this book under the Ægis of pataphysical good humor. I want to explore the liberating joy of transgressing rule, order, convention & boundary. & so the constellation of muses I have decided to attempt to channel into the spirit of these margins within the first part of this book will include the following:

Claire Wyman (Baltimore)
Mother of Opal (Beth Downey?) (Baltimore)
Gary P. Letteron (Baltimore)
Joan E. O'Brien (Baltimore)

BB (Bonnie Bonnell rcv'd an honorary degree wch she didn't deserve) (Baltimore)

<u>2nd graduating class:</u>

Giddy Gayle 26/x "Bunny Luv" EnHANSENment (Gayle Hansen) (Baltimore)
Sheila (Baltimore)
Alfred E. Nudeman (Baltimore)

Lowrence Sterne— whose proliferation of digressive anecdote challenges the orderly progression of linear thinking.

Alfred Jarry— whose science of the exceptional inverts, unbes, & turns upside down Cartesian, Newtonian, Comptean, & Etceterean conceptions of the universe.

Lewis Carrol— whose elastic bending of literary reference into a symbolic fantasia of encrypted meanings wends its way through the labyrinthine duck-rabbit hole in which possibility & incomprehension replace likelihood & certainty.

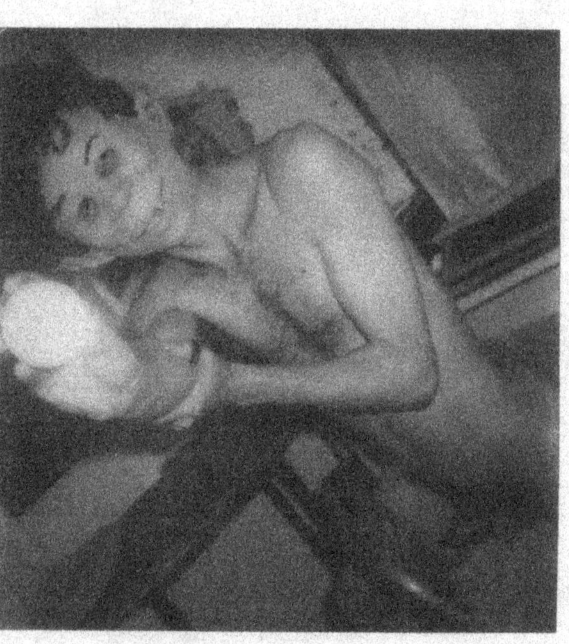

polaroid of Alfred taken by Miss Stepp around the time of his graduation

Cecil (Peema & Alex's Cat) (Baltimore)
Peema Tartoon (Amy Plamondon) (Baltimore)
Svexner (Alec Bernstein) (Baltimore)
id entity (myself) (Baltimore)
Some Queer (Lee Warren) (Baltimore)
Charles (Brohawn & his girlfriend) Robin (Baltimore)
Robin (& her boyfriend) Charles (Brohawn) (Baltimore)

3rd graduating class:

John Berndt (Baltimore)
Laura Adele Trussell (Baltimore)
tENTATIVELY, a cONVENIENCE (Baltimore)
Michelle Estebon (North Carolina/Baltimore)
Brian Gentry (North Carolina/Baltimore)
Nathan Long (Baltimore)
Wendy Lee Parker (Baltimore)
Peter Zahorecz (Baltimore) (alas, deceased)
John Sheehan (Baltimore)
Janet Axe (Saxe) (Baltimore)

4th graduating class:

Mark Pawson (London)
Scott Larson (London)
Pete Horobin (Dundee, Scotland)
Stewart Home (London)
cris cheek (London)
Sianed Jones (London)
Shaun Caton (London)
Elisabeth Morcellet (Paris)
Reinhardt U. Sevol (Paris)
tENTATIVELY, a cONVENIENCE

5th graduating class:

Rebecca Barten (Baltimore)
tENTATIVELY, a cONVENIENCE (Baltimore)
Jake "Unclean" True (Baltimore)
John Berndt (Baltimore)
Jeanine Helene Farrall (Baltimore)
etta cetera (Pittsburgh)
Brainpang (Peter Warner) (NYC/Pittsburgh)
Warren Burt (Melbourne/Pittsburgh)
John Anders Evans (Pittsburgh)

Jack Spicer-- whose wit in the rearchitecture of nomadic & mesmerizingly metastable ambiguity throws a baseball through the window of language's clarity, leaving its uptight apologists & doxologists to pick up the pieces.

Kenneth Patchen-- whose ridiculously simple & insightful drawings recapture the playful spirit of children who have not yet learned that games have rules.

Maya Espy (Michelle Dunn) (Baltimore/Pittsburgh)

Never heard of *most* of these people?
That just goes to show how *important* they are.

Tristan Tzara-- who revealed that dada was a more interesting word when it referred to hobby horse & to nothing at all than when it stuttered toward an identification with the male parent.

Interlude from method:

"Epiphenomenon-- an iteresting choice which implies a relationship to an object or phenomenon which is possible & exceptional rather than directly causal, consequential, & determinate-- the middle term between determinate & indeterminate between law & exception between causality & independent adjunct event."

Jarry's use of the term clearly at odds w/ what Huxley made of it in 1874.

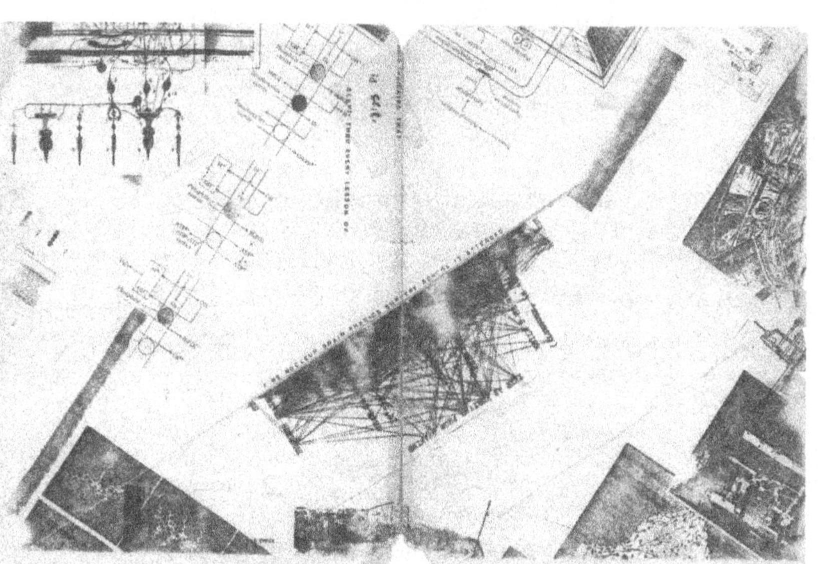

the other side of my 1st Nuclear Brain Physics Surgery's cool diploma

Understandably, such details might be of limited interest to a general readership & may seem tangential to a bk hypothetically about some sort of number theory *BUT* I include this info in order to map a developmental path. Readers who've followed a similar trajectory may be more engrossed as a result. When I read mention of a record that I know, eg, being played by characters in a novel (Julio Cortázar's Hopscotch, let's say) it increases my identification.

Back on track: In Simon Watson Taylor's English translation (from the French) of Jarry's truly 'pataordinary Exploits and Opinions of Doctor Faustroll Pataphysician - A Neo-Scientific Novel, Jarry describes 'pataphysics as an "epiphenomenon" ("that which is superimposed upon a phenomenon"), as the science "extending as far beyond metaphysics as the latter extends beyond physics." Furthermore, "Pataphysics will examine the laws governing exceptions" & "is the science of imaginary solutions".

In the same bk, Jarry has Faustroll *born* at age 63, taking sponge-baths w/ wallpaper, using a boat that's a sieve, becoming smaller than himself; sending a telepathic letter to Lord Kelvin in wch it's explained that:

"The sun is a cool, solid, and homogeneous globe. Its surface is divided into squares of one meter, which are the bases of long, inverted pyramids, thread-cut, 696,999 kilometers long, their points one kilometer from the center. Each is mounted on a screw and its movement toward the center would cause, *if I had the time*, the rotation of a paddle at the top end of each screw shaft, in a few meters of viscous fluid, with which the whole surface is thinly covered..."

The final chapter provides a mathematical 'proof' "CONCERNING THE SURFACE OF GOD" ending w/:

"-Let us note, in fact, that, according to the formula

$$\infty - 0 - a + a + 0 = \infty$$

the length a is nil, so that a is not a line but a point.

Therefore, *definitively*:

GOD IS THE TANGENTIAL POINT BETWEEN ZERO AND INFINITY. Pataphysics is *the* science..."

Earlier, Jarry states that Pataphysics "will explain the universe supplementary to this one; or, less ambitiously, will describe a universe which can be - and perhaps should be - envisaged in the place of the traditional one, since the laws that are supposed to have been discovered in the traditional universe are also correlations of exceptions, albeit more frequent ones, but in any case accidental data which, reduced to the status of unexceptional exceptions, possess no longer even the virtue of originality." Hardly the way science is ordinarily referred to!

Raymond Queneau, the Transcendent Satrap of the *Collège de 'pataphysique* (founded May 11, 1948 in Paris), wrote, in 1937, a novel entitled Odile. In it, the protagonist is a mathematician named Roland Travy. He's interrogated by a magistrate after being exonerated in connection w/ a murder (English translation by Carol Sanders in the Dalkey Archive 1988 edition):

"But I might tell you that what interested me the most were your personal papers. I had to look through them, that's my job, and you'll understand how interested I was when I tell you that I'm something of an amateur mathematician too. You won't be too surprised because you're an admirer of Fermat, who was a magistrate, although of course I am not comparing myself with that famous and obscure genius. In the old days I spent long hours trying to prove his famous theorem but in the end I realized it was futile. I have to say, however, I have to say that I think his

Return to method. In part two, I will follow a different course, investing a different sort of energy borrowed from two of the muses whose names do not appear in the index of this book --> Jean Genet & Antonin Artaud. Their presence chosen because of the intensely marginal course of their lives & experiences.

I expect that the transition between the two sections will be jarring & disturbing. I hope that it will not reify the distinction, the boundary, between the two explorations of the margin.

theorem is either right or wrong. I'm not a follower of Brouwer. And what about you, my dear sir, are you a believer in the validity of the law of the excluded middle?"

Fermat & Brouwer will be 'explained' a bit more later (in case you don't already 'know' who they are). Shortly thereafter, in a conversation w/ a friend, Travy explains the magistrate's question:

"'To put it another way: are there some propositions that one can never prove or disprove? Some people say there are; some people even think that there might be some propositions that you could prove that you could never prove true or false. Between true and false the middle is not excluded.'"

Near the end of the bk, Roland confesses to his love-interest, Odile, that: "For years I've been mistaken about myself and I've lived under a misapprehension. I used to think I was a mathematician. Just recently I've realized that I'm not even an amateur mathematician. I'm nothing at all. I don't know anything about it. I don't understand anything about it."

I actually read this novel *after* conceiving of P.N.T. but Queneau's words make for an appropriate epigraph - even though they're not quite used that way here. Where was I? I don't *know* where I am but I won't let that stop me from navigating thru the chaos of believing in *nothing* at the same time I believe in *everything*.

****

NAVIGATION

<u>Fermat's Enigma - The Epic Quest to Solve the World's Greatest Mathematical Problem</u> - Simon Singh. Fermat was a 17th century French mathematician now credited by some w/ being the father of modern number theory. He tantalized 3 centuries' worth of thinkers w/ the (unsubstantiated) claim that he'd discovered a proof that there are no whole number solutions for the equation:

$$x^n + y^n = z^n \text{ for } n \text{ greater than } 2$$

Attempts to prove this were futile, despite the efforts of many great minds, until Andrew Wiles did so in October of 1994. *But that's neither here nor there for the purposes of this bk.* I mention it only to clarify the thread begun in the 1st <u>Odile</u> quote. My intention is to pick out odds & ends of the math bks I've read as spikes for the meandering

Interesting to note, here that Fermat's Last Theorem is considered to be the single most famous example of the marginal note. According to historians, the theorem was written by Fermat in 1637 in his copy of Claude-Gaspar Bachet's translation of the Arithmetica of Diophantus. Apparently, Fermat was not the first mathematician to inscribe his copy of the Arithmetica with marginalia. The Byzantine mathematician Maximus Planudes wrote in his: "Thy soul, Diophantus, be with Satan because of the difficulty of your theorems." The statement sheds some light upon a certain type of determinate thinking: "That which cannot be solved belongs to the devil."

tracks of this narrative.

How much to explain & how much to assume as already known? What's probably familiar to the dilettante reader I'll explain cursorily or not at all. Pythagoras was an ancient Greek philosopher & mathematician whose philosophy centered around a universe constructed of 'perfect' rational numbers. Legend has it that he sentenced his student, Hippasus, to death by drowning because Hippasus introduced the notion of an irrational number.

This story might well be far-fetched (& my telling of it poor) but it's a 'perfect' place for me to start a Pantheon of Perverse Number Theory Outlaws. My position, such as it is, being rooted in sympathy for the heretic.

Then along comes varying degrees of infinity! Now that's *sexy*! Page 93: "However some infinities are larger than others. For example, any attempt to pair every rational number with every irrational number ends in failure, and in fact it can be proved that the infinite set of irrational numbers is larger than the infinite set of rational numbers." Hippasus' infinity was *bigger* than Pythagoras'! So there!

& the possibilities increase. While physical 'reality' may seem to be bounded every wch way, THOUGHT EXPERIMENTS work toward knowing *no* bounds. The previously alluded to "imaginary numbers" complicate the previously 'simple'. Page 116:

"For example, if we restrict ourselves to real numbers then the number 12 can be factorized only into 2 x 2 x 3. However, if we allow imaginary numbers into our proof than 12 can also be factorized in the following way:

$$12 = (1 + \sqrt{-11}) \times (1 - \sqrt{-11})$$

Here $(1 + \sqrt{-11})$ is a complex number, a combination of a real and an imaginary number."

Ignoring the "loss of unique factorization" that this example is used to demonstrate in Singh's bk, I'm struck by the incredibleness of taking a 'real' number (1) & adding an 'imaginary' number ($\sqrt{-11}$) to it & then multiplying *that* times the same real number w/ the same imaginary number *subtracted* from it! It's a sortof borderline tautological nihilism, a startling example of what I'll later discuss as "anything is anything".

Alan Turing: The Enigma - Andrew Hodges. On page 83 (of my 2000 edition paperback) I find mention of G. Frege's "*logistic* view of mathematics, in which arithmetic was derived from the logical relationships of the entities in the world, and its consistency guaranteed by a basis in reality." But is "reality" itself axiomatic or an assumption to be challenged? "Frege's task was to abstract the ideas of 'any', 'thing'".

Even more interesting if we consider the etymological significance of another name for Satan -- Lucifer -- "bringer of light." For Maximus the Planudes, "light of the human soul" could only shine in certainty & solution -- but perhaps the light which Lucifer introduced was one of doubt & indetermincy. That certainly makes sense if we also associate this figure with the tempting serpent represented in the myth of the garden of Eden.

This work was "overtaken" by Bertrand Russell's work w/ sets - something referred to in the "Haircut Paradox" section of this bk.

Topology: "This branch of mathematics could be described as the result of abstracting from geometry such concepts as 'connected', 'edge' and 'neighbouring' which did not depend upon measurement." Now this idea of abstracting concepts to create analytical disciplines fascinates me.

What if we were to abstract something like: {[insert name of fingernail technician here] in relation to the 1st part of a meteorite to burn upon entering the Earth's atmosphere at an unspecified precise point over a gyre used by the aforementioned [insert name of fingernail technician here] to reach an area of more precision tools for painting the entire Warren Commission Report on a pinky nail}? Wd we end up w/ something 'pataphysically related to a bk like <u>A Stress Analysis of a Strapless Evening Gown</u> (wch I haven't read) - edited by Robert A. Baker?

But I digress. Let's backtrack to AXIOMS & then segue to.. (but I fast-forward). If an axiom is: "1. a self-evident truth, 2. a universally accepted principle or rule, 3. *Logic, Math.* a proposition that is assumed without proof for the sake of studying the consequences that follow from it." (The Random House College Dictionary, 1975) then we're immediately presented w/ some obvious philosophical problems, eh?

How is anything ever "self-evident"? (Anything is [Self-Evident]) P.N.T. accepts *no* "universally accepted principle or rule". If logic & math are dependent upon axioms as their bottom line then that's a demonstration of logic & math's most self-conflicted dilemma insofar as they require absolute irrefutable proof *w/o assumption.*

One of the axioms of arithmetic is the *Commutative Law of Addition*:

For any numbers m and n,

$$m + n = n + m$$

That probably seems to most people to be a "self-evident truth". But, 'pataphysically speaking (perhaps), there must be an exception to this. Finding it & describing it precisely is the problem. Therefore, in P.N.T. we have:

$$(m + n \neq n + m) \text{ -> } x$$

((m + n does not equal n + m) is isomorphic to x)

The problem is, therefore, what is x?

What's being postulated here (w/o any substantiation) is a thought experiment providing an alternative to a "self-evident truth" that then, hypothetically, calls into question *any* "self-evident truth"s. This is, of course, certainly not unprecedented. Let's take the example of Non-

It is also interesting to note that a symmetric encryption algorithm developed by Horst Feistel in the 1970s was designated by the name "Lucifer". Cryptology derives its name from the Greek word Kryptos, meaning "hidden". It would be "natural" or intuitive, according to convention to associate the hidden with invisibility & darkness & yet here it is named after the bearer of light! Remarkable! Compare this rendering to the term the Nazis used to describe their cipher during World War II: "Enigma". I suppose it is Alan Turing's work toward decoding this particularly malevolent cipher which Andrew Hodges writes about.

Euclidean Geometry.

Euclid was a Greek mathematician from 300BC now known as the "Father of Geometry". The 4 postulates of his "absolute geometry" were followed by a 5th one, more problematic, commonly called the "parallel postulate". In this, it's stated that for every line w/ a point not on that line there is only *one line* running thru that point on the same plane that doesn't intersect the 1st line. Over 2,000 yrs later, Non-Euclidean Geometry was conceived of - in wch this postulate no longer holds true. Hyperbolic Spaces are modeled on the possibility of at least 2 lines passing thru the point that don't intersect the 1st line.

I was fortunate enuf to've been the A/V technician for a presentation at the Andy Warhol Museum in Pittsburgh by artist/mathematician Margaret Wertheim from the Institute for Figuring. We talked briefly afterward & I showed her my knuckle tattoos (pictured on the cover of this bk & explained elsewhere). Her presentation was on "An Exploration of the Intersection of Higher Geometry and Feminine Handicraft" - more specifically on the depiction of Hyperbolic Space thru CROCHETING. I loved it!

Margaret was generous enuf to give me a copy of her bk, A Field Guide to Hyperbolic Space, that her lecture reiterated. We struck up a very brief email correspondence. An edited version of what was probably the last email I sent her is presented below. Alas, she never replied. I suspect that this non-response indicates, as is so common in the history of math, that I've been relegated to the lowly status of 'crank' (then again, maybe not!) - not surprising but still disappointing!

From: anon@fyi.net
Subject: A Field Guide to Hyperbolic Space
Date: March 16, 2007 11:14:23 AM EST

Hi Margaret,

I'm reading your book now. I'm very glad to have your lecture presentation in book form to refer back to. I find it inspiring & think it'll be a great launching point for what I call P.N.T. (Perverse Number Theory). [..] These days I think of myself as a COGNITIVE DISSIDENT - wch, I think, sums it up nicely. Reading yr bk makes me wonder what is to hyperbolic space what hyperbolic space is to Euclidean geometry.

The main quality of Pittsburgh that I publicly extol is its topography. It's too hilly to be turned into a grid-like urban sprawl. There are a plethora of nooks & crannies - just as I like it. It was wonderful when a woman in yr audience pointed out that the brain & intestines (etc) are more hyperbolic than Euclidian. Well, Pittsburgh is a much closer analog to my brain than flatter, more grid-controlled cities like LA. How can you stand it there? I know you sd you like it.

Back to the question: What is to hyperbolic space what hyperbolic space is to Euclidean geometry? Perhaps the way the brain thinks is to the brain's shape what the brain's shape is to Euclidian geometry.

I'm reminded of Duchamp's 3 Standard Stoppages. Are you familiar w/ that piece? According to Arturo Schwarz's THE COMPLETE WORKS OF MARCEL DUCHAMP (bottom of page 131), Duchamp's 3 Standard Stoppages was "a work Duchamp considers to be one of the main turning points of his life." As I recall, Duchamp took 3 one-meter long threads & dropped them. The resultant curve became a new unit of length in his playful physics, "the meter diminished" - as Duchamp is quoted by Schwarz as saying.

SO, when I read at the bottom of yr bk's page 45 about the measuring of angles between stars whilst searching for evidence of non-Euclidean space on a cosmological level: "All measurements revealed the standard sum and for most of the past century the evidence has pointed to a Euclidean framework." I imagine those measurements redone using Duchamp's 3 Standard Stoppages. While this is obviously a 'fanciful' idea, I think it also points in a similar intuitive direction as the crocheted hyperbolic spaces.

[..]

In PNT, as perhaps w/ Duchamp & Jarry (see 'pataphysics), whatever is NON-AXIOMATIC can be embraced as liberatory. As I write in my (admittedly somewhat overwhelming) email signature: "(m + n does not equal n + m) is isomorphic to x". In other words, I'm interested in looking at such an "x" - partially 'because' such a notion is another step outside the restrictive 'accuracy' of Euclidean space & the 'narrow-mindedness' that has followed it. Don't misunderstand, I have a great respect for Euclid. He had a fantastic clarity. I, on the other hand, prefer a fantastic ambiguity. It's COGNITIVELY DISSIDENT to what I call "reality maintenance traps".

[..]

THANKS EVER SO MUCH FOR YR BK!

yr pal,

tENTATIVELY, a cONVENIENCE
COGNITIVE DISSIDENT

Turing is mainly known for his invention of the "Turing Machine", a prototypical computer used to crack the nazi Enigma code during (the so-called) "World War II". As Hodges podges it:

"The Service chiefs were highly indignant at:

...the condition of creative anarchy, within and between the sections, that distinguished G[overnment]C[ode] and C[ypher]S[chool]'s day-to-day work"

May P.N.T. continue in the spirit of creative anarchy!

A particularly brilliant insight of Turing's was that the ACE (Automatic Computing Engine) was *not* doing arithmetic or playing chess but was simply being activated by electronic pulses: "one might regard pulses as representing numbers, or as representing instructions. But it was really all in the mind of the beholder. The machine acted, as he put it, 'without understanding'".

I return here to Raymond Queneau's character Roland Travy's saying "I don't understand anything about it" in his revelation of himself as *not* a mathematician. Turing was noted for his "Turing Test" wch he asserted cd demonstrate that computers are as much qualified to be considered as 'thinking' as people are. Perhaps Travy, & by extrapolation *all* supposedly sentient beings, cd no more be *proven* to be THINKING than a computer can be.

Supporting such a parallel might be the ideas of McCullough & Pitts "which suggested that the logical functions of electronic valves bore some similarity to the structure of neurons in the human nervous system." In other words, if the Turing Test implies that computers 'think'

"Fantastic" — again a provocative word choice. In Tzvetan Todorov's account of Fantastic literature he describes the essence of this genre as being the liminal state between the supernatural & the psychological/perceptual. The fantastic can be read as one or the other, as both, or as something uncertain within the flux & doubt of Niether/Either. More indeterminate than either the uncanny — which has a rational explanation, or the marvellous in which supernatural forces themselves emerge as real & measurable. The telos of the fantastic is the creation of doubt & uncertainty.

as much as humans do, then perhaps humans aren't thinking *at all* & are merely responding to "electronic pulses" (neurons firing) & outputting whatever their 'hardwire' (or soft machine) channels it into.

For Turing, "it was obvious that a symbol had no intrinsic connection with the entity that it symbolised". Wch may just bring us back to Anything is Anything again. Or, more obviously, to the debate in linguistics about whether such a relationship exists. In the "Adamitic" or magikal school of thought, there *is* a definite & vigorously important relationship between signifiers & signified but this relationship may be muddied or lost. Such a belief/theory may just be wishful thinking. Or it may not be.

Turing's definition of the symbol seems to reiterate Saussure's claim that the relationship between the signifier & the signified is simply arbitrary. Seimiology has wasted quite a bit of time coming to terms with this oversimplified postulation. Peirce had already begun to explore the issue in greater depth -- defining some of these relations as congruent & mimetic, others as causal (the index) & still others as strictly conventional. Of course, the great twentieth century

In July, 1997, I met w/ my friend & fellow neoist Florian Cramer in Berlin where we discussed such issues & then conducted a relevant experiment or exercise. Below are some transcribed excerpts from both:

Excerpt from the Interview:

FC: Adamitic language..
t,ac: Adamitic? I think that the idea of an Adamitic language is interesting but I'm wondering, you would know much more about this than I do because I know nothing about it since I know nothing about everything & everything about nothing, etc, etc.. - but, is there any sort of theory amongst linguists, or whatever the appropriate field of study would be, that you know of, that tends to trace language back to common roots of any sort?
FC: Yeah, there is, um, for example in Chomskian linguistics you have this idea that you have something like semantics & patterns in a language which are common to all languages.
t,ac: Does he develop this theory in great detail? In other words does he have a technical description of it?
FC: Yeah, it's called [unintelligible] schematic transformational grammar.
t,ac: Could you say that again, please?
FC: Generative transformational grammar.
t,ac: Ok.
FC: But actually I'm not that familiar with this kind of linguistics because linguistics in this century has very much split into various fields. You could say, from something like literary linguistics, which is mainly from the structuralist tradition; from Ferdinand de Saussure over Roman Jakobson to post-structuralism, deconstructionist approach as well as Roland Barthes, Umberto Eco - but, on the other hand, you have this

kind of technical linguistics, Chomsky, for example, which is, actually, more the kind of linguistics which you study if you study linguistics properly, which is, for example, also important for computer linguistics if you generate speech recognition or speech systems & then you, mostly [unintelligible] to this kind of scientific linguistics - & then you also have philosophical linguistics like, uh, for example, speech act theory by Austin & Searle..

t,ac: Which is what?

FC: Well, uh, this is actually something where you could say that modern linguistics have an approach which is closer to the idea of Adamitic language because, well, the primary assumption of modern linguistics is that language is arbitrary - that a linguistic sign has no absolutely whatever organic relation to the thing which we represent.

t,ac: So no onomatopoiea? or whatever?

FC: Yes, that would be, actually this is a different [unintelligible] which has been introduced by Charles Sanders Peirce who differentiated between the iconic, the indexical, & the symbolic sign where you actually have these possibilities of the onomatopetic relationships but, um, no, the question's rather, to quote Austin, how to do things with words. There is 1 problem - if you have arbitrary language, it just means that, for example, if I say the word "cassette" or if I write it down then it has no relationship whatsoever to a cassette & by saying the word "cassette" I'm not manipulating the matter of the cassette in a way. So, it's a purely arbitrary relationship..

t,ac: So that's..

FC: Somebody has just decided just to call this piece a cassette.

t,ac: Which is opposite to Adamitic language.

FC: Which is opposite to Adamitic language because in Adamitic language you will have an organic relationship between the word & the thing so that by uttering the word you would, for example, invoke or manipulate the thing so like the classical example is of the Genesis where god says, uh "It shall be light" & then it's light. This is Adamitic language. & the theory, the theory of Adamitic language as it's notably present in the Kabala & in Jewish mysticism is that in the paradise, before the expulsion from the paradise Adam actually possessed a language which was similar to that of the divine language - where he was capable, for example, of naming animals. & that this original language where you could invoke & manipulate things with was lost when

proponent of the mystical connectedness of all symbols & referents is Velimir Khlebnikov.

Searle, of course, never clearly defines the difference between what is immanent in the power of words & how language simply functions in the service of facilitating certain types of human ritual. In god's language, described here, the power of transformation is immanent — in the priest's declaration of the marriage rite this power is extrinsic to language & resides in the network of the social organization. It is dependent upon agreement, promise, & submission to a socially organizing set of status codes. In neither type is there any fantastic ambiguity?

humanity was expelled from the Garden of Eden. So, um, the whole, um, occupation of Kabalism, or also you could say magic in general, is to, sortof, regain command over things by the means of language. & you could say that, in a way you could use it as a critique against modern linguistics because, for example, if Bill Clinton, today, says, uh, "Drop the atom bomb over Moscow" then the atom bomb would actually be dropped because he has the power & the possibility to do so. & just by saying this & by, maybe, having a few codes, or whatever, this would be made to happen today. So you could say that modern linguistics in defining language as arbitrary is actually missing some aspects. It cannot answer the question of how language is actually capable of directly invoking things or making things happen. & this is, for example, a matter which has been discussed by speech act theory - that's exactly the question of speech act theory, how you..

t,ac: Speech act?

FC: Speech act theory, yes, by, notably by Austin & um..

t,ac: Austin's spelled A,u,s,t,i,n?

FC: Exactly, yeah. He was an Oxford linguist, I think in the 1930s.

t,ac: So is the concept of Adamitic language mainly supposedly originating from Kabalists or from who?

FC: I would say it's probably related in all kinds of magical or even metaphysical notions of language. I have thought about, for example, what, how 1 could locate multiple names as they are used in Neoism - in, uh, in either Adamitic or arbitrary language. I think this is extremely interesting because my theory is that they are both - or neither of them, in a way - because, when you say, you have a multiple name, an open situation, everybody can use that name & share this identity there was an extreme case of an arbitrary name - because the name is not naturally given to you - you know, it's not like somebody's born & he has, uh, he gets a name & the name is stamped on the passport but, it's, it's, it's a name, say, Monty Cantsin, Luther Blissett, Karen Eliot. &, um, uh, as you wrote, the name is fixed, but the people using it aren't. So this would be like the classical definition of arbitrary language in a way - the same way as I say, for example, if I take beer, then the notion, the word beer, b, double e, r, is fixed, but, for example, the meaning may change over the centuries - something like this..

t,ac: Let's make a projection right now. Am I interupting your train of thought too much?

But again -- because this power is not immanent w/in language -- it's just an other ordinary act of communication -- an order by an individual socially authorized to give this order -- We needn't allow Searle down this garden path of misdirection

FC: A little bit. Ok, so 1 could say, on the 1 hand, the use of multiple names is a use of language as extremely arbitrary - where you've got an extremely flexible signifier-and-signified or sign-and-thing relationship. It's the highest possible flexibilization of the sign-and-thing relation. On the other hand, as soon as you participate in that multiple name, you are immediately, since there is no fixed referent, say there is no fixed referent for Luther Blissett because there is no person Luther Blissett - or, also, Monty Cantsin - it's a fiction, it's a fiction created by those using the name. So, you could say that by sharing this identity, by adopting this arbitrary name, you, you get the immediate power to, to change it. Yeah? Which is like Adamitic language. Because you are now able to do something in the name of Monty Cantsin, Karen Eliot, Luther Blissett, & so on & actively participate in the shaping of the identity & you can, sortof, directly invoke the character of Monty Cantsin by using the name. So that would be an extreme example of Adamitic language. So, so that, that's, uh, that multiple names, sortof, a kindof flip-flop thing, you know? where you..

t,ac: What d'ya think about the idea of extending that type of thinking so that, for example, beer, the word beer, would be an open concept that could refer to any object? etc, I mean, this obviously refers back to my interest that anything is anything or anything as anything, etc, etc.. Or just taking all words & making them open contexts which can be used freely by the people who choose to use those words in this manner. So, for example, I might say to you "Pass the beer" but I could mean anything by that & you could respond in whatever way you felt appropriate.

FC: Yeah, this would actually be the, exactly match post-structuralist or contemporary linguistics. That you say there is no fixed meaning for any word & the meaning actually.. the, the - this is justified by the use or by the difference - that you say "beer is not wine", for example. Yeah, that you have a purely relational definition & usage but there is no actual referent to the word.

t,ac: What do you think, um.. Can you imagine an experiment in which there were a group of people living together for a month who agreed to not use any word with any fixed reference? but who would try to interact with each other using language? I mean, would you have any prediction about what might happen in that circumstance?

FC: Foo.. I think, um, 1st of all it would require a kind of deprogramming

exercise on behalf of the people who participate in that experiment - because the whole use of referential language is something that is part of the whole metaphysics & history of metaphysics of all culture. So, um, I think it's hard to get rid of referential usage of language - because even if you don't use it in a fixed sense, you, you might, just, for example, use a metaphorical operation where you, for example, say "beer" instead of "wine" but you still refer to something. So, actually, actually come to a complete - to a point where you totally detach signs from any referents would be something really hard to achieve in everyday situations.
t,ac: But imagine it as possible & imagine that the grammatical structure would stay intact & that people would be extremely disciplined & would try to apply this type of thinking. So, for example, instead of, perhaps ordinarily I would say "Pass the beer" or "Pass the salt" - instead I would say..
FC: "Go home.."
t,ac: "Eyesocket"
FC: Yeah..
t,ac: "Eyesocket table"..
FC: Yeah..
t,ac: ..or whatever. Can you imagine any sortof interesting consequence that, interesting state of mind that would result from that, that.. might relate to magikal desires for Adamitic language? - for example?
FC: Well, no, I would simply think of it as an allegory of what's already happening. I think that's the state we have right now. I mean, who knows that what you perceive as beer is the same as what I perceive as beer. Nobody has any idea, I mean you might be a fiction & I might just be a brain somewhere hooked up to some electrical impulses. Even if you actually exist your, your brain might be programmed in such a way that it has a completely different perception of the entire outside world & that, that the reason that we believe that what is beer is actually the same, yeah, when I speak about as when you speak about it, might be pure coincidence..
t,ac: But for you personally, if you imagine that type of experience. How would you speculate you would be effected by it? - for a month, in a closed situation, with a group of 10 people.
FC: I still, I still slip away from your question in saying that for me it's not so much a thing of creation but rather a thing of reading. Yeah, that it's.. I don't think.. you could say that it's already the situation we have & 1

could follow that in, sortof, practically trying to apply it, 1 would make things more honest - but I think that's a complete delusion. It wouldn't, perhaps it wouldn't change anything. Yeah..
t,ac: I think it would change things - but I don't.. I'm not really thinking of it in terms of creation either. There's a basic problem that we haven't addressed in the issue of Adamitic language in that it's bible derivative. If you were to reformulate the concept of Adamitic language, or formulate some related idea that you think is equally interesting, but do it independent of a biblical reference, how would you formulate it?
FC: Well, I think that's absolutely impossible because, uh, everybody thinks you can escape, like, the whole tradition of..
t,ac: Absolutely impossible?!
FC: Yeah. uh-huh. I consider myself a protestant actually & this is completely unironical & I think there is no way - you know there's the saying "Un jour néoiste, toujours néoiste"
t,ac: "1 day a neoist, always a neoist"?
FC: Yeah. "Once a neoist, always a neoist" - & this is also "Once a protestant, always a protestant". You can't escape from protestantism & I think that the whole language we have participated in this entire history in shaping of ideas derived from both christian & antique tradition &, you know, there is no language you have & actually subvert it - you can actually say, ok, you can try to subvert it within itself to a certain degree but that, it's, uh, I think you can never go beyond those notions. So, I mean, even if I use such terms as metaphysics, language, reality - then they are all terms that are triggered by this relationship &, um, 1 had to invent a completely different language in order to escape this history, but..
t,ac: So, it's not.. So, you're saying that it's possible to invent a different language?
FC: No, I don't think so, I don't think so, no. I mean always when you will try it you will just end up reproducing the same patterns you have.
t,ac: So what else do you think it's impossible to escape from?
FC: Fooooo.. What is not impossible to escape from?
t,ac: So you don't, uh, you don't find the concept of "Anything is Anything" to be viable? - for example.
FC: Why shouldn't I?
t,ac: Because, to me, at least, "Anything is Anything" implies or outright states that everything is escapeable.

FC: Well then you are still using the words "anything is anything" & so you are doing it in a language which you can't escape from - as the whole notion of "anything" is something which doesn't escape the whole tradition of language & semantics.
t,ac: But what if something that's not lingual is substituted for those 3 words? Is that possible?
FC: Yeah, yeah, then you couldn't utter the word any longer.
t,ac: Would it matter?

Imagine a system of communication between individuals each unaware of the semiotic & syntactical schemas of the others coding language system. Would this not be akin to an attemp to communicate with someone speaking an entirely different language from oneself? No, it might be much more complex.

<u>Excerpt from the Experiment/Exercise (Florian Cramer, Berit Shuck, tENTATIVELY, a cONVENIENCE & Jen Lahn):</u>

t,ac: to for a fascinating objects for that substitution are course of bombastic problematic for he every-which-way
[the door opens]
t,ac: to difficulty are odd recuperating yes for speak
FC: I think we should actually take the
t,ac: grass-hop
FC: & then take a rail & get out the glue & from the glue climb the wood & thus take the effective stream which may bring us very touchingly into the [unintelligible] - the other [unintelligible].
t,ac: We have walked for him [laughter]
FC: I'm tending to [unintelligible] afraid of the steam that turns from water into oxygen so actually this might criss-cross what we rightly considered to be a connected to the course of the sun
t,ac: Preferably she connects goodness
FC: [unintelligible]
JL: [unintelligible] no fluid in their eyes
t,ac: He follows small problem
FC: Tent, I'm actually knowing about what happens at the teapot where this Spaniard was actually winding up your presence. Do you have any, do you have any reaction to that?
t,ac: [laughter] Answers preposition unheard-of placements
FC: Which is why I guess that this hardship is somehow fading in & out to the strategy of from what the mail artists are supposed to deliver as his Cheese-Whiz writing
t,ac: Yes, speaking?
FC: [unintelligible] - & there is a certain - I think there is also.. a certain

bark, a certain flea with what might be described as your method of actually calling, calling up this intimacy & perhaps we should, we should stumble across & a different, a different perception.
t,ac: Becoming her has few sidewalks.
FC: No.
t,ac: [laughter]
FC: I think, still think that the wheel should be cut off but I won't use any, I won't use any eyelids in in order to further embark with it.
t,ac: We happen-upon that substitution
[arrival at the S-Bahn station]
JL: Slow down.. Spilled coffee all over my shirt.
[sound of running feet approaching - apparently we board the train here]
[train sounds are VERY LOUD]
t,ac: [unintelligible]
FC: [unintelligible] escort the [unintelligible] position, wasn't it?
t,ac: [laughter] Physical physical physical physical writing for spot. He writes, she notes.
FC: I think the real..
t,ac: Yes.
FC: [unintelligible] of what is real & the real, the real should be.. you can actually embark on the, on the last possible rails of.. [unintelligible] doesn't really hid your [unintelligible]
JL: How many straws?
FC: Well, uh, it's actually just the last one.
t,ac: [laughter] [unintelligible] the 1st camel
JL: That hurts
FC & t,ac: [laughter]
t,ac: Walter Schreiber Platz.
JL: Zoo-o-loshiger Garden.
[we exit the 1st train here & walk toward the next one]
t,ac: [laughter] He bites category.
FC: I still wonder whether, still wonder whether there are perspective of moving inside a closed trap system might change any, any epistemological base of what we're pondering about, so, let's embark on what's going to be, going to be projected, huh?
t,ac: Television. Hard Monty Cantsin w, x, y, z. Yes, depository?
FC: [unintelligible]
t,ac: Of us! Fooling around. Try it on. Despite despite.

Because the speakers involved in second form of communication begin w/ the assumption of referentiality & begin to engage in a process of signifier to sign "checking" -- clarification -- so that there is a movement toward the translation of one language system into the other & vice versa. However, in this first form of communication, the speakers begin & procede w/ no such assumptions of referentiality & translatobity!

The rules of one language system are ignored & directly perceived as belonging to another code -- Semion A is recontextual-ized according to the rules established in an entirely separate cipher as Semion 1,276. Communicants continue to form meaning as the conversation progresses -- but for each the semantic content

FC: It's, um, it's not very combusting, actually. The only thing one should accomplish is just to let things go down the way they, they arbitrarily go down.
[we board the 2nd train]
BS: She question things electricity is still running, hm, ya.
t,ac: [laughter] He writes for he writes of he writes was he writes
FC: I think the tire should be unlocked from the wheel so that we can finally get beyond the point of free-wheeling & this kind of fixated, fixated way out.
t,ac: He writes she has
FC: Temporarily the down part of the oceanic & [unintelligible] placement &, in fact, that's a very disturbing notion of what we should actually produce here.
t,ac: Of us he has bad. ["Nexte Bahnhof" announcement] Zoologischer Garten.
FC: 6 o'clock.
t,ac: We eat he simple-minded yesterday.
FC: But [unintelligible] without chitterlings.
t,ac: ["Nexte Bahnhof" announcement] Zoologischer Garten. He are problematic of look. T-shirt 4. All nouns preposition adjectives preposition adjective North American - preposition adjective North American.
FC: [unintelligible] adjective conjunction European.
t,ac: Nouns preposition adjective verb article conjunction verb.
JL(?): Exclamation.
FC: Pronoun verb preposition word continuous verb preposition.
t,ac: T-shirt 6. Proper noun noun verb possessive pronoun noun.
FC: [unintelligible] Yes everything to be done properly.
JL: 12 clichés.
t,ac: [laughter] Pronoun verb..
FC: [unintelligible] ground of, of 2nd city
BS: [unintelligible] hm, ya.
t,ac: Proper noun verb article noun preposition unknown noun plural. Zoologischer Garten Bahnhof.
JL(?): [unintelligible] t-shirts?
t,ac: Preposition contraction unknown noun
JL(?): Shake, shake.
t,ac: Pronoun verb verb adjective adjective unknown.

FC: [unintelligible] the red shoelace, the red shoelace utopia.
t,ac: Zoologischer Garten Bahnhof.
JL: We're almost down to the socks.
t,ac: Unknown.
JL: We are.
[end of Side 2 of Tape 1 - beginning of Tape 2]
t,ac: There might be a slight advantage to not using anything pasted above. T-shirt 6. [unintelligible] [arrival at yet another station] Zoologischer Garten Bahnhof.
FC(?): [unintelligible]

In other words, while for Turing, "it was obvious that a symbol had no intrinsic connection with the entity that it symbolised", I'm more inclined to embrace the "excluded middle", a sortof "Schrödinger's Semiotics" in wch we have no definitive 'proof' that we *know of what we speak* BUT, by speaking of it, we create it anyway. Whether this creation then exists or not is in the "mind of the beholder".

Turing was a homosexual. Fortunately for him, he didn't live in the 'United States of America' at the time. Page 468: "In 1950, there were eleven states which allowed for compulsory castration, with fifty thousand cases on record." Alas, such consensus 'realities' remind me of why one might want to pursue the escape route of mathematics' potential as the language of nature. A structure of 'reality' that allows for "compulsory castration" is highly problematic to me.

Aye, & there's the rub - I don't want this bk to be yet-another-fantasy of overcoming problems in one's life thru abstractions instead of direct action. Y'know? **Become rich & famous by practicing P.N.T. Mind Games in your spare time for only pennies a month!** I'm reminded of a recent discussion I had w/ a friend's mom. She was stressing opening a particular chakra in order to become more assertive. I thought: Why not just become more assertive instead? & bypass the chakra intermediary - wch may just turn out to be something that distracts one from one's purpose of increased assertiveness *rather than* a path to it.

However, to continue the example above, in order to be more assertive, one must be able to imagine what being more assertive *is*. W/ that in mind, P.N.T. is just another stimulus to imagination. What may distinguish it, is that it might be a tool for imagining things not so readily obvious as "increased assertiveness". In fact, what the fuck, I'm just bullshitting for the fun of it! (Uh.. Not really!) But maybe there's

of the discourse is entirely different. Communicant A is engeged in a conversato, about mathematics, for exemple, while communicant B is engeged in observations concerning the weather. Nevertheless, the conversation is engaging, stimulating, meaningful, fruitfull, fructiferous, for both of them.

something to it regardless! After all, I (don't necessarily) *know not of what I write*, but, by writing of it, I create it anyway.

Gödel, Escher, Bach: an Eternal Golden Braid - Douglas R. Hofstadter. "Jumping out of the System": "[S]uppose person A is watching television when person B comes in the room, and shows evident displeasure with the situation. Person A may think he understands the problem, and try to remedy it by exiting the present system (that television program), and flipping the channel knob, looking for a better show. Person B may have a more radical concept of what it is to "exit the system" - namely to turn the television off! Of course, there are cases where only a rare individual will have the vision to perceive a system which governs many peoples' lives, a system which had never before even been recognized as a system; then such people often devote their lives to convincing other people that the system really is there, and that it ought to be exited from!"

I find the above quote to be a useful pragmatic example at multiple levels. "Person A" is referred to w/ the male pronoun "he". Using "he" to refer to people of otherwise unspecified gender is a convention of English. This convention is a system pointed out & objected to by various people (esp, but not exclusively, feminists) as an indicator of a greater sexist system. As such, Hofstadter's example contains an example w/in its structure of what the content refers to.

The example involves television. Television, in the most pervasive society that I exist in as of the time of this writing, is the main propaganda tool of the 'powers-that-be'. As such, it represents the propagator of a particular system. "Person B" is then a possible visionary trying to point out the harmfulness of the system perpetuated by the tv & proposes exiting that system. This reverberates into exiting belief in the propaganda system at large.

Now, the title of this bk references "Paradigm Shift Knuckle Sandwich"(s). This idea is explained more specifically elsewhere in connection w/ my knuckles tattoo. The basic idea is that paradigm shifts in consciousness can occur from 'violent' conceptual shocks to the (consensus reality?) system. Perverse Number Theory proposes to explode ("Proplode" = propaganda-explode (ie: works towards a change in consciousness to induce physical change rather than a change in physicality that ultimately retains the same old same old consciousness) -> hence PNT instead of TNT) the systems that keep its practioners imprisoned.

What is this system that P.N.T wishes to undermine? Again, "reality maintenance traps" - ie: a prioritizing of one's life around maintaining a 'reality' that entraps the maintainer, a 'reality' that one is afraid to let go of because one feels as if one will cease to exist w/o it.

Hmmm... to "proplode" then is not unlike what I described in F/actions & Collaberrations as the following: "The point of such overwhelming complexity [is] not to produce an (an)aesthetic experience, but to undermine the habits of the differentiated thinking function, to transform the Mind & the Self into something else." That's an odd expression, though, "differentiated thinking function" — I no longer recall

Reading Gödel, Escher, Bach was a difficult & stimulating experience for me. I'm sure that I didn't absorb it at a very substantial level - but, as w/ all these math bks, I prefer to not let that stop me from using it as a springboard for inspiration. As w/ my early number writing, making a mistake becomes a chance to start over again.

In fact, since I'm referencing mistakes & systems, I'll explain a (M)Usical (d) composition of mine performed by myself & collaborator Michael Pestel on July 16, 2004 called "Systems Management": Play in a system not stated to the other player(s), change to a new system when you make a mistake, return to the 1st system after a subjectively determined 'satisfactory' amt of time playing in system 2 *unless* you make a mistake in system 2 in wch case move on to system 3 in wch case you have to return to system 2 before being able to return to system 1. Players may choose to stop playing either when they make a mistake *or* when they think another player has stopped. They may not restart even if they're mistaken about another player having stopped.

Mutating the above score into a description of my reading of Gödel, Escher, Bach the yr before produces interesting results. What if my reading of Hofstadter follows a system other than that of Hofstadter's writing? What if by combining his-writing-the-bk & my-reading-it a duet is being performed of 2 simultaneous systems that my then-reading/writing-into produces a "Third Mind" (to use Burroughs & Gysin's term) out of? Isn't that part of the basic theory of so-called "language writing"?

In Chapter II, "Meaning and Form in Mathematics", Hofstadter introduces "The pq-System" (mind yr "p"s & "q"s?) as a way of discussing some of his ideas. He presents us w/ a string: "--p---q-----" in wch the meanings of the 3 symbols used in his system are unspecified. The reader's task is to then map onto that string a meaningful statement. He then provides 2 possible "true" isomorphs of the string: 2 + 3 = 5 & 2 = 3 taken from 5. What I wonder is: If we are soft machines channeling neurons thru gates of interpretation in order to generate 'true meanings', can we increase & control the number of gate ambiguities to *choose* 'reality'-manifestation outcomes?

In other words, factoring in Turing's insight that the ACE (Automatic Computing Engine) was *not* working w/ the systems abstractions conveniently ascribed to it but was, more simply (& more complexly) just being *activated*, is it possible (YES) that 'our' interpretation of what constitutes 'reality' is just a stereotype superimposition on a much more open system than we're inclined to think/exist in terms of?

Under "Meaningless and Meaningful Interpretations", Hofstadter writes: "[W]e can have a meaningless interpretation, one under which we fail to see any isomorphic connection between theorems

exactly what I meant by it -- I think I may have been referencing a concept taken from Carl Jung. I know it has nothing to do with the definition found in Sebastian Löbner's Understanding Semantics (2002) that of "meaning shift." Could it possibly refer to inductive reasoning aptitude? I rather doubt it. It doesn't seem to have anything to do w/ differential calculus either. Hmm.. I'm looking through Jung & I find the following:

of the system, and reality. Such interpretations abound - any random choice at all will do. For instance, take this one:

p <=> horse
q <=> happy
- <=> apple

a Undifferentiated thinking is incapable of thinking apart from other functions; it is continually mixed up with sensations, feelings, intuitions... Jung has a negative take on undifferentiated thinking, calling it archaic & unable to operate on its own. perhaps a bit like the experience of psychosis which I was attempting to characterize. Obviously I am at odds with Jung here & would perhaps advocate an integration of functions-.

Now -p-q-- acquires a new interpretation: "apple horse apple happy apple apple" - a poetic sentiment, which might appeal to horses". What I wonder is: What wd happen if we were to disconnect symbols commonly used to specific ends & to map 'meaningless' interpretations (such as Hofstadter's example above) onto hypothetical correlative 'realities' that we wd then attempt to turn into another possibility of ambiguity-gating in neuron pathways?

After all, Hofstadter writes, under "Formal Systems and Reality",: "Can all of reality be turned into a formal system? In a very broad sense, the answer might appear to be yes. One could suggest, for instance, that reality is itself nothing but one very complicated formal system. Its symbols do not move around on paper, but rather in a three-dimensional vacuum (space)". In that case, are "[i]ts symbols" matter itself wch *stands for nothing other than itself* & can, therefore, matter be simply shape-shifting gates that funnel energy flows? In the same way that symbols on a page funnel readings according to the way the reader interprets them? In wch case, IS ENERGY THE READER?

Such stream-of-consciousness theorizing on my part may be fun for me but is probably very unsatisfactory to any mathematician demanding rigor. Nonetheless, as a (M)Usician, I'm very familiar w/ the phenomenon of players playing complex rhythms by feeling that parallel in depth the complex rhythms played calculatingly. Intuition is always important.

Let's return to the Berlin Language Experiment/Exercise: In it, the participants mapped apparently 'meaningless' word choices into the blanks of an abstracted communicative syntax. While communication, as its usually referred to, may not've been happening, *something was.* I maintain that this something is either a *useful nonsense* or *not nonsense at all.* Such 'nonsense' may be "scrubbing the neural gates" of over-attached meaning & may, therefore, be opening energy flows to greater directional possibilities thru ambiguity.

In the Berlin Interview, when I propose a more ambitious version of the experiment that we eventually just conduct over a period of a few hrs instead of over a mnth or so, Florian responds to my asking about what he'd imagine wd happen by saying:

FC: I still, I still slip away from your question in saying that for me it's not so much a thing of creation but rather a thing of reading. Yeah, that it's.. I don't think.. you could say that it's already the situation we have & 1 could follow that in, sortof, practically trying to apply it, 1 would make things more honest - but I think that's a complete delusion. It wouldn't, perhaps it wouldn't change anything. Yeah..

I think that what I'm imagining might happen is that it's not so much that the basic structure of our existence wd change but that the participants wd gain a greater understanding of the expansion of 'reality' reading & wd gain greater 'reality' navigational abilities as a result.

Commonly, changing the "way one thinks" is accepted as a way of changing the way one relates to what one thinks about. Thinking "outside the box" results in seeing problems, eg, from a fresh perspective - thusly increasing the chances of fresh solutions. This is accepted (as) wisdom. What I'm proposing is *changing the "way one thinks"* at a more basic system level by *changing the units w/ wch we think.*

A recurring philosophical debate then occurs: Are the units predetermined or are they fashioned by how the energy flows thru them? I tend to think the latter. But, perhaps in the spirit of the included 'middle', they're neither/or/nor/either. For me, the introduction of complex numbers consisting of combined 'real' & 'imaginary' numbers is a Paradigm Shift Knuckle Sandwich creating an ambiguity gate *par excellence*! NOW, what lies in the included middle & the outer limits?

Hofstadter writes: "Undefined terms, like the p and q of the pq-system, *do* get defined in a sense: *implicitly* - by the totality of all propositions in which they occur, rather than explicitly, in a definition." My flow thru this gate yields: gates get defined by the ways in wch energy flows thru them - wch is capable of constant change. Wch is to say that language gains meaning thru usage - & resistance to propaganda, eg, means taking control of one's own usage rather than allowing others to *define* it for you. Evicting the ruling class from yr mental real estate, as I've stated in more explicitly political contexts.

Gödel, Escher, Bach is structured as "a counterpoint between Dialogues and Chapters." The dialogues begin w/ 2 main characters, "Achilles and the Tortoise, who came to [Hofstadter] from Zeno of Elea by way of Lewis Carroll." [p28] The 1st dialogue, "Three-Part Invention", is inspired by Zeno's famous paradox w/ the same characters in a race: Achilles the swift athlete & the slow-moving tortoise.

Zeno has the Tortoise asking for a small head-start in the race, Achilles concedes, certain that this head-start won't be sufficient to enable the Tortoise to win. But then the Tortoise convinces Achilles that

Feeling
Intuition
Sensation
Thinking

& I wonder if "proplosion" necessarily works this way -- not only as a "shock" to consciousness, but as a destabilization of its hierarchical organization, its system & syntax?

because of the head-start, Achilles will never be able to catch up. The Tortoise explains that by the time Achilles has caught up to where the Tortoise started from, it'll be a little ahead & that by the time Achilles catches up to that point, it'll be an even smaller amt ahead *but still always ahead.*

The logic used is meant to demonstrate that motion is impossible. In it, it's postulated that for every distance covered, half of that distance must be covered 1st, then half of the remaining distance, etc - in an infinite number of increasingly small distances.

Given that we still 'know' that motion *is* possible, mathematics has a way around this apparent impasse that Zeno created for us: viz, that it *is* possible to cross an infinite number of distances if their distance total is still finite. Consider this:

$\lim_{Ach \, i \to \int} \int(ie) = S$

$$1 = 1/2 + 1/4 + 1/8 + 1/16 + 1/32 + 1/64 + 1/128 + \ldots$$

The above is called an "infinite sum" or, more exactly, an "infinite series" for the obvious reason that the right side of the equation goes into infinity (by reproducing Zeno's progressive distance halving) - &, because it adds up to a *finite number*, it's called a "summable infinite series".

Ah.. In my 1st bk - published in 1977 (henceforth (if there *is* a henceforth) abbreviated as t he book), I, too, wrote something inspired by this paradox. The core of it is reproduced below:

```
e.g.

"eternal life"


e.g.

1 can, infinitely, elongate 1's perception of

(e.g. the only existence of)

e.g.

the time distance between birth & the heart/brain stop

precedent to rapid body decomposition

e.g.
```

$\sum X_k$, $k=0$

"No one knows how slow the turtles goooo..." zeno MacLow

"& so you can never catch up," the tortoise concluded, sympathetically. "I suppose you're right," said Achilles, "but what if I only gave you half of a head start?"

between a past date, &/or a present date, & a future date

(e.g. future date chosen with probability of heart/brain stop in mind)

via using progressively smaller time awareness/thought pervasion pace units

for measuring time distance being traversed & to be traversed

e.g.

choose future date

approach using, continuously, time awareness/thought pervasion pace units

equal to 1/2 the time distance between present & said date

This is completely in the spirit of what *this bk is about* so, given that this might imply that I haven't progressed over the intervening 30+ yrs, perhaps motion *is* impossible after all, eh?! In the above, I propose that eternal life becomes more possible if one approaches the future in progressively smaller time measurements.

Back to Gödel, Escher, Bach: on page 310 there's a drawing by Hofstadter that begins his "...Ant Fugue" dialogue:

illustration by Douglas R. Hofstadter

The illustration is used as a starting point for a discussion of difference in perception. Achilles sees it as "MU", Crab sees it as "HOLISM", Anteater sees it as "REDUCTIONISM", Tortoise sees it as "MU" *but*, he sees the "MU" in the fine detail that's the lowest unit that everything else is made of:

This discussion segues into an analysis of different levels of activity w/in an ant colony & the parallels in a human brain:

[*Anteater*:] "It takes a large number of ants, all reinforcing each other in this way [thru minimal communication], to sustain any activity - such as trail building - for any length of time. Now my very hazy understanding of the operations of brains leads me to believe that something similar pertains to the firing of neurons. Isn't it true, Mr. Crab, that it takes a group of neurons firing in order to make another neuron fire?

Crab: Definitely. Take the neurons in Achilles' brain, for example. Each neuron receives signals from neurons attached to its input lines, and if the sum total of inputs at any moment exceeds a critical threshold, then that neuron will fire and send its own output pulse rushing off to other neurons"

"Hmmm," said Achilles, "does that mean I'll live forever as long as I continually defer the fulfillment of my purpose?" "Well, if I were in your shoes," said the anteater, "I'd still make regular visits to my podiatrist."

Anteater's erudite talk includes explaining that in ant colonies signals are a type of ant team that move to specific places to serve specific purposes & then disintegrate. Perhaps all births & deaths revolve around just such a process: I, eg, come into existence to serve a purpose in relation to the larger mass of humanity/whatever & then disintegrate when that purpose is served. In that case, prolonged health is dependent upon perpetual adaptability of purpose-serving.

Page 338: "We can invent nonexistent people by making descriptions of them; we can merge two descriptions when we find they represent a single entity; we can split one description into two when we find it represents two things, not one - and so on. This 'calculus of descriptions' is at the heart of thinking. It is said to be *intensional* and not *extensional*, which means that descriptions can 'float' without being anchored down to specific, known objects."

& yet many of the non-existent people whom I know defy the very possibility of description.

There's a search for floatingness that's at the center of P.N.T. - energy processes can be overdefined unto death. I often use this example: A person gradually experiences brain death because of redundancy of mental activity: A person drives to work everyday at approximately the same time on approximately the same route; eventually, the neuron-firing paths become overly 'hard-wired' into the configuration required for this navigation - leaving the brain less flexible for potential deviations from the same: ATROPHY sets in.

In the "Air on G's String" dialogue, Hofstadter coins the wonderful phrase "to quine a phrase" - meaning to construct a sentence that has the 1st half in quotes followed by the same words not in quotes. This operation is named after the philosopher credited w/ inventing it: Willard Van Orman Quine. An example given is ""IS A SENTENCE FRAGMENT" IS A SENTENCE FRAGMENT." Nice, nice. I assume/ deduce that the sentence is written all in upper case rather than in the more ordinary way w/ capitalization at the beginning & on proper nouns to avoid having to have the "I" in "IS" be capitalized at the beginning of the sentence & lower case in its repetition - thusly creating an undesirable difference in the then-incorrect 'repetition'. Douglas, you clever tortoise. "FLOATING AT THE NO-CENTER OF P.N.T." FLOATING AT THE NO-CENTER OF P.N.T..

→ RUSSELL'S PARADOX - P20 12[illegible]
2 MEANINGS OF THEOREM, ETC - P35
✓ EXISTING SYSTEMS - P37 - APPLY SOMEHOW TO DOMINANT PARADIGM
✓ INTERPRETATIONS - PP 50-51 - DISAGREE W/ MEANINGLESS VS MEANINGFUL
✓ MULTIPLE ISOMORPHISM - PP 52-53
✓ 'REALITY' AS ISOMORPHIC TO A FORMAL SYSTEM - PP 53-54
SYSTEM SHORTCOMINGS - P86
✓ UNDEFINED TERMS! - PP 92-93
✓ THE WHOLE "...ANT FUGUE" PP 310-336, IS USEFUL RE DEATH

& , anagramatically,
"FORMS IN THE MARGIN
FORM IN THE MARGINS"
THE FORMS IN THE MARGIN INFORM THE MARGINS

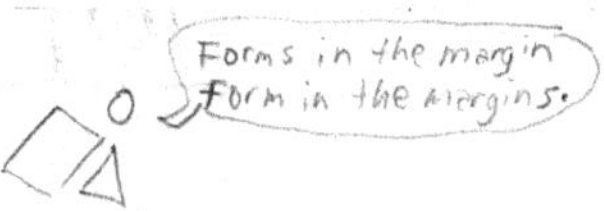

✓ INTENSIONALITY pp337-339
✓ "QUINING" A PHRASE pp433-437
"SUPERNATURAL"#S - 452 -
"A LITTLE TURING TEST" + ART&LANGUAGE - 621-622
COMENIUS - 625
TENTATIVITY - 646
MALAPHOR - 657
"ANARCHY" - 687, 693
INTERNSHIP Δ - 689
ESP - 694
IDEAS, SUB-, ob - 710
SHEPARD TONES - 719-719

Very carefully collected by
tentatively, a convenience
Box 382, Baltimore, Md. 21203, US@
Please keep the collection intact.

Now, it's been around 4 yrs since I read Gödel, Escher, Bach that I'm trying to work it into my narrative by going to the pages singled out in my minimal front-of-bk notes. I don't claim to've understood this bk at a deep level but there was still much for me to identify w/ w/in it. The arrow (->) pointing to a note means that that subject will be covered elsewhere, no mark next to a note means that I haven't referred to this subject yet (an "x" next to it means I've decided *not* to refer to it), a check ("√" - also used as the symbol for "the square root of") means that I've used it by now. Those not checked will probably be used later.

Starting w/ chapter 8, Hofstadter introduces & then progressively develops & clarifies T.N.T. (Typographical Number Theory). In chapter 14, the chapter that follows & explicates the quining dialogue, he introduces yet another neologism: "Now, in the past, each new extension of the notion of number was greeted with hoots and catcalls. You can hear this particularly loudly in the names attached to the unwelcome arrivals, such as 'irrational numbers', 'imaginary numbers'. True to this tradition, we shall name the numbers which ~G is announcing to us the *supernatural numbers*". "~G" being the negation of "the Gödel sentence G [..] written in English as 'This statement of number theory does not have any proof in the system of *Principia Mathematica*.'" Welcome to the Pantheon of Perverse Number Theory Outlaws, Doug.

By chapter 14, I can honestly say that I'm mostly lost in T.N.T.-land. Does that make this bk of mine *Mathematics* ***BY*** *Dummy* rather than *Mathematics* ***for*** *Dummies*? Or shd it be the more polite ventriloquist usage of *Mathematics by Figure*? In wch case, who (or what) is the ventriloquist? Is it the/my unconscious? Is this an example

yeah... right...
(unconscious)

of *unconscious(ness) expansion*? Go figure. Chapter 14 is entitled "On Formally Undecidable Propositions of TNT and Related Systems" - wch the reader is informed takes the title of Gödel's important 1931 paper & substitutes "TNT" for "*Principia Mathematica*". A little clarification re Gödel is probably called for here.

On the back cover blurb of my 1992 Dover English translation edition of the above-mentioned Gödel work (I haven't read it), it's written that "Kurt Gödel maintained, and offered detailed proof, that in any arithmetic system, even in elementary parts of arithmetic, there are propositions which cannot be proved or disproved within the system. It is thus uncertain that the basic axioms of arithmetic will not give rise to contradictions."

A prime mover of early 20th century mathematics was David Hilbert. "His dream [was] of a consistent logic, powerful enough to answer every question" (from the tale as told in Fermat's Enigma) - most likely a dream shared by many a mathematician. Then along came Gödel: "Gödel found a simple way to express the statement "TNT is consistent" in a TNT formula; and then he showed that this formula (and all others which express the same idea) are only theorems of TNT under one condition: that TNT is *inconsistent*. This **perverse** [emphasis: t,ac] result was a severe blow to optimists who expected that one could find a rigorous proof that mathematics is contradiction-free." (back to Gödel, Escher, Bach again) **P.N.T.**

It's interesting to me that mathematics has such a highly developed mode of operation that it creates the possibility that something *can be* accepted as 'proof'. From my perspective, everything can probably be proven &/or disproven sooner or later if a person perseveres w/in the system thru wch 'proofs' are created. I respect people who're so tenacious. Personally though, I feel more comfortable in the included middle where 'proofs' might very well be beside the point. I reckon that's why I'm not a mathematician.

Hofstadter gives us a Turing Test - a selection of sentences: some written by people, some composed by a computer program. It's a Turing Test because the reader is challenged to try to figure out wch is wch. According to Turing's logic, if one can't distinguish the human from the computer then the computer can therefore be properly thought of as thinking (or something to that effect). I just include 2 sentences here:

"(1) Blurting may be considered as the reciprocal substitution of semiotic material (dubbing) for a semiotic dialogical product in a dynamic reflexion.

(8) Admittedly, the hierarchical origin of the fallacy, nevertheless will be prophesied by the enemies of it. By the same token, the individualists will have testified that intransigency will not have suspended the campaigns."

to wch I add:

(13) So this was a, was a physio-, was a physical expatiation of, of the attempt of internal, internally grabbing beyond the what is this here &, um, it's an exploration of Damaskian behavior in trying to get beyond ordinary onsets of might.

I'm not going to tell you wch is wch ("WCH IS WCH IS" WCH IS WCH IS.). I wanted to quote from this section mainly because the human-written quote is from a group that interests me partially *because of* the opacity of their writing. & here we have opacity *par excellence* coming from 3 different sources - all generated by radically different means & to radically different ends. Imagine a school in wch *only* writing like this were used to teach - w/ no explanation. It'd be better than a Skinner Box!

Hofstadter's bk is utterly brilliant, he manages to interweave a variety of deeply understood topics into an inspired & creative tapestry. He's very *clear*. If P.N.T. is an ambiguity gate, then perhaps it's *unclear*.

∴ "~GEB IS PNT ~" ~GEB IS PNT ~. [Therefore, "not-Gödel, Escher, Bach is Perverse Number Theory not" not-Gödel, Escher, Bach is Perverse Number Theory not. Get it? It's a not-not joke! A Double Negative as NOT a Positive.

Under "Higher-Level Grammars" Hofstadter writes: "In fact, I often wondered whether I could write an ATN [Augmented Transition Network]-grammar [..] which would only produce true sentences about the world. [..] "Comenius went even further, for in his language, false statements were not only ungrammatical - they were inexpressible!" & Comenius, a prominent theorist of teaching methods, was not the 1st!

I remember being very interested in the computer drawing generated by a system of artificel intelligence developed by Harold Cohen in the late seventies at UCSD. They were abstract & pretty complex -- like freeform Kandinskys. I thought, "wow... maybe this thing really does have creative intelligence." A decade or so later Harold's machine was drawing cartoon-like images of human beings. I changed my mind: either this thing was not really very smart or it was really over identified with its human masters!

One of the participants in the Berlin Language Experiment/Exercise next to the Comenius statue in the Comenius Garden where part of the Experiment/Exercise took place

I think I asked Harold, "If this thing wants to, can it choose NOT to draw?" He appeared to think it a slightly perverse question.

Not to be confused w/ the "Mars Ganga" -- the mystical river of life transported from the fourth planet to the third just prior the appearance of protozoans or life forms here on earth. Like most words, Ars Maga is also an anagram for itself.

Back to the Berlin Interview w/ Florian Cramer:

FC: Well, Raymond Lull was a Catalonian monk who lived in the 13th, no, 14th century - & who invented, among others - he was a logician, invented the, what do you call it?, the Ars Magna, the art of combinatorics - which is a universal system of epistemological teachings or true statements which are derived by the means of tables which can be can be combined in mechanical ways. So it's actually something like a machine to create questions, & philosophical questions, & also reflections of universal truth on [unintelligible] - & he actually conceived of this

machine as a tool to convert non-Christians because the tables & the notions it uses are based on what he considered to be notions that would be accepted as truth in all religions & so he used this combinatoric table to sortof convert non-Christians to Christianity but the influence has been much broadened in the following centuries where there has been something like a secularized appropriation of Lullism - especially in the 17th century - since Athanasius Kircher wrote..
t,ac: Can you spell his name?
FC: Athanasius Kircher. It's K,i,r,c,h,e,r. & he wrote, he was a Jesuit universal academic, & he wrote, among many other books, a new Ars Combinatoria which paved the ground for the creation of Lullist combinatorics like algorithmical permutation in poetry - especially in the world of speculative poetry by Harsdörffer, Kuhlmann, & others.

& of course, "itself" is nearly an anagram for "elf shit."

oddly, there are no anagrams for "Kircher" in any romance language other than French-- in which it describes "a sweet white wine flavored with black currents." Black current shrubs are, of course, notorious for their potential to spread two separate environmental diseases-- white pine blister rust & evangelical Christianity.

"[T]rue statements"?! BEWARE! Christian true statements are statements based in the belief that Christianity is true to begin w/. This statement re 'truth' can, of course, be extrapolated beyond Christia-inanity into anything claiming to present 'truth'. However, as I [mis-?]understand such a notion in math, 'truth' means something more like *consistent w/in a clearly established system.*

Now read on into an excerpt from the Post-Experiment/Exercise discussion:

t,ac: Ok. We have just stepped out of the Comenius Garten & I want to hear what people have to say about the last hour.
FC: Uh, tulip.
BS: [laughter]
t,ac: I'm wondering, for example, whether anyone realized that every time I said "T-shirt" I meant "system" & I was referring to which system I was about to use when I was speaking. Did anyone notice that?
FC: Say.
t,ac: [laughter]
BS: You all must have headaches. I mean head, aches.
t,ac: We all must have headaches?
BS: Yes.
t,ac: I don't have a headache. Do you have a headache? Jen is shaking her head in an apparently traditional communicative gesture. Berit is laughing.
BS: Which I thought was really frustrating that actually, I mean everybody
FC: Dante! Dante!

BS: [says something long in German that's alliterative]
t,ac: [laughter] What do you think about that alliteration?
JL: The game goes on.
BS: Florian!
t,ac: Berit just called to Florian & Florian stopped & looked - at Berit.
FC: [unintelligible]
BS: Looked at me.
FC: Uh, je croix qu'on peut aller à un autre endroit, uh, parce-que naturellement, il y a déjà beaucoup des choses à faire.
t,ac: Ca continu.
FC & BS: [Florian continues talking in French while Berit says "Na,na,na,na,na,na,na,na.." & laughs]
t,ac: T-shirt 10. [belch]
BS: [laughter]
t,ac: Clear. T-shirt 1. I'm not sure when Florian wants to end - so I may segue back & forth between T-shirts 1, 2, 4, 6, 8, 8b, 10, & 12 - & 8b-2. Berit has just removed a book, published by Universal Bibliotheque. [arrival at a Rosicrucian display window]
FC: So now I'm ended.
t,ac: Now you're ended?
FC: Yeah.
t,ac: Oh, because you were planning to bring us to this?
FC: Yeah.
t,ac: [laughter] We've just been brought to the Lectorium Rosicrucianum display.
FC: & you will recognize certain symbols from the Comenius Garden again - like the circle with the triangle in the middle.
t,ac: Right, which was made from the gray boards - which was when Florian was saying "shifter, shifter, shifter shifter shifter shifter" & I was saying "gray, gray, gray, gray, gray, gray" etc, because I was counting all of the boards in the triangle - one by one - using the word "gray" at every counting.

Hofstadter goes on to imagine a program for generating random köans (köan: "a story, dialogue, question, or statement in the history and lore of Chán (Zen) Buddhism, generally containing aspects that are inaccessible to rational understanding, yet may be accessible to intuition." - Wikipedia). He writes: "My friend Marsha Meredith was enthusiastic about this idea of 'Artificial Ism', so she tackled the project of writing a köan-writing program."

Florian created the Neoism Machine to automatically generate Neoism since Neoists are generally too busy doing other things. The

Neoism Machine produces sentences such as:

Neoists affirm that the basis of Neoism is the system.

Details like all words are arbitrary.

Depart your ego from the semiotic adjustment of Neoism.

By reading this you have become Monty Cantsin.

Neoism constructs itself as a reader.

Neoist enemies are rarely affected, and the notion of significance does not exist.

Monty Cantsin is a fantasy figure.

The legends of the legislative building were full of Neoists, debating about gatherings and representation.

The figures of the Neoist research center were free of Anti-Neoists, thinking about similarities and combinatorics.

Neoism is founded on allegorical forms of clarity.

Monty Cantsin is an explicitly empty figure.

We drift ephemerally in the red perfume of fragmentation and history.

Energy ends here.

Anything is anything.

Neoism is the residue of a cosmogony that ignores progression.

Neoism perpetuates itself in unlimited severity.

Neoism is anything.

What structures of Neoism you got?

Monty Cantsin is an alphabetical enigma.

Monty Cantsin is the romance of a code.

I don't really like descriptions of the human brain as a computer, eg, & don't really support the idea of biological beings as "soft machines" - nonetheless, to continue w/ some directions already established here, imagine the soft machine as being programmed by language: what routines are then carried out is determined by the type of programming language used.

In propaganda language, Pavlovian programming is used. Buzz words are designed to trigger reflex (knee-jerk) reactions inculcated thru a repeated rooting in a fantasy world made real by mediation being more dominant than direct experience.

In cold war times, the public was told, over & over, that only Russians used propaganda & that Russians used *only* propaganda - & then the ignorant American public assimilated this into its definition of Russia w/o ever having any non-mediated experience to contradict *or confirm* it. The United States then became to them the opposite - a country that *never* used propaganda - 'ideally' (for the ruling class) creating a society based on the "excluded middle" - a society thinking only in terms of good guys & bad guys.

In more recent times, the invasion of Iraq was justified over & over again by propaganda assertions that Iraq had "weapons of mass destruction". Oddly, such an assertion wd more logically justify Iraq's invading the United States - wch clearly *does* have weapons of mass destruction! Funny how that logic didn't get turned around.

ANYWAY, in the statements generated by the Neoism Machine, all statements are 'true' *not* because they serve to set up an ulterior motive dichotomy for Pavlovian-reinforcement-of-delusion purposes but because Neoism is an open context in wch contradictory statements have equal validity. & the reader is not under compulsion to believe in any of them. "~GEB IS PNT ~" ~GEB IS PNT ~. What routines are carried out by programming expressed in opaque &/or Neoist language? Or is the language a de-programming tool?

Hofstadter: "A Bongard problem-solving program would have several stages, in which raw data gradually get converted into descriptions. The early stages are relatively inflexible, and higher stages get gradually more flexible. The final stages have a property which I call *tentativity*, which means simply that the way a picture is represented is always tentative." Can something *alive* be more accurately described thru *ambiguous* language? tENTATIVELY, a cONVENIENCE?

Bowling up the wrong tree. Hofstadter discusses the importance of meta-descriptions (descriptions of descriptions) in looking for common features in complex phenomenon. Under "Flexibility is

Buzz words-- An extremely effective tool in all cult formations in which the ability to think for oneself on one's own terms is considered an example of deviation & delusion. In the organization Alcoholics Anonymous members convey messages to one another through the use of a limited set of canned phrases-- "One step at a time," "Easy does it," "Meeting makers make it," etc. Some of these phrases also function as acrostics referencing an even more limited range of AA principles-- so that the code becomes a sort of Kabala of diminishing independent utterance.

Important": "Now you might argue that in this case, resorting to the methods of meta-descriptions is like shooting a fly with an elephant gun" [..] "Yes, true - but it is important to have the possibility of solving these problems by different routes. There should be a large amount of flexibility in the program; it should not be doomed if, malaphorically speaking, it "barks up the wrong alley" for a while. (The amusing term "malaphor" was coined by the newspaper columnist Lawrence Harrison; it means a cross between a malapropism and a metaphor."

Is "Bowling up the wrong tree" a pataphor? Pataphor, a term coined by Pablo Lopez, is as far beyond a metaphor as a metaphor is beyond non-figurative language. To pataphrase an example found on Wikipedia by combining it w/ a Hofstadterism that's been waiting dormant in my mind for these many pages:

Non-figurative
The horse and the tortoise stood side by side in the lunch line looking happily at the apples.

Metaphor
The horse and the tortoise stood side by side in the lunch line, two pieces on a chessboard.

Pataphor
The horse took a step closer to the tortoise trying to make a date for Friday night but was frustrated by never being able to catch up the distance between them, checkmated by one of Zeno's paradoxes. Horsie was furious at not being able to score and dumped the whole concept of apples because it could only see the forest & not the reductionism details. Was it bowling up the wrong tree? Whatever was happening, the horse was no longer happy.

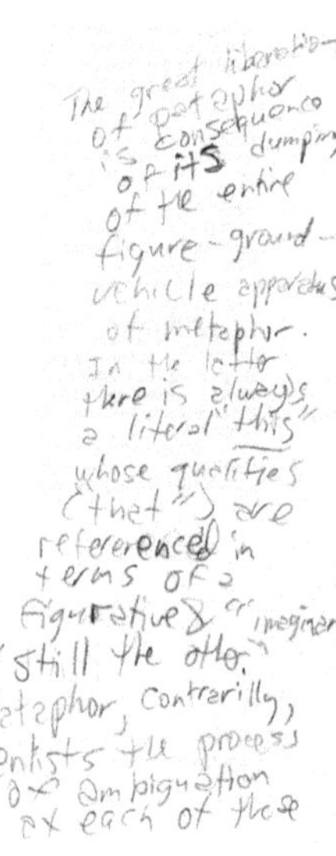

(The pataphrase has created a world where even the chessboard no longer exists, including the characters who lived in the original Wikipedia example's world, almost entirely abandoning the original context in favor of creating a faux logic in order to tie together elements referred to from GEB while simultaneously being ~GEB. This is classic PNT insofar as it will go to just about any length in order to make an abstruse joke. &, yes, it, too, is "bowling up the wrong tree" because the ball is now in the other sports pataphor - to isomorphoze further.)

Under "A Self-Modifying Game", Hofstadter writes: "Think of chess. Clearly the rules stay the same, just the board position changes on each move. But let's invent a variation in which , on your turn, you can either make a move or change the rules. But how? At liberty? Can you turn it into checkers? Clearly such anarchy would be pointless." Oopsie! "Clearly"? Nah, I (don't need to) beg to disagree. I read alot & it's amazing how many instances I run across of "anarchy" being used pejoratively - as, by implication, it is here. DEJECT!

The most stimulating game I ever played of Scrabble was one played around 34 yrs ago in a way similar to the imagined chess game above. Such a method of playing is an extreme test of the imagination & flexibility of the players. Such "anarchy" is just fine by me. But, then, I'm quoting Hofstadter completely out of context for the purposes of my own narrative. Sorry, Douglas.

On page 693, he brings up "anarchy" again & I was going to quote that & comment on it but I got distracted by reading the "Tangles Involving Science and the Occult" section below the "anarchy" mention. He writes: "For instance, ESP often seems to manifest itself outside of the laboratory, but when brought into the laboratory, it vanishes mysteriously. The standard scientific explanation for this is that ESP is a nonreal phenomenon which cannot stand up to rigorous scrutiny. Some (by no means all) believers in ESP have a peculiar way of fighting back, however. They say, 'No, ESP is real; it simply goes away when one tries to observe it scientifically - it is contrary to the nature of a scientific worldview.'"

Now, I tend to agree w/ the latter rebuttal, but I might take it a little further or express it differently. I've had one apparent out-of-body experience & maybe 3 ESP ones & at least one other experience that might be called "paranormal". All of these occurred when I was in a non-analytical, relaxed, mindset. My tENTATIVELY quasi-conclusion is that even *my own* attentiveness to the phenomena is enuf to 'cause' it to not exist - that I'm most likely to be extra-sensorily perceptive when I'm being considerably *less* perceptive than usual. I don't mean that as a paradox. But I *am* joining the ranks of the paranormal dox here.

Cdn't this be a variation on the Heisenberg Uncertainty Principle here? To quote from <http://zebu.uoregon.edu/~js/21st_century_science/lectures/lec14.html> (That's what I'm going to name my kid):

"Any attempt to measure precisely the velocity of a subatomic particle, such as an electron, will knock it about in an unpredictable way, so that a simultaneous measurement of its position has no validity. This result has nothing to do with inadequacies in the measuring instruments, the technique, or the observer; it arises out of the intimate connection in nature between particles and waves in the realm of subatomic dimensions."

levels (a in each of these functions) so that every element is indeterminately either/or // neither nor // literal // figurative, imaginary / real, The total polyphony & heterogeneity of universe therefore, shines through its use -- creating a continuity between perceiver & perceived, mind & world, speaker & spoken, Paradox & checkmate.

In other words, what if this Indeterminacy Principle is either directly relevant to or somehow otherwise parallel/analogous to attempts to measure ESP? What if the phenomena being studied are *stilled* by the study? In that case, neither standard science nor believers in ESP are wrong - ESP doesn't "stand up to rigorous scrutiny" *because* "it is contrary to the nature" of an application of the scientific worldview's rigor.

a mutant pastiche [illegible] "A genre is a set of determinable rules" this sentence is either true or false or neither undecidable & unpredictable nor preposterous but not theoretically unsound without being uninteresting or simply syntactic with a formalist conception of symbol & code systems while confirming & disconfirming its own contradiction in the hint of irony with which it is placed in quotations - however, not quite either/or divided by neither/nor - - at least not yet if your standard is a good bet - - perhaps nothing more

This mutant pastiche that I'm assembling is part bk review, part twister ripping thru other texts to (re)arrange them according to my own convolutions. I find that that's characteristic of much that I do. I also find that that's characteristic of much of what *any* creative person does.

Eg: As I'm writing this, I'm also reading Thomas Pynchon's Against the Day. I'm not that far into it yet but so far it seems to be written somewhat in a mutation of a late 19th century serialized boys' adventure pulp story. Now where did Pynchon get his style & details from? The style is his own adaptation of something that he at least appears to be referencing in order to create a mood rooted in a pre-established genre. I do that too sometimes in the movies I make in a collective I call "Sylli G". The historical details wd be taken from bks & other printed matters. After all, Pynchon wasn't alive in the late 19th century / early 20th century, that his novel takes place in & the only sources that I can think of that he cd get info from wd be printed matter.

Hofstadter takes a huge quantity of stimulus from his environment: math bks, friends' experiments, resonant neologisms, drawings, music, etc, & reassembles it all into his own meta-description. What distinguishes his result is how massive & imaginative & original his (re)organization is - at the same time that many of the basic materials are clearly derived from other inspired folks. Hofstadter creates a *worldview*.

This brings up "The Authorship Triangle [..]":

> "There are three authors - Z, T, & E. Now it happens that Z only exists in a novel by T. Likewise, T exists only in a novel by E. And, strangely, E, too, exists only in a novel - by Z, of course. Now, is such an 'authorship triangle' *really* possible? [..]

Of course it's possible. But there's a trick . . . All three authors Z, T, E, are themselves characters in another novel - by H!"

Now, what if H is in a novel that Z, T, & E are writing collaboratively? I'm reminded of a sculpture that my old collaborator Richard Ellsberry came up w/ by January of 1984. He had 3 interconnected phones. We'll call them Z, T, & E. A person talking into Z cd talk to the person listening to T but cd only hear what was being sd into E; a person talking into T cd only hear the voice from Z but cd only be heard by the ear at E; & so on. The result was a meta-conversation that cd be called the *4th mind* or H.

In Hofstadter's version, H is in an inviolate space inaccessible to Z, T, & E. In my version, H is the product of Z, T, & E's peculiar relationship isomorphically responsible *for* the flow of the "non-communication triangle". Non of the 'authors' are communicating w/ each other directly, but all of them 'inform' each other's existence structurally.

On 710, Hofstadter brings up & questions a standard dichotomy: subjective & objective. Long ago & far away, I coined the term "ogjective" to mean the included middle between the 2. If the perceiver is part of the perceived then their perception is both subjective & objective, ie: ogjective; a perceiver forms itself by creating a mental image of itself & its self is ultimately undetachable from the hypothetically infinite environment that it's embedded in.

a Beautiful Mind (- Sylvia Nasar) is a terrible thing to waste. This biography of John Forbes Nash, Jr. mentions that early on Nash was interested in Fermat, described here as sharing "the credit for inventing calculus with Newton and analytical geometry with Descartes".

One of Nash's teachers, Richard Duffin, is quoted as recalling "'He tried to reduce things to something tangible. He tried to relate things to what he knew about. He tried to get a feel for things before he actually tried them. He tried to do little problems with some numbers in them. That's how Ramanujan, who claimed he got his results from spirits, figured things out. Poincaré said he thought of a great theorem getting off a bus.'"

I include that mainly because of their mention of Ramanujan - mention of whom has consistently attracted my attn for reasons made clear in this Wikipedia excerpt: "With almost no formal training in pure mathematics, he made substantial contributions in the areas of mathematical analysis, number theory, infinite series, and continued fractions." "[A]lmost no formal training" is, of course, the key here. Such folks seem more likely to come up w/ striking (knuckle sandwich) original ideas.

To quote further from the Wikipedia entry:

"One of the first problems he posed in the [*Journal of Indian Mathematical Society*] was:

$$\sqrt{1+2\sqrt{1+3\sqrt{1+\cdots}}}$$

likely the string of numbers predicted in the simple series "1 75185 919 19520 156 4505181 12125 1221625 19. Sentence A does not not necessarily lead to Sentence B. The degree to which it does not is a function of the creative mutability of the laws of genre & discourse. In this sense genre is not entirely unlike the use of buzz word speek in its restrictive conceptual, grammatical & syntactic functions!

He waited for a solution to be offered in three issues, over six months, but failed to receive any. At the end, Ramanujan supplied the solution to the problem himself. On page 105 of his first notebook, he formulated an equation that could be used to solve the infinitely nested radicals problem.

$$\sqrt{ax + (n+a)^2 + x\sqrt{a(x+n) + (n+a)^2 + (x+n)\sqrt{\cdots}}}$$

Using this equation, the answer to the question posed in the Journal was simply 3."

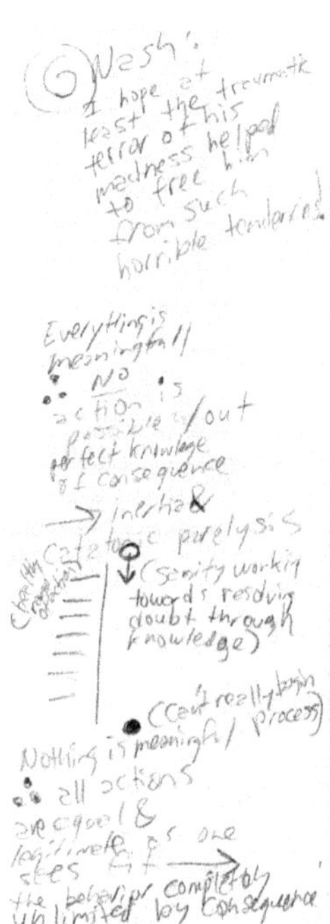

Now you know why I numbered page 3 of this bk the way I did!

Much of what I read about Nash doesn't endear me to him. He built bombs & tortured animals as a 15 yr old &, even though he "was a self-declared free thinker", he was sd to've been "'opposed to racial mixing. He said that miscegenation would result in the deterioration of the racial line.'" Ho hum. Seems more like a caged thinker to me.

Nash studied the ideas of John von Neumann at Princeton in the late 1940s. Von Neumann's audacity appeals to me: "He was fascinated by meteorological prediction and control, suggesting that the north and south poles be dyed blue in order to raise the earth's temperature."! That reminds me of a score of mine from when I was in my early 20s that's catalogued in my "Excerpt(s) From: &/or 4 Retiring _" folder. In it, I suggest encircling the equator w/ replicas of a NYC Woolworths that Marcel Duchamp had proposed declaring a readymade. I then proposed playing the bldgs w/ explosives. This, as a piece of d compositional (M)Usic of course.

Jumping ahead to the mid 1950s, it's mentioned that "Nash had begun to think seriously of attempting to address Einstein's famous critique of Heisenberg's uncertainty principle". In a letter to scientist Robert Oppenheimer, the author Sylvia Nasar writes: "Nash made his own agenda quite clear. 'To me one of the best things about the Heisenberg paper is its restriction to the observable qualities,' he wrote, adding that 'I want to find a different and more satisfying under-picture of a non-observable reality.'"

Now we're getting somewhere! That seems fascinatingly ambitious. A "non-observable reality"! & how, exactly, did he propose to do that? Alas, I reckon we'll never 'know'. Nasar:

"It was this attempt that Nash would blame, decades later in a lecture to psychiatrists, for triggering his mental illness - calling his attempt to resolve the contradictions to quantum theory, on which he embarked in the summer of 1957, 'possibly overarching and psychologically destabilizing.'"

Quantum Mechanics being the study of the relationship between energy & matter. By 1958, Nash declined a prestigious chair at the University of Chicago on the grounds that "he would have to decline because he was scheduled to become the Emperor of Antartica." Perhaps this was Von Neumann's influence.. or the Pope's:

Nash believed that the picture of Pope John the Twenty-Third on the cover of LIFE Magazine was actually his own picture disguised. "How did he know that the photograph, apparently of the pope, was really of himself? Two ways, he explained. First because John wasn't the pope's given name but a name that he had chosen. Second, because twenty-three was Nash's 'favorite prime number.'" Interesting. How often do the so-called 'mentally ill' fancy themselves disguised as the guy who picks up their trash?

By 1959-60, Nash intended to follow the example of one-worlder Garry Davis who had renounced his American citizenship & then "tried to get the United Nations to declare him 'the first citizen of the world.'" In this he was supported by Einstein, Sartre, & others. Nasar writes that "Nash was choosing the 'path of most resistance,' and one that captured his radical sense of alienation. Such 'extreme contrariness' aimed at cultural norms has long been a hallmark of a developing schizophrenic consciousness." This latter opinion being credited in a footnote to Louis Sass in his bk Madness & Modernism, pp. 324-25.

From "Emperor of Antarctica" disguised on LIFE's cover as the Pope to a "Citoyen du Monde"! I'm beginning to like this guy. Around this time, he managed to visit "Liechtenstein, where he considered requesting citizenship, on account of the fact that Liechtenstein didn't levy income taxes on foreign residents." Liechtenstein is the world's smallest German-speaking countries w/ one of the world's highest standards of living.

Nash & his wife Alicia were living in Paris. One of their most striking visitors "was Alexandre Grothendieck, a brilliant, charismatic, highly eccentric young algebraic geometer who wore his head shaved, affected traditional Russian peasant dress, and held strong pacifist views." Sounds like the Unknown Neoist to me!

By 1961, Nash was back in the US in the care of Trenton State Hospital. By 1970 he was back in the vicinity of Princeton, 'haunting' it & leaving mysterious messages on blackboards for students to find. Nasar provides these examples: "'Mao Tse-Tung's Bar Mitzvah was 13 years, 13 months and 13 days after Brezhnev's circumcision'", "'I agree with Harvard: There is a brain flat.'" Another example:

"Shortly after Hironaka won a Fields prize for his brilliant proof of the resolution of singularities, one of Nash's messages read:

The diagram on the previous page illustrates my own personal experience attempting to work w/ two extremes of the empathetic process. At the top -- these folks make the ethical connections but get caught in the sublime impossibility of acting in a world where each moment entails a no-win situation -- a counselor can help them -- these people are generally labeled "schizophrenic". At the other end are those who have zero degree empathy -- they tend to be beyond help -- these people are generally labelled "sociopaths"

$$N^5 + I^5 + X^5 + O^5 + N^5 = 0$$

Can Hironaka resolve this singularity?"

One part of this bk will be about my graffiti using this self-inking rubber-stamp:

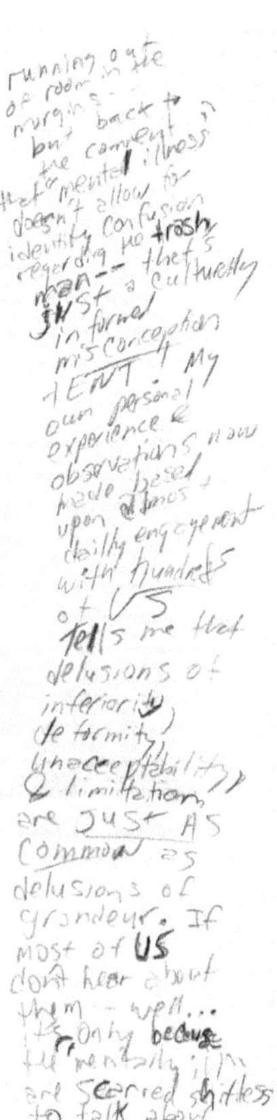

$$(\forall x)x = (\forall x)x$$

wch I call my "Anything is Anything" stamp. Reading these stories about Nash makes me feel like I've been in good company! a Beautiful Mind is certainly a thoroughly researched bk but I'd love to see an *entire* bk of *just Nash's blackboard graffiti* - w/ detailed & serious analysis of it from multiple perspectives.

Commenting on Nash's schizophrenic use of numbers, Nasar writes: "Margaret Wertheim, author of *Pythagoras' Trousers*, a history of numerology, has pointed out that 'people look to the order of numbers when the world falls apart.'"

By the early 1990s, Nash is sd to've recovered from his schizophrenia. "He wrote in 1995:

Gradually I began to intellectually reject some of the delusionally influenced lines of thinking which had been characteristic of my orientation. This began, most recognizably, with the rejection of politically-oriented thinking as essentially a hopeless waste of intellectual effort."

Tell that to John Von Neumann or Robert Oppenheimer! Or Einstein & Helen Keller! Or any politician! Hhmmm.. methinks Nash was still delusional, eh?! In other words, while his own "politically-oriented thinking" may've been "a hopeless waste of intellectual labor" because of his own personal ineffectualness along those lines, that doesn't mean that politically-oriented thinking in general shd be rejected. Thank goodness, the history of mathematics is full of outspoken pacifists & (free) radicals!

Somewhat tangentially, mention is made of Gödel's "well-known paranoia and terror of decision-making". Critics of his 2 incompleteness theorems cd have a field (theory) day w/ that!

By 1994, Nash was awarded the 'Nobel Prize in Economics' - strictly speaking, not really "a Nobel Prize, but rather 'The Central Bank of Sweden Prize in Economic Science in Memory of Alfred Nobel.'" I don't get it, I mean shdn't such a prize at least go to someone who works for a living? Nash certainly didn't for most of his life. Still, it's kindof heartening, maybe I'll get a Nobel Prize in literature for my bathroom graffiti (JUST KIDDING, OK!!).

Nash's receiving the prize had to do w/ his accomplishments in game theory. a Beautiful Mind explains that before 1994 the US government gave away business licenses for free, under the guidance of regulators, to companies for scarce public resources such as tv broadcast bands. This was considered inefficient, etc.. *SO*, auction rules were created by game theorists to revise the licensing situation - & a development of Nash's was at the center of it. I think I prefer Nash's blackboard messages.

The Man Who Loved Only Numbers - The Story of Paul Erdös and the Search for Mathematical Truth - Paul Hoffman. We're back to Pythagoras of Samos again. "[H]e was a vegetarian who refused to eat beans because they reminded him of testicles" - you have to wonder about a story like that! After all, he lived in the 6th century BC - did someone really chronicle his aversion to beans & did that story actually survive? Maybe so, if so, that's PDF (Pretty Damned Funny). I think I'll make up a story about Pythagoras here just for the fun of it:

Pythagoras's sandals were always coming untied. One day, as Pythagoras was seated lecturing members of his Brotherhood on the "Harmony of the Spheres" (some not very believable sources say that actually this notion is a malapropism for "Hominy on the Spears" - a reference to a threshing practice often used as a euphemism for "cornholing").

During the talk, Hippasus, always a mischievous lad, tied Pythagoras' laces together. The great philosopher, being wholly engrossed in deep discourse, didn't notice the boy's nimble fingers at work. At the conclusion of the lecture, Pythagoras stood up & started trying to walk & promptly fell flat on his face. Hippasus' friends' laughter immediately gave away who the culprit was & Pythagoras pissed on him in retaliation. Hippasus, deeply humiliated in front of the Brotherhood, then complained that Pythagoras was "acting irrationally" & sd he'd "get his number someday".

Pythagoras "considered 220 and 284 to be friendly. His notion of a 'friendly' number was based on the idea that a human friend is a kind of alter ego." "These numbers have a special mathematical property: each is equal to the sum of the other's *proper divisors* (divisors other than the number itself). The proper divisors of 220 are 1, 2, 4, 5, 10, 11, 20, 22, 44, 55, and 110, & they sum to 284; the proper divisors of 284 are 1, 2, 4, 71, and 142, and they sum to 220.

Delusions of grandeur are expressed more easilly because their communication offers a pay off & reward -- what would you rather admit to -- being worthy of the respect that's continuously denied you OR deserving being treated like the piece of shit that the others already think you are? (Well, that having been said -- I don't work with any folks who participate in Nash's circle of class priviledge -- still let's not generalize from him to US, eh?)

"A second pair of friendly numbers (17, 296 and 18, 416) was not discovered until 1636, by Pierre de Fermat." OOOOH!! Those wd be nice numbers to have tattooed on the tips of my thumbs & fingers! Back to Fermat again! What a guy for Paradigm Shift Knuckle Sandwichs *he was*!!

"Pythagoras saw perfection in any integer that equaled the sum of all the other integers that divided evenly into it. The first perfect number is 6. It's evenly divisible by 1, 2, and 3, and it's also the sum of 1, 2, and 3. The second perfect number is 28. Its divisors are 1, 2, 4, 7, and 14, and they add up to 28." I can certainly see the appeal of all this! What a neat & tidy system! What a feeling of divine revelation the Pythagoreans must've felt. Why, I'll bet Pythagoras didn't really pee on Hippasus after all!

Erdös pioneered an aspect of **combinatorics** (see the interview w/ Florian Cramer at the bottom of page 41 of the NAVIGATION: Language Experiment section re Raymond Lull to see why I'd **embolden** this) called Ramsey theory. "The idea behind Ramsey theory is that complete disorder is impossible. The appearance of disorder is really a matter of scale."

"Erdös was born in Budapest on March 26, 1913". "Throughout his life he would fearlessly defy 'Fascist' authorities of every stripe, be they armed thugs, mindless university bureaucrats, the U.S. Immigration Service, the Hungarian Secret Police, the FBI, Los Angeles Traffic Cops, or the SF [Supreme Fascist = God] himself."

"With anti-Semitism on the rise in the early 1930s, Erdös [who was Jewish], Vázsonyi, and a few other young Jewish mathematicians got together weekly". They "liked to meet in the city park at a bench by the hooded bronze statue of *Anonymous*, a historian who chronicled twelfth-century Hungarian kings. 'When we got together as a group,' Vázsonyi recalled, 'we were always concerned that the police would come and question us. Group meetings were prohibited during the Horthy dictatorship. We could not speak freely. We thought there were spies everywhere. That's when Erdös started developing his private language. Many of us were Communists in the sense of what it meant at that time: that we were against the Horthy regime.' But it wasn't safe to use the word *Communist* out loud, so Erdös started referring to Communists as people 'on the long wavelength,' because in the electro-magnetic spectrum the red waves were long. He said that Horthy supporters and other Fascist sympathizers were 'on the short wavelength.' That's when he started calling children and other small things 'epsilons' [the 5th letter of the Greek alphabet &, In mathematical notation, the minuscule open e symbol - its size probably being Erdös's reference here]".

the author on the Anonymous statue's lap in Budapest - 1997 - photographer: etta cetera

in 1934 Erdös left Hungary & went to Cambridge in England where he met G.H. Hardy, "the father of modern analytic number theory". Hardy had previously been the British sponsor of our old friend Ramanujan & much of the next section of The Man Who Loved Only Numbers is filled w/ wonderful stories about the "legendary Srinivasa Ramanujan Aiyangar, a poor, sickly, self-taught Indian".

Hardy tells of Ramanujan's being able to "'remember the idiosyncrasies of numbers in an almost uncanny way. [..] I had ridden in taxi-cab No. 1729, and remarked that to me the number seemed a rather dull one, and that I hoped it was not an unfavorable omen. 'No,' he [Ramanujan] reflected, 'it is a very interesting number; it is the smallest number expressible as the sum of two cubes in two different ways." Ramanujan had seen that

perhaps equivalent to the semantic operators "this" & "that" - whose degree of semantic differential allows them to refer either to differences of perspective or differences of observed object.. making them wonderful words to use in poetry.

$$1729 = 12^3 + 1^3 = 10^3 + 9^3"$$

Ramanujan, sometimes w/ Hardy, worked on so many things that I hesitate to do more than list them here lest this section just turn into a giant descriptive quote: asymptotic formulas, highly composite numbers, round numbers, partitions, etc..

By 1938, Erdös moved to the US; by 1943, he attempted to join the Los Alamos secret atomic weapons project. "'Unfortunately, he refused to sign a paper saying that he would not talk about the A-bomb after the war." He was rejected. I admire his honesty.

Right in the midst of my notes re Erdös, I find a P.N.T. proposal of Special Instance Subtractive Commutativity. I have no idea of what use this cd be (or to what degree such a thing has already been discussed elsewhere) but the basic idea is this:

In certain special instances,
$(m - n = n - m)$ -> to x

Now, I can only think of *one* instance when this wd hold 'true': when m & n are the same number. However, my intuition tells me that there're probably complex numbers that cd be substituted for m &/or n. But, then, what do I 'know', eh?! & is that "-> to x" really necessary? Nope. *Unless*, what $(m - n = n - m)$ is isomorphic to is some sort of exception, some sort of 'untrue' or *qualified special* statement. Eg: (6 - 5 [1] = 5 - 6 [-1]); 1 = -1 -> to x. x then becomes some sort of number-line mirror image in wch things can only be 'equal' if they're equidistant in opposite directions from zero. Hence, the "-> to x" is useful *after all* insofar as it requires that a special instance be conceived of.

Gödel pops up again. Math bks are a regular soap opera of fascinating characters. This time, I got to read that "Gödel was not a poster boy for mathematical sanity. Obsessed with ghosts and demons and an imaginary heart ailment, [..he refused to take] food from anyone other than his wife Adele, fearing that other people were secretly trying to poison him. [..] In the middle of 1977, when Adele was hospitalized for major surgery, he stopped eating altogether, and by the following January starved himself to death at the age of seventy-one." Sheesh! & to think that this guy & Nash were both married & I haven't gotten laid for over a yr! Life ain't fair.

Still, Gödel had a profound effect. Erdös didn't much care about incompleteness - he had plenty of things to work on w/o getting upset about hypothetical ones that, in theory, cd never be solved. Bertrand "Russell, on the other hand, was crushed: 'I wanted certainty in the kind of way in which people want religious faith.'"

<u>The Man Who Loved Only Numbers</u> is alotof fun. It's racy. I learned from it that Fibonacci didn't figure out the numbers that he's associated w/ - that didn't happen until 6 centuries after his time! He just posed a simple calculating problem about rabbit reproduction in an enclosed space that generated the series now known after him! I wonder what Fibonacci Numbers complicated by a rabbit-proof fence are.

careful [illegible] the delusional are not lovable? certainly a lot of baggage there — most of my coworkers in the mental health profession inevitably raise their eyebrows when they find out a client is engaged in a relationship. My response is invariably: "That's part of being human isn't it? So what are you implying?" (see my note on the Conservative function of humor in the Cognative Dissidents

Page 204: Ronald Graham, a mathematician at AT&T: "'Some areas of mathematics are so obscure or so new that people in those areas have no one to share their results with.'" Indeed. & what do we make of that? Just as in other creative endeavors, there must be amazing things being accomplished by some people out there that're going completely unnoticed. Not only do people working in unpopular & obscure & new fields have to develop the field & create something substantial w/in it but they have to also *market* it in order for it to not fade out w/o ever being significantly noticed. Flashy & urbane mathematicians (& d composers & what-not) that're good self-sales(wo)men (or just plain lucky) are in the minority.

Carl Friedrich Gauss, born in 1777 in Germany, is another frequent character in the math soap/soup. A story told about him is that when he was 10 he was posed a problem by an intimidating teacher that he, & his classmates, were expected to very slowly solve. The problem was to find the sum of all the whole numbers from 1 to 100. Gauss realized that he cd get a correct result by mentally vertically stacking the numbers in forward order over them in reverse order & then adding the columns:

$$\begin{array}{l} 1 + \;\; 2 + \;\; 3 + \;\; 4 + \ldots + 97 + 98 + 99 + 100 \\ 100 + 99 + 98 + 97 + \ldots + \;\; 4 + \;\; 3 + \;\; 2 + \;\;\; 1 \end{array}$$

Since each vertical stack produced a sum of 101, he 'simply' needed to multiply 101 times the number of pairs (100) & then divide it in half (since each pair is repeated). The gist of it is that Gauss's insight is the insight of pattern recognition - pretty much the insight of every mathematician/analyst. Recognize a pattern, recognize its significance, apply it. Most importantly, perhaps, one needs to be able to be flexible enuf to *play*, to *not* accept things in their given order. & *that*, Nash be damned, is *political*.

Apparently, I 1st ran across the number *e* in this bk. *e*'s a number that eventually factors into my Paradigm Shift Knuckle Sandwich. I read an entire bk about it. That'll be navigated thru later. But, before I explain *e* (much later), I shd explain a few other things: "Real" numbers are numbers that can be represented in infinite decimal form. They include 0, 1, 2, 3, ... in both negative & positive; "Rational Numbers": any number that can be expressed as a ratio - such as 1/2; & "Irrational Numbers": numbers that *can't* be expressed as ratios but only as non-repeating decimals.

Now, not being even an entry-level amateur mathematician, what I wonder is: What is a negative number divided by a positive one & vice-versa? Then, in this mod-a-go-go day & age, I ask WikiAnswer online &

Discussion groups Just because he's got ghosts & demons doesn't mean that they don't serve some useful purpose — however indirect. My Constructivist approach would look at both what the delusions are doing against him & what they are doing for him. Without being a proponent of madness, I do recognize it as a form of internal dialog & attempt at resolution -- even if mediated by unconscious forces which appear on the surface to be singularly destructive.

online & I'm told that the number is negative in both cases. I'm disappointed. I also don't get it. The explanation of this must be intriguing. I can understand dividing an absence by a presence: in other words, 1/2 of negative 22 = negative 11. Fine. But how do you divide a presence by an absence?

If we have 10 apples being divided between 10 horses, each horse gets one apple (in a hypothetically functional communist society) or one horse gets 9 apples & the other 9 horses fight over the remaining apple (in a simplified version of a typical capitalist society) or whatever. the 1st solution being the "Happy Solution" (if we can have "Friendly Numbers" why not "Happy Solutions"?) [That doesn't seem to be a pre-existing term so I reckon I can take provisional credit for this one!].

BUT, [Yeah, yeah, don't begin sentences w/ "but" - that's ass-backwards - *BUTT*, in math, isn't ass-backwards forwards? Aren't 2 negatives a positive?] what're 10 apples divided by an absence of 10 horses? 1 rotten apple per horse? 10 apples that get eaten by 10 something-elses? & what's the Happy Solution? & is there an "Unhappy Solution" (the "Final Solution", of course, is a prime candidate for the Unhappy [or, as its otherwise unknown, the "Miserable"] Solution)?

Right, the operation itself seems intuitively "illegitimate" & the transposition of the act of division into multiplication seems to be merely a rhetorical move that confuses negative & positive with the inverse operations of these two processes. Somehow it seems more conceptually legitimate in terms of addition & subtraction — perhaps because we are all so familiar & used to the idea of debt.

Let's try again: Using the analogies of commerce that much of this math is rooted in, a horse pays a grocer for 10 apples because it's having a party for its 9 friends; unfortunately, the grocer only has one apple & promises to deliver the other 9 apples to the stable before the party starts; as such, the 9 apples are *negative* apples until the grocer does something positive about them. Fine.

Let's ignore all the problems - such as: How does the horse communicate its need? Does it use Francis the Talking Mule as an intermediary? What does the horse pay w/? Is the grocer a crook? What court will recognize the horse's demand for reparation? What court will recognize Francis the Mule as an attorney? In what countries will Francis & the horse be imprisoned as dissidents for shitting in the courtroom? Will the US government convince its peoples thru the Fox TV network that torturing Francis & the horse is the only way that the good guys can find out whether they were planning to throw the apples at the president? Right. Ignore all that.

Ok, I can't resist, I go to WikiAnswers & ask: "If Francis the talking mule represents a horse in a courtroom dispute with a grocer over non-delivery of apples will they be imprisoned?" & I get this response:

Step 1: Is one of these your question?

button: No, skip to step 2

If you dispute a debt from a collection agency and send it with delivery confirmation does the 30 days start from the date they received the letter?

button: NO, but show me this one; button: YES, this is my question

Why does the movie name refer to 'Two Mules' in the Clint Eastwood classic Sister Sara gets rid of her mule for a burro. Hogan has only one mule?

button: NO, but show me this one; button: YES, this is my question

What is the difference between a jack mule and a john mule?

button: NO, but show me this one; button: YES, this is my question

Western comedy where he threatens to trade his mule for a real horse and at the end comes riding into town on a little pony do you know the name of the movie?

button: NO, but show me this one; button: YES, this is my question

Can a horse eat too many apples?

button: NO, but show me this one; button: YES, this is my question

There! Wasn't that fun?! Better than actually trying to solve the problem?! How can mathematicians resist going off on *tangents* like this *in perpetuity*? But I'll knuckle down here:

How do I even express this? 10 nonexistent horses divide up 10 apples? 10 horses owed to the stable owner eat 10 actual existing apples in the mind of sd stable owner? Is that like counting yr chickens before they get to eat their own shit on the assembly line? Was that a "pataphor"? No, because the sentence included the word "like" - making it a simile. As far as I'm concerned, if the horses don't exist (or, more appropriately perhaps, are the *absence* of 10 horses that *shd've* existed), then the apples are still positive because the horses can't eat them!

I'm serious, I really don't get it. If only this bk had some ***MARGINALIA*** explaining it to me (this is where I really start to get tricky). The stable owner buys 10 horses from the grocer, who's now expanded his business to being a supermart chain store of some sort. The supermarketeer takes the stable owner's money & promises to deliver the horses before the 10 apples rot. At this point, for the stable owner at least, the 10 horses are represented by *negative numbers*. For that matter, so are the 10 apples; after all, the horse that's going to buy the 10 apples for his or her 9 friends (*see*, we don't even know their sex yet!) doesn't even exist yet so how can it be ripped off by the supermarketeer? This whole situation is getting very *unstable*. We'll call this the "Unstable Problem".

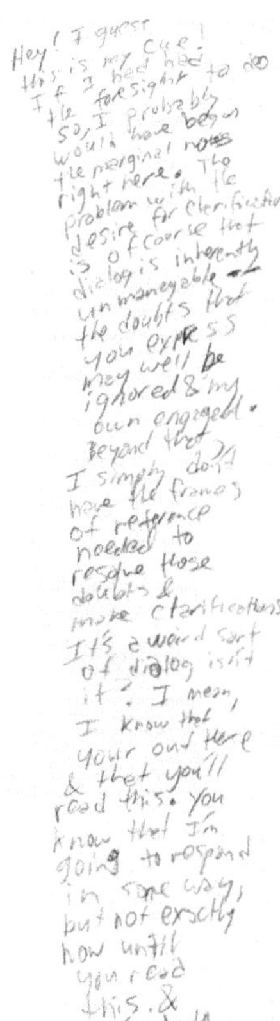

One version of the Unstable Problem: A stable owner buys 10 horses from a supermarketeer. The supermarketeer doesn't have the horses at the time of sale so the horses are *represented by negative 10.* The stable owner really does have 10 apples for the horses - these are represented as *positive 10.* What is 10/-10? According to mathematics, the answer is -1. The apples have been whisked out of existence by some horses that aren't even there yet. Is this where Hofstadter's *supernatural numbers* come in?

To quote a quote quoted on page 38 of <u>Paradigm Shift Knuckle Sandwich</u> (the bk you're reading/writing, *dummy*, unless you're too busy watching some Bruce Willis movie or something!): "True to this tradition, we shall name the numbers which ~G is announcing to us the *supernatural numbers*". "~G" being the negation of "the Gödel sentence G [..] written in English as 'This statement of number theory does not have any proof in the system of *Principia Mathematica*.''"". I shd mention that I'm a bit confused here because I understand Hofstadter's use of the tilde ("~") to mean "not" but ordinarily ~ means "equivalent to" *except* in Propositional Logic where it means "not". Even though I'll be phrasing statements as propositions (by using the "if" framework ("<>") I'll use "¬" as my symbol for "not". Now we're getting *deep*, huh? I'm writing over my own head again (& getting myself in *deep shit* w/ people who actually understand this stuff, like Hofstadter).

Let's say that "¬H" = "not-Horse" (meaning the negative of a horse). This ¬H is *supernatural* or, at least, *announces the supernatural*. ¬H is the negation of 'This statement of Perverse Number Theory does not have any proof in 'Pataphysics." That means it's unexceptional since 'Pataphysics is the science of exceptions. Let's say that the *unexceptional* is a *negative concept* represented by ¬E - ie: not Exceptional.

¬H is the negation of ¬E. When I asked WikiAnswers: "What is the mathematical symbol for is the negation of?" one of the possible questions I got in reply was: What does the warning symbol on your hoda intigra means it is under the oil symbol rectangle with like fumes three letter s's like on top?

Naturally, I just *had* to find out the answer to *that* one even if it wasn't the question I was asking & the answer *isssssssssssssss* (drum roll please): a girl named sara rocks hahahahahaha and is better thatn u

Another possible question that I got back from WikiAnswers was: **What is negation logic?** That seemed potentially relevant so I clicked for the answer & got: **This Question Has Not Been Answered Yet**.

IN HONOR OF THIS, I HEREBY ANNOUNCE, THE "cONVENIENCE'S 1ST & LAST THEOREM PRIZE":

I WILL TAKE OUT TO (A FAIRLY CHEAP BUT NOT TOO SHABBY) DINNER WHOEVER ANSWERS THE QUESTION:

Does negation logic help solve the problem of whether the warning symbol on your HODA INTIGRA flashes when supernatural horses are about to eat Sara [Sara being the positive number here]?

Lardy! being an 'almost completely unknown writer who doesn't even have more than 5 friends who'll ever read his bks' [I wonder what the mathematical expression for *that* is?!] is fun! But, I digress.

It's time to bring out the heavy hitters: let's add Special Instance Subtraction Commutativity to the Unstable Problem. In order to express "¬H is the negation of ¬E", I'll harken back to "(m - n = n - m) -> to x".

aware that there will be other readers-- that we are writing for them... but it's not quite a dialog... I don't even know how much of this marginalia will be included in the final outcome. & I have the advantage as the second writer-- "the post-textual writer"?-- in that I can address these three entities-- I can recognize you, our audience, & myself-- whereas your address is confined to the first two of those relationships.

In case you don't remember where this was explained, subtract $\sqrt[5]{3{,}125}$ from $\sqrt[61]{8.03748056 \times 10 \text{ to the 108th}}$ to find the relevant page number. [C'Mon! That's an *easy* one!] Thus we have:

$$\neg H \; x \; \neg E$$

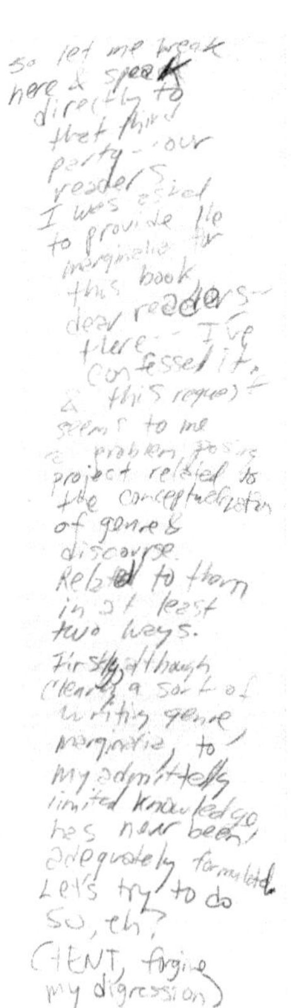

Elegant, eh? To recapitulate: in English, "$\neg$H *x* $\neg$E" might mean: "Not-Horse is the mirror image on the negative side of the number line of the positive representation of the unexceptional." Simplifying that in standard formulaic reductionist manner by equating the negative reflection of the unexceptional w/ the *exceptional*, we end up w/: "Not-Horse is *exceptional*". *Therefore*, while under usual circumstances a non-existent horse cannot negate an apple, this *supernatural horse* is an *exception* & can do whatever it damn well pleases - such as eat a 'real' apple. Get it? That's why a positive number divided by a negative one is negative.

Alright, alright, I realize that I've fudged all sorts of things to get this result. Eg: I started w/ 10 horses *dividing up* 10 apples, because the initial problem is a division problem, but then I use alotof dubious intermediary steps to change that to *subtraction*: I have the 10 horses divide the apples so that they can eat them. This leads to the final result of the ~H being able to *eat/negate* the apple but it still doesn't really explain the division. Back to the drawing board. To reinterpret:

"$\neg$H *x* $\neg$E" translates into English as: "A negative horse is the mirror image on the number line of a positive apple. The negative horse goes into the positive apple a negative amt of times because the positive apple is inviolate by the negative horse regardless of whether it's its mirror-image or not. SO THERE!fore, the result of the division is a negative result because the total doesn't exist." Does that work for ya?!

The Slap-Happy Solution:

$$<(m - n = n - m) -> \text{to } x \supset x \neg E/\neg H \supset \neg E \text{ € } \{\neg H\}>$$

If we apply the Special Instance Subtraction Commutativity assertion that a negative number is the mirror image of a positive number on the number-line by specifying it as the relationship between an unexceptional apple & a negative horse with the negative horse as the divisor & an unexceptional apple as the dividend then the unexceptional apple belongs to the set of negative horse.

In other words, the damned horse eats the apple & brings it over to its side of the fence (w/ the "fence" here represented by the slash ("/") of the division & otherwise representing the zero ("o") of the number-line division between negative & positive numbers) & into its "set" ("{}") - otherwise known as its stomach. There we have Perverse Number Theory in a nutshell.. or a stomach lining.

But I've been neglecting The Man Who Loved Only Numbers. Much of what my notes in this bk refer to are things that're great stories but that don't fit into my stream-of-(un)consciousness (a 20th century mathematician getting trepanned, Erdös getting off on a jaywalking charge because he had one of his math bks w/ him) or mathematical territories that are interesting to me but that I have no use for here & that're better explained elsewhere so why mention them? (algorithms, complexity theory, regular & irregular primes, place-value systems, calculus, etc..)

It is a kind of writing that goes unread— except by it author & excepting exceptions — which of course there always are — this piece being an example of such. It constitutes a formal heterotopia — I can write in any mode or manner as I see fit as I shall be the only reader of these Marginal notes (sic.) Heterological, fleeting, & unpredictable reactions (Note my comments on tENT's comments regarding the "mentally ill.") to the engagement of text, unbound by conventions of narrative writing or expository analysis—

There is one bit though that *has* to be here & that's the subject of grouping infinite series. An infinite series is an expression like this: IS = 1 + 1/2 + 1/4 + 1/8 + ... The dots mean that infinitely many terms in the implied progression follow. Therefore, one 'knows' that after 1/8 there's 1/16 & then 1/32 etc because the denominator (divisor) is progressively doubling. Since the infinite extension of this can't be written, it's just notated by 3 dots. (Parenthetically, everytime I see the word "can't" I try to imagine how it *cd* be possible.) Hoffman provides this example of how grouping can work:

"And then there are simple-looking infinite series that seem to misbehave entirely. Take a look at the series

1-1+1-1+1-1+1-1+1-1+...

What does it add up to? Grouping the terms one way,

(1-1)+(1-1)+(1-1)+(1-1)+(1-1)+...

the sum is definitely 0. But if the terms are grouped another way,

1+(-1+1)+(-1+1)+(-1+1)+(-1+1)+...

the sum is definitely 1."

There's a very slight & 'perfectly valid' shift taking place here. In the 1st grouping, parentheses just reorganize the already existing notation. In the *2nd grouping* the notation is restated: "1-1" becomes "1+(-1)". It means the same thing but enables an entirely different outcome. *Therefore,* it *doesn't* mean the same thing. It means the same thing. "But" enables an entirely different outcome. A whole branch of P.N.T. shd look into this.

Wch brings me to varying degrees of infinity. What a delicious idea. To determine the degree of infinity of a group of numbers, a test of one-to-one correspondence is tried. Georg Ferdinand Ludwig Philipp Cantor, a German mathematician born in 1845, pioneered this. To demonstrate that there's a one-to-one between counting numbers & their squares, he presented this (or some variation thereon):

1	2	3	4	5	6	7	...
↕	↕	↕	↕	↕	↕	↕	
1	4	9	16	25	36	49	...

See the logic? You can imagine that for every number on the top there's a corresponding number on the bottom. Therefore, each of these infinite sets (counting numbers, the squares of counting numbers) is equal in size. Cantor called this size *aleph-null* - aleph being the 1st letter of the Hebrew alphabet. The symbol for this being $\aleph_0$. The $\aleph_0$ size is also called *denumerable* or *countably infinite*. Aleph-numbering is used to refer to the different sizes of infinities. A larger infinity than *aleph-null* is, therefore, *aleph-prime* aka *aleph-one*. Cantor got trickier w/ his demonstration that there's a one-on-one between counting numbers & fractions:

unfettered by the rigor of scholarship, unhindered by concern that anyone might give a damn, take offense, or retract publication (sic again of course on my part here). So marginalia is always already a genre (?) which has much in common with tENTS "mutant pastiche" approach to violating the strictures of generic & discursive codes here. Marginalia is perverse discourse theory in question -- well.. maybe to every

1/1	2/1	3/1	4/1	5/1	...
1/2	2/2	3/2	4/2	5/2	...
1/3	2/3	3/3	4/3	5/3	...
1/4	2/4	3/4	4/4	5/4	...
1/5	2/5	3/5	4/5	5/5	...
...	...	...	...	...	

1/1→2/1	3/1→4/1	5/1→...
↙ ↗	↙ ↗	↙
1/2 2/2	3/2 4/2	5/2 ...
↓ ↗	↙ ↗	↙
1/3 2/3	3/3 4/3	5/3 ...
↙ ↗	↙ ↗	
1/4 2/4	3/4 4/4	5/4 ...
↓ ↗	↙ ↗	↙
1/5 2/5	3/5 4/5	5/5 ...
... ...	... ...	...

In the organization on the left, it appears that there are infinite rows of infinities w/in the fractions &, thusly, no one-on-one between the top row (the counting numbers) & those rows beneath it. Cantor figured out a way around this. By skipping over the ratios that repeat numbers (2/2 is one again, 3/3 is one again, 4/2 is two again, 9/3 is three again, etc..) & following the arrows, a linear path that enables one-on-one is created. 1<->1/1, 2<->2/1, 3<->1/2, 4<->1/3, 5<->3/1, 6<->4/1, 7<->3/2, 8<->2/3, 9<->1/4, 10<->1/5, etc..

Ingenious.

Of course, we here, the dummy at the HQ for PNT have to ask: What happens if we allow the repeating expressions to remain? If we don't skip over them? What if: 1<->1/1, 2<->2/1, 3<->1/2, 4<->1/3, ***but 5<->2/2***?, etc..? We're still following a navigable one-on-one path between counting numbers & fractions ***but*** now the same numbers are repeated in different expressions. Isn't this akin to the grouping tricks a few pages back that led to an infinite series equalling one & *also* equalling zero?! "It means the same thing but enables an entirely different outcome. *Therefore*, it *doesn't* mean the same thing. It means the same thing. "**But**" enables an entirely different outcome."

In this new allowed correspondence, five then corresponds to one - a member of its own group but then we already have *one* corresponding to *one* - a member of its own group! So if one can correspond to 1/1 w/o our being bothered by its just being a different expression instead of a new number altogether, why can't three, eg, correspond to 3/1 (ordinarily skipped over) *instead of 1/2*! What happens if we do this? Then we no longer have a one-on-one correspondence between the counting numbers & *all* fractions.

If we have 1<->1/1, 2<->2/1, 3<->3/1, 4<->4/1, etc; then we have a new one-on-one that eliminates all the other fractions that aren't just different expressions of the counting number. Of course, a way around this is to just not allow any repetitions from one group to the other. As such, 1 can no longer correspond to 1/1, 2 can no longer correspond to 2/1, 5 can't correspond to 3/1, 6 can't correspond to 4/1, etc.. Instead we just have 1<->1/2, 2<->1/3, 3<->3/2, etc..

That seems more correct to me. Checking online (<http://www.jimloy.com/math/trans.htm>) I found this instead:

```
1/1  1/2  1/3  1/4  1/5  1/6 . . .
2/1  2/2  2/3  2/4  2/5  2/6 . . .
3/1  3/2  3/3  3/4  3/5  3/6 . . .
4/1  4/2  4/3  4/4  4/5  4/6 . . .
. . .
```

We can count these merely by (1) starting in the upper left corner, (2) move right, (3) move down, (4) move left, (5) move down, (6) move right, (7) move right, (8) move up, (9) move up, (10) move right, etc. We get this list of fractions: 1/1, 1/2, 2/2, 2/1, 3/1, 3/2, 3/3, 2/3, 1/3, 1/4, etc. And so the rational numbers are countable, merely because we can count them.

exceptional genre there is always a non exceptional exception which disproves the exceptional rule ... if you follow my drift. I suppost Fermat -- who instigated practically eg. entire tradition of solutional problem solving within mathematics would be such an exceptional except to the exceptional exceptionality of the genre of marginalia. His was hardly a marginal textuality at all. Anyway, & before I lose my train of thought.. on to the Secondly Do you recall that a "secondly"?

See the difference? The rows & columns are switched. He *doesn't* go diagonally (he sortof zig-zags) & *doesn't* eliminate repetitions. Maybe this guy is 'wrong'. Looking at another web-site's explanation (<http://www.homeschoolmath.net/teaching/rational-numbers-countable.php>), Hoffman's version is repeated.

The point here is, if grouping in infinite series can lead to different conclusions, I wonder if different arrangements of the path thru a one-to-one correspondence chart & different rules about excluding or including fractions that equal the counting numbers makes any difference in the infinity size classification or in any other aspect of it?

As is, we have 3 $\aleph_0$s involving fractions - one that includes repetitions (the Jim Loy one), one that doesn't (the Hoffman one), & one that excludes anything not expressed in its lowest form (mine). Of course there can be an infinite amt more. We cd, eg, create correspondences like these: 1<->1/1, 2<->2/2, 3<->3/3, ... or 1<->1/2, 2<->1/3, 3<->1/4, ... or 1<->1/2, 2<->2/3, 3<->3/4, ... All of these are valid infinite correspondences in wch the progression is clearly recognizable. It's just a matter of giving each group of fractions a name & designating them all *countable infinities*.

However, it's the 1st 3 mentioned that I'm concerned w/. What happens if we represent reducible fractions in their lowest forms? I'll modify & expand Loy's diagram to make this clearer:

001	**1/2**	**1/3**	**1/4**	**1/5**	**1/6**	1/7	1/8	1/9	1/10	1/11	1/12	. . .
002	**001**	**2/3**	**1/2**	**2/5**	**1/3**	2/7	**1/4**	2/9	**1/5**	2/11	**1/6**	. . .
003	**3/2**	**001**	**3/4**	**3/5**	**1/2**	3/7	3/8	**1/3**	3/10	3/11	**1/4**	. . .
004	**002**	**4/3**	**001**	**4/5**	**2/3**	4/7	**1/2**	4/9	**2/5**	4/11	**1/3**	. . .
005	5/2	5/3	5/4	**001**	5/6	5/7	5/8	5/9	**1/2**	5/11	5/12	. . .
006	**003**	**002**	**3/2**	6/5	**001**	6/7	**3/4**	**2/3**	**3/5**	6/11	**1/2**	. . .
007	7/2	7/3	7/4	7/5	7/6	**001**	7/8	7/9	7/10	7/11	7/12	. . .
008	**004**	8/3	**002**	8/5	**4/3**	8/7	**001**	8/9	**4/5**	8/11	**2/3**	. . .
009	9/2	**003**	9/4	9/5	**3/2**	9/7	9/8	**001**	9/10	9/11	**3/4**	. . .
. . .												

The non-underlined **bold** numbers are 1st appearances; the underlined **bold** numbers are repeated appearances reduced to their lowest form. Numbers like "005" & "10/5" (wch wd be reduced to "002") wd, of course be **bold** if we were to continue the diagram further. I only point out patterns w/in the diagram's limits. In all 3 1st one-to-one's presented, particular paths result. In the following diagram all numbers reducible to non-fractional form & all repeating numbers are removed:

Wes coming? If not, please refer back to page $\sqrt[62]{1.3464565 \times 10 \text{ to the 111th}}$ Secondly, simply by asking me to write this marginalia, tENT has created a situation & context in which the "rules" of the genre of marginalia cannot apply, challenging its sense of self-unawareness, its strictly personal responsiveness, its fleetingness as unreproduced textual object, its "aura" to use Benjamin's phrase, its unique material essence.

1/2	1/3	1/4	1/5	1/6	1/7	1/8	1/9	1/10	1/11	1/12	. . .
	2/3		2/5		2/7		2/9		2/11		. . .
3/2		3/4	3/5		3/7	3/8		3/10	3/11		. . .
	4/3		4/5		4/7		4/9		4/11		. . .
5/2	5/3	5/4		5/6	5/7	5/8	5/9		5/11	5/12	. . .
			6/5		6/7				6/11		. . .
7/2	7/3	7/4	7/5	7/6		7/8	7/9	7/10	7/11	7/12	. . .
	8/3		8/5		8/7		8/9		8/11		. . .
9/2		9/4	9/5		9/7	9/8		9/10	9/11		. . .
. . .											

In the next diagram, lines are drawn thru adjacent empty spaces & dots are marked in empty spaces w/o any adjacent partners:

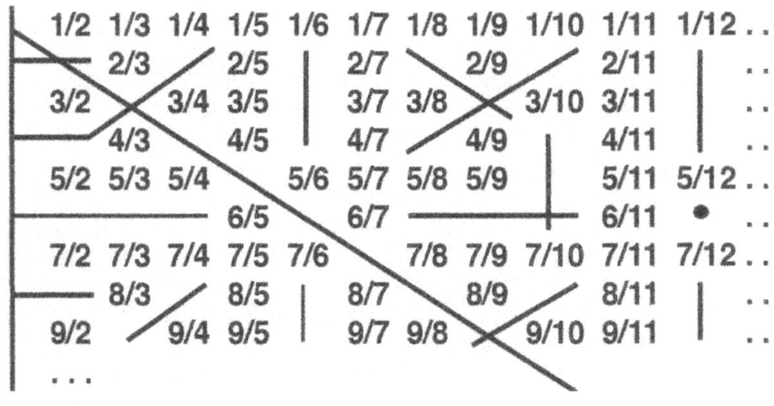

Obviously, certain predictable patterns appear. A continuous vertical line runs down the left side where the counting numbers reduced from fractions wd've been. A continual diagonal line runs from the left corner where all the "1"s wd've been. In the 2nd column (now the 1st w/ numbers), every other fraction is removed. In the 3rd column (now the 2nd w/ numbers), every 3rd number is removed. The reducibility (&, therefore, removability) of the numbers depends on whether the numerator (dividend) & denominator are divisible by the same number. A fraction in lowest terms with a prime denominator other than 2 or 5 (i.e. coprime to 10) always produces a repeating decimal.

In effect, it is a challenge to produce a discourse genre outside the range of its limits. Paradoxically, these limits where already within the limited of what is marginal to genre to begin with. Still following? Good! Then on to the task -- how does it unfold? what level of dissimulation will it seek? Another paradox: If it decides to be "outside" the limits of marginalia, will it once again return to the center? To return to tENT'S thinking once again . . .

Then, we remove the numbers altogether:

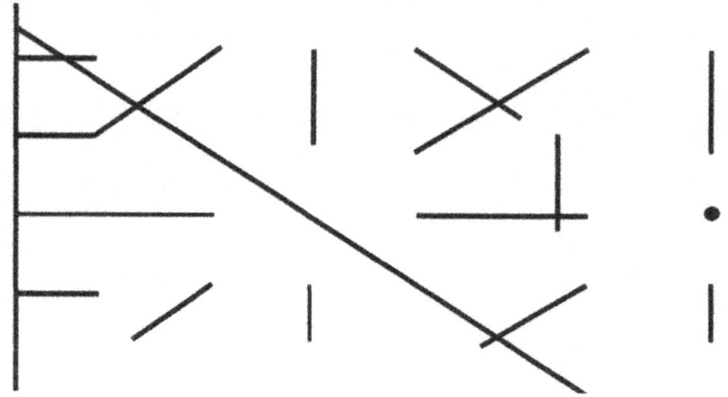

Keep in mind that the Fibonacci Numbers (numbers that add together the previous 2 numbers in their sequence: 0, 1, 1, 2, 3, 5, 8, 13, 21, 34, 55, 89, 144, 233, 377, 610, 987, ...) have parallels in nature. Their relationship to the reproduction of rabbits has already been noted. According to <http://www.mcs.surrey.ac.uk/Personal/R.Knott/Fibonacci/fibnat.html#petals>, "On many plants, the number of petals is a Fibonacci number:buttercups have 5 petals; lilies and iris have 3 petals; some delphiniums have 8; corn marigolds have 13 petals; some asters have 21 whereas daisies can be found with 34, 55 or even 89 petals." &, of course, this is *not* a coincidence but is, instead, a clear clue as to the structural underpinnings of nature.

Notice that when the above pattern, wch I call the "Flatland", is superimposed on this random selection of photos from the internet, that the only place where text appears has lines running thru it wch, when read by connecting the letters crossed over from left-to-right & then from top-to-bottom (since this is in English), produces the word "SEER". Now notice that in the 2nd random example, the letters crossed over spell out "DEAD ED". There's no doubt in my mind that this is a reference to Edwin Abbott Abbott.

Is a double negative not not necessarily a positive? What I'm wondering right now is how you, dear reader would take these marginal notes if they at least had the appearance of something belonging inherently to this publication? What if for instance they looked like the following (perhaps ignoring my use of scotch tape):

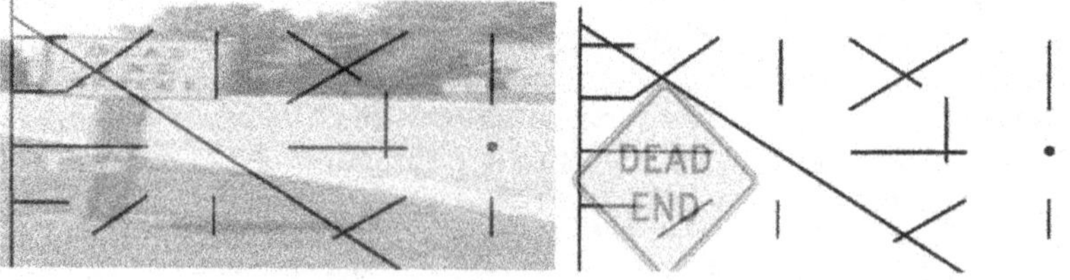

"Cantor was also taken with the weird arithmetic of aleph-null, $\aleph_0$ the countably infinite. For example:

$$\aleph_0 + 1 = \aleph_0 \text{ and } \aleph_0 + \aleph_0 = \aleph_0"$$

Why stop there? **Cantor and Hilbert sitting in a tree; CAL-CU-LATE, I, N, G; First comes $\aleph_0$; Then comes the Poisson boundary of random rational affinities; Along comes baby in different degrees of infinities!** Ok, ok, Hilbert came after Cantor - I just drop his name because he was very enthusiastic about Cantor's work. As for the "Poisson boundary"? I have no idea what it is but I wanted some math thing to rhyme w/ "infinities" & "affinities" was close enuf.

Cantor started working on "real" numbers (numbers expressible as decimals) & decided that they constitute a larger size infinity that he called *aleph-one.* As a demonstration of the non-correspondence between reals & counting numbers, Cantor proposed "listing all the reals between 0 and 1 in no particular order":

.12146789 . . .	1
.32769234 . . .	2
.71234568 . . .	3
.43567233 . . .	4
.645946784 . . .	5
. . .	. . .
. . .	. . .

"According to Cantor's diagonal argument, construct a new number from the digits in boldface, .12264..., by replacing every single digit by a different digit of your choosing. For example, you could add one to each digit to produce the number .23375... This new number is clearly a real number so it should appear somewhere on the list. But it can't be anywhere on the list because, by its very construction, its first digit differs from the first number on the list, its second digit differs from the second number on the list, its third number differs from the third number on the list, and so on. Thus, the list of reals isn't complete, there are more reals than counting numbers to pair them with. Cantor had found a larger infinity" What's the sound of me clapping for Cantor in text version?

Thanks to Wikipedia, I then learned a little about Beth two: $\beth_2$:

⌐A genre is a set of determinable rules.⌐ This statement is either true or false or neither indecidable & unpredictable nor prepostorous but not theoretically unsound nor uninteresting or simply syntonic with a formalist conceptualization of symbol & code systems while affirming or disconfirming its own contradiction in the hint of irony set within the simplicity of its syntax--however not quite either/nor

C.F. Kramer,
Edna. E.
Princeton 1970
PP. 332-333

also referred to as 2c (pronounced two to the power of c).

Sets with cardinality [**cardinal numbers**, or **cardinals** for short, are a generalized kind of number used to denote the size of a set, known as its **cardinality.**] Beth ["Beth", of course, being the pronunciation of the 2nd letter of the Hebrew alphabet (*alpha-beth*)] two include:

- The power set [given a set *S*, the **power set** (or **powerset**) of *S*, written *P*(*S*), or 2*S*, is the set of all subsets of *S*.] of the set of real numbers, so it is the number of subsets of the real line, or the number of sets of real numbers
- The power set of the power set of the set of natural numbers

etc, etc.. It seems so predictable that there'll be an infinite amt of infinities that I *almost* wish that someone wd prove that there *aren't* an infinite amt but, rather, something like an infinitely repeated decimal expression's amt (or whatever). *&/OR* that something even more mind-boggling will come along - like proof that infinity is really only embodied in the navel of a woman named Beth who had an oven named after her at the steel mills & subsequently a street named "Bethoven" wch people assumed was a misspelling of a famous composer's name.

"One of Erdös's most fundamental contributions to mathematics was to come up with a powerful new form of an existence proof ["demonstrating the existence of something without being able to display that something"] called the *probabilistic method.*" [..] "Erdös in effect flipped a coin to decide whether each pair of people are friends or strangers." We remember Friend(ly) Numbers, right? Has anyone ever heard of "Strange Numbers"? I found this post online by someone named "Turtle": "*Begin with the idea of perfect numbers;OK? Now if a number's divisor sum equals that number it's perfect, if the sum is less it's called deficient, & if the sum is more it's called abundant. So all Strange numbers are abundant by exactly 12(twice the first perfect number six) The set of Bizarre numbers abundant by exactly 56 (twice the second number 28). The sets' discovery I believe is original to me; I have never seen them described*" Are Stranger Numbers *stranger* than Strange Numbers? Are there Perfect Strangers whose "number's divisor sum equals that number"?

"The idea of flipping a coin to help solve mathematical problems was disconcertingly radical in a field prized for its precision, but it is now commonplace in computer science. The irony is that Erdös, who shunned computers, made a major contribution to the theory of computing."

not yet if your standard is a good bet (the string of niethers & nors, identities, etc. etc.-- perhaps all more than likely no more than the simple series ┘1 75185 919 19520 156 452051813914 12125 182112519.┐

Sentence A does not necessarilly lead to Sentence B. For which we are sentenced to Sentence C-- once again etc. or ad infinitum. (One cannot avoid the temptation to seek the pun: Add Infinitum-- a sentence which cannot be generated in certain

Originally, I was going to quote regarding the *continuum hypothesis* - another notion re degrees of infinity, etc, but when I went to what I thought was the correct page number that I'd jotted down, I found this material about tossing coins instead. That was because I cdn't read my own handwriting. The number I went to was "229", but the number I wrote was "225". By this point I was a bit 'bored' by infinity (especially since so many other people have commented on it so much better) so I liked the coin-tossing more. I thought it tied nicely into my mistake-making.

I'm happy to say that I'd never heard of Marilyn vos Savant before reading The Man Who Loved Only Numbers. She writes math brain teasers (&, I perhaps incorrectly assume, other things) in a column called "Ask Marilyn" in *Parade* magazine. Why am I glad I'd never heard of her? Because, supposedly, her writing is read by millions of people every Sunday (or was as of 1998). In my worldview, if something's that popular there's probably something unforgivably 'wrong' w/ it. Thank goodness I'll never be 'wrong' in that way!

According to Hoffman, "She bills herself as the person with the 'Highest IQ' ever recorded, a whopping 228, according to the *Guinness Book of World Records*." For the record, I've been called a "genius" now & then - mainly by friends who probably don't have very strict standards for such things - but I think that most people who're egomaniacal enuf to consider themselves such are too oblivious to criticisms of standards used to measure such things. IQ measurements have been demonstrated over & over again to serve ulterior ruling class interests. Perhaps she shd read Stephen Jay Gould's The Mismeasure of Man (perhaps she probably already has). Maybe *I* shd read it! Never trust anyone who considers themselves to be a genius. They're probably severely lacking in self-criticism. For the flip side of the record, *my* IQ is:

71

grammatical systems according to our first statement: ⌟A genre is a set of determinable rules.⌝ Whether true or false or indecidable, statement A can & should in fact be expanded to the following: ⌟A genre is a set of determinable rules... which can be expanded to encompass other sets of rule ordered & non-rule ordered codes within the limits of recognizability-- which itself is a subjective variable of interpretation.⌝ In this way a ⌟genre⌝ is different from a ┘grammar┐ because, if you like, its process is not only a matter of socio-pragmatic intra-evolution but of heteroreferential inter-combinatroiesis.

I begin with a few brief observations concerning genre rather than mathematics for myriad reasons most of which will be left unnaccounted for here-- I have no intention of adding them up or dividing them by zero, of determining their function, or subtracting them from the sum of what

(enlarged here to the largest font size that doesn't drive it onto another page - does the font enlargement indicate a parallel degree of IQ enlargement? In other words, does not only the *numerical size* of the IQ count but the *font size* of its expression as well? In this case, I think the font size is around 125. If we allow the *font size* & the *numerical size* of my IQ to be added together, my IQ *still* isn't as big as vos Savant's! Darn!)

"71" is a Gaussian Prime. I had to consult a Jim Loy site again to learn more about this. Thanks, Jim. I'm sorry I sd "Maybe this guy is 'wrong'" about you earlier. I filched this definition from<http://mathworld.wolfram.com/GaussianPrime.html>:

Gaussian primes are Gaussian integers $z = a + b i$ satisfying one of the following properties.

1. If both a and b are nonzero then, $a + b i$ is a Gaussian prime iff [If and only if (i.e., necessary and sufficient). The terms "just if" or "exactly when" are sometimes used instead" - nIFFty, huh!] $a^2 + b^2$ is an ordinary prime.

2. If $a = 0$, then $b i$ is a Gaussian prime iff $|b|$ is an ordinary prime and $|b| \equiv 3 \pmod 4$.

3. If $b = 0$, then a is a Gaussian prime iff $|a|$ is an ordinary prime and $|a| \equiv 3 \pmod 4$.

So, you see? Having an IQ of 71 isn't so bad. One might even say that it's Gaussian Prime (mental) 'Real' *e* State. (&, yes, a google search for "e + the Gaussian prime 71" did yield some results *but* ""e + the Gaussian prime 71"" didn't - it's all in the way you group it)

I like this term "iff". Naturally, being the homonyphonemiac that I am, it makes me think of a pun: If "iff" means "If & only if" (being, therefore, a fairly stringent qualifier), what does "iffy" mean? In 'natural' language (ie: language not so tied to a specific discipline) something that's "iffy" is "uncertain" or "dubious" (&, therefore, an ambiguous qualifier), then "iff" & "iffy" appear to be opposites. Think of it like this: "I like Marilyn vos Savant *iff* she considers an IQ of 228 to be an *iffy* concept".

I hope that none of these folks that I quote who I'm relying so heavily on for explanations ever get mad at me for doing this. After all [is there a strict mathematical expression for "After all" in wch "all" is precisely defined as the set of all infinities?), they're the mathematicians. I'm just a wise-ass.

To prove my point further that an IQ of 71 doesn't necessarily indicate a lack of cleverness & skill on my part (even, dare I say it?, *genius*), I took a plot of the complex plane that shows the Gaussian primes as filled squares off the Wolfram MathWorld site again (Hey! I bet they took it from somewhere else!) & combined it w/ my own Flatland diagram:

Okay.. so that is clearly NOT NOT NOT what kind of thing I have been producing here.. but if it were, so what? Well, it would change your relationship to both parts of the text, wouldn't it? To mined to + ENTS.. & to the interface between them.

Here's still another formal arrangement which would alter all of these relationships:

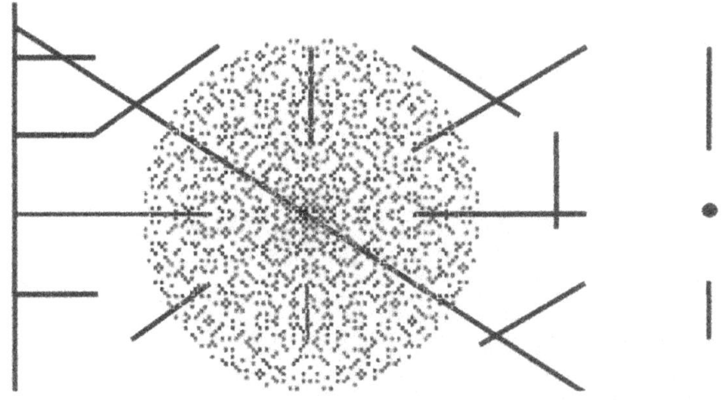

See how pretty it is? Note that Flatland easily nullifies Complexity. Story of my life. Ron Graham wasn't *really* talking about me when he was quoted in <u>The Man</u> as saying:

"'There's a premium in mathematics on doing something new, something different. It can carry over to the rest of your life. There's a tendency to make your own rules. $A + B$ is not equal to $B + A$. Well everyone knows it is, but suppose it isn't. Now what have you got? Why should you drive on the right when you can drive on the left?'"

[..]

"'There are problems in graph theory called *extremal problems*. They are problems about extremes, like 'What's the largest number of edges a graph can have?' Erdös was into questions like that. I had heard about a European aristocrat who bought out an opera house and gave tickets to his friends and put all the bald ones in a certain position so that when he looked down from the balcony their bald heads spelled out something spectacular, unbeknownst to them.'"

Whoever that aristocrat was, he's another honorary addition to the Pantheon of Perverse Number Theory Outlaws! If only I had a photo taken from his operatic vantage-point. I'd love to superimpose the Flatland overlay on it. "Logicians joke about a black theorem, which if you discover it, drives you mad." But, you know the saying? "I don't get mad, I get even"? As if that isn't twisting Hoffman's words around enuf already, consider my favorite SubGenius alternative saying: "I don't get even, I get odd."

e: The Story of a Number - Eli Maor. Oi! Here I am, sitting at my computer, not sure how much longer I'll have to write this bk in uninterrupted alone-time before having to do something practical for bucks, & I'm starting to lose steam. I need my pipes cleaned or something. I have 5 more bks to write about in this "NAVIGATION:" section & I'm looking at e w/ trepidation: Do I have the energy &/or inspiration to go on?! I'm on a roll but the slope's beginning to feel like it's flattening out.

Not only did this bk clarify *e* for me, but, even more importantly, it probably introduced me to the idea of **INDETERMINATE FORMS**. These are so much up my alley that I don't know whether I'm bowled over by them, down w/ them, or cornered in them. 1st:

A *limit* is something that 'bounds' an infinite series by having the series *approach* it w/o ever being able to *reach* it. *SO*, if we create a series like 1, 1/2, 1/3, 1/4... & designate that as *a* then the limit of *a* is 0 (zero). The progression will never reach 0, just as Achilles wd never reach the Tortoise in Zeno's paradox, because it just keeps getting infinitely smaller, but it can get infinitely close to it - such as just .0...1 nanoseconds away from scoring! This is expressed as: $1/n \to 0$ as $n \to \infty$ or as:

$$\lim_{n \to \infty} \frac{1}{n} = 0$$

The limit of one divided by *n* as *n* approaches infinity is zero.

e stands for a specific non-repeating, non-terminating decimal number that begins: 2.7182818284590452... It's generated by an infinite series: $e = 1 + 1/1 + 1/(1x2) + 1/(1x1x2x3) + 1/(1x2x3x4)$... *e* is the base

Yeah... right.

for the natural logarithm & is used in many calculations involving limits. The choice of the letter "e" honors a mathematician named Leonhard Euler (1707-1783).

Wikipedia: "In mathematics, a logarithm of a given number to a given base is the power to which you need to raise the base in order to get the number. For example, the logarithm of 1000 to the common base 10 is 3, because 10 raised to a power of 3 is 1000" "In simple terms, the natural logarithm of a number x is the power to which e would have to be raised to equal x — for example the natural log of e itself is 1 because $e^1 = e$, while the natural logarithm of 1 would be 0, since $e^0 = 1$." If a logarithm uses e as its base, it's written as "$\log_e$".

e^0? Hhmmm... I reckon that that's the sortof expression people use w/o questioning because they know it produces certain results. For me, it just hovers there mysteriously. I don't know what making something to the power of 0 does. Multiplying *e* by itself *no* times?! To make it 1? Is everything to the power of 0 = 1? Well, I tried googling that last question & got links about the power of prayer & some self-help stuff or some such, etc.. So I tried "zero as a factorial". That was the ticket.

Now, I *thought* I found that any factorial number (call it **n!**), may be written, **n! = n x (n-1) x (n-2) x (n-3) x ...** (understanding the ellipsis to mean: repeat this process until all positive results are exhausted). In other words, **n** is defined as the product of all positive integers from 1 to **n**. The factorial number is the number of possible set arrangements made w/ a set of numbers - eg: "1" can only be arranged one way: {1} so its factorial is "1". 3 can be arranged 6 ways so its factorial is 6: {1,2,3}, {1,3,2}, {2,1,3}, {2,3,1}, {3,1,2}, {3,2,1}. To re-express the latter example, 3! [the factorial of 3] = 3 x (3-1) x (3-2) = 3 x 2 x 1 = 6. Now, of course, to all of you who remember basic math, such a description is probably so simple that you're irked w/ me.

Note that in these examples, the number 0 is left out: It's *not* "3 x (3-1) x (3-2) x (3-3) = 3 x 2 x 1 x 0 = 6" *because* if you multiply by zero you get **zero**. So we stop when the positive integers stop. Why do you get zero when you multiply by it? Imagine a horse is fed an apple every day, the amt of apples it has per mnth is equal to the number of days in that mnth. If it's February in a leap yr, then it gets 29 apples; if it's a mnth that I just made up called Starvation wch *has zero days* then it gets zero apples - even though it's getting one apple per day.

As I understand what I'm reading on the net to try to explain this zero factorial thing, we use the form 1! = 1 x (1-1) = 0. Uh.. Why do we get down to *n* minus itself here & not elsewhere? I'm sure it's all very simple (**not**?) but looking at sites isn't helping.

The 1st website I looked at seemed to express the factorial equation incorrectly. No doubt they didn't, but it made no sense to me. The 2nd one <http://www.zero-factorial.com/whatis.html> expressed the equation as, in the example of 3,: 3! = 1*2*3. That seemed fine because it only uses the positive integers & the use of the asterisk ("*") seems to just mean "times" in the same way that "x" or "•" typically does so I just figured it's an alternative usage. *Then* they threw me for a loop:

Same text... more or less-- significantly different form... which author is more interesting? which editor is an author? Curious. It would seem that the idea of the margin depends quite a bit on the style of presentation regardless of content. In textual production the last word concerning legitimacy is simply visual presentation within the limits of a recognized code. I can write almost anything as long as it looks good. Take the following:

n! = 1*2*3*...*(n-2)*(n-1)*n
and so on.
Logically, n! can also be expressed n*(n-1)! .

Therefore, at n=1, using n! = n*(n-1)!
1! = 1*0!
which simplifies to 1 = 0!

Now, it's no longer clear to me *what* "*" means. Neither of the 2 expressions used seem to equate the one I *thought* (apparently incorrectly) I understood. *So*, I decided to look up the "*" & see whether it has a meaning I cd slot in. In the Wikipedia Table of Mathematical Symbols, it's related to "convolution" & "conjugation" - neither of wch seemed to be relevant. Given that the above description uses the word "Logically", I tried the Wikipedia Table of Logical Symbols - where it isn't listed *at all*. SO, I'm *still* confused.

Fortunately, they express it a different way by explaining, again, that the factorial of a number is the number of set permutations that're possible w/ its components. One can only have 1 arrangement, a set w/ "1" in it; & zero can only have 1 arrangement, an empty set. Fine. But is that what the above says? Not as far as I can tell!! I mean it says that the factorial of 1 = 0, *not* that the factorial of zero = 1. Right?! So I'm still confused..

But I can run w/ 0! = 1 easily enuf. It helps me understand that e^0 = 1 means that there's only 1 set of *e* multiplied by itself zero times - wch is an empty set. In other words, 2.7182818284590452... horses multiplied by themselves zero times is an empty stable - similar but not identical to 2.7182818284590452... horses eating 0 apples in the mnth of Starvation = starvation. Ok, I know that's a bad comparison because the mnth of Starvation has no days & the 2.7182818284590452... horses can't starve in no time at all! I mean it takes some time!

Whew! Remember long ago (page $\sqrt[74]{2.10449191 \times 10 \text{ to the 138th}}$) when I was about to discuss the luscious **INDETERMINATE FORMS**?

∞ is not a number. Simply explained, ∞ + ∞ = ∞. Compare that to 1 + 1 = 2 & you'll see what I mean. As to who "you" are, I have no idea. I mean, this bk isn't marketed at anyone. That's a math problem in itself. I'm writing it for myself & hoping that there's someone else out there who'll find it interesting. It's like reading the autobiography of someone trying to absorb some info that they no longer feel that much affinity for but who get inspired by it *anyway* & decide to mutate it at their leisure just for the sake of having an active, *creative* relationship to the world.

The readership then becomes, at a meta-level, people who share that approach to experience & who enjoy seeing it applied by someone else. The math becomes somewhat secondary. & I like to think that people who *do* share that approach are in an extreme minority, are *exceptional*, & that this shared experience then becomes a 'pata-level.

In e̲, I find:

"Expressions such as ∞/∞ or ∞ - ∞ are known as "indeterminate forms." These expressions have no preassigned value; they can be evaluated only through a limiting process. Loosely speaking, in every indeterminate form there is a "struggle" between two quantities, one tending to make the expression numerically large, the other tending to make it numerically small. The final outcome depends on the precise limiting process involved. The indeterminate forms most commonly encountered are 0/0, ∞/∞, 0 • ∞, ∞ - ∞, 0^0, ∞^0, and 1^∞."

Such a notion as "indeterminate forms" excites me in the same way that Heisenberg's Uncertainty Principle does, or Schrödinger's Cat, or Gödel's Undecidability. They all point to a *flexible* 'reality'. A 'reality' that evades definition by its very nature. For me, as a person trying to have a frequently creative relationship to 'reality', I find myself *limited* by those "reality maintenance traps" I keep coming back to.

The limit of "tENTATIVELY, a cONVENIENCE" as it reaches for infinity is undermined by what it doesn't have control over by virtue of not understanding it. That's not quite right: I don't necessarily want to *control*, I want the "Happy Solution of a Friendly Cooperative Relationship with [almost?] *everything*" is more like it. There must be a mathematical way of expressing that & it's precisely as a result of creating such an expression & understanding fully the meanings & relationships between the symbols that go into making such a thing that one is enabled to have a more detached perspective on whatever the expression symbolizes. One can then play w/ those symbols as an aid to imagining different relationships. & *that's* a sort of magik. What primitive expression can I make to represent this - if only for humor's sake?

(discourses, genres, grammars) may be trivial &/or phallacious (who truly believes that our ultimate purpose is to ┘express ourselves clearly┐?).

but tl
signi
of the
at lec
attemp ne
trespc fance
bounda e need
the li ist to
of ┘on it to
own fa ss the
langua ries,
into t mits
margin e^{L}s
anothe millar
ge┐.
he
s of .
r--

that is what drive^{L}s both tENT^{L}s project & the journey

however & therefore who can tell?).

of this marginalia (or not or both not & yet both

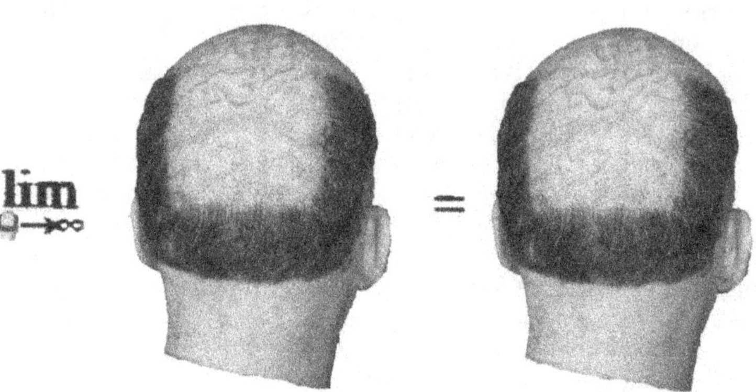

barber & photographer: Julie Gonzalez

I know, I 'know', that's pretty cutsie. Nonetheless, it serves my primitive purpose. It's also an example of "Folk Math" - in this case a rebus-like image that combines pictures & symbols w/o necessarily having a rebus-like purpose insofar as the picture of my head isn't meant to be suggestive of a word in a typical rebus-like way - eg: by substituting an image of an eyeball for the word "I". The expression reads as: "The limit of "tENTATIVELY, a cONVENIENCE" as "tENTATIVELY, a cONVENIENCE" approaches infinity is "tENTATIVELY, a cONVENIENCE". So the sky's not the limit, I am. I *did* sneak in a complication there that's *oh-so-typical* of Perverse Number Theory. I use a cropped photo of myself as a symbol of "tENTATIVELY, a cONVENIENCE". It's cropped just to show my head (on my neck). I'll allow such a reduction/over-simplification to symbolize "the thinking of 'tENTATIVELY, a cONVENIENCE'" or some such.

Shaved into my hair is *another* symbol. That's where I get tricky. The symbol is a "Blissymbol" for "open". Blissymbols are the creation of a man named Charles Bliss. He was born in 1897 in the Austro-Hungarian Empire about wch he is credited as having sd that the "20 different nationalities hated each other, mainly because they spoke and thought in different languages." He was a survivor of 2 concentration camps, who created an international picture language somewhat rooted in the logic of math symbols. His ideas were rejected & he had to work as a laborer to support his family. I'm w/ ya there, Charlie!

By the time he was around 68, children w/ cerebral palsy at a particular center in Canada were being taught to communicate w/ his symbols. Alas, he didn't learn about that until he was around 74 so the satisfaction of that success came late.

lim(inal) Bliss
HEAD → ∞ = (eg: "egg")
this & or that
margin(al) symbol
FORM(ULA)

Fig. 78 "Obulata mandala"

As first formulated by Victor Turner in his essay "From Rebusal to Rebusoid" which, by means of perverse noematic abstraction can be "reduced" to: (al) = (al)(LA) [bearing NO relation to the deity of similar name]

Since my head tattoo of a brain is somewhat visible in the enclosure by the Blissymbol for "open", I call the overall combination my symbol for an "open mind". Of course, that makes "mind" synonymous w/ "brain" wch is pretty questionable.. but I'll let that slide.

Therefore, a restatement of the above expression is: "The limit of "the open mind of tENTATIVELY, a cONVENIENCE" as "the open mind of tENTATIVELY, a cONVENIENCE" approaches infinity is "the open mind of tENTATIVELY, a cONVENIENCE". NiffyTY, huh!

It just brings me back to what I've been saying all along: *Increasing flexibility in the neural pathways thru wch one accomplishes what one's striving for (thru ambiguity gates) leads to an increasingly open mind wch decreases the limits (or increases the limit - to, perhaps, express it a little more mathematically correctly) to approaching infinity.* Does that seem like the worst self-help drek you've ever read? Nah.. I'm *sure* there must be worse!

Redefining "ambiguity gate" might be in order. A neuron receives pulses from other neurons firing, this either triggers or doesn't trigger it to fire in turn - ultimately contributing to some sort of signal resulting in some sort of triggered action: the blinking of the eyes, the thinking about defining "ambiguity gates", the tattooing of one's fingertips w/ 1, 7, 2, 9, & 6 on one hand's tips and 1, 8, 4, 1, & 6 on the other hand's & a subsequent ritual of friendliness-expressing thru hand-shaking, etc..

An "ambiguity gate" is a flexibility of neuron firing response patterns in wch options are increased by different paths being used to achieve the same result & by the same path being used over again to achieve different results & analogies, similes, metaphors, & pataphors used toward the same or related purposes of flexibility & creative open-mindedness. Or something like that.

In the chapter entitled "*Some Curious Numbers Related to* e", the 1st example regards e^{-e} wch equals 0.065988036...: "Leonhard Euler proved that the expression $x^{x^{x^{x^{\cdots}}}}$, as the number of exponents grows to infinity, tends to a limit if x is between e^{-e} ($=1/e^{e}$) and $e1/^{e}$." Yes, there are people who sit around & figure these things out. I'm not one of them. I sit around & figure *other* things out that might very well be even *more* obscure & of even *less* interest to most people. Still, this Leonhard Euler character was quite a feller.

In the same chapter we find several numbers about wch "it is not known whether this number is algebraic or transcendental: $e + \pi$; $e \times \pi$; e^{e}; π^{e}; &, climactically, y about wch it's also not even known whether it's

maybe bearing some relation on second thought— "AI = Allah" or "This everyday guy = god." kind of makes sense. I remember misremembering an Albert Einstein quote. He actually said: "Whether it's the janitor or the president of the University, I speak to everyone the same." What I like better is: Everyone is the same -- whether the janitor or the president of the Universe." I thought he was using a euphemism for the g-word.

rational or irrational.

In "The Function That Yields Its Own Derivative" (snappy title) we read that "The rate of decay of a radioactive substance - and the amount of radiation it emits - is at every moment proportional to its mass *m*" [..] "We see [..] that *m* will gradually approach 0 but never reach it" In other words, Zeno's paradox embodied in decaying radioactive material. Achilles is radioactive & will never catch up to the Tortoise because their half-lives ("the time it takes a radioactive substance to decay to one-half of its initial mass") are different.

Achilles might be Uranium-238, the Tortoise might be Radium-220. If so, the half-life of Achilles is "about five billion years" & the half-life of the Tortoise is "only twenty-three milliseconds". Achilles doesn't stand a chance. If he gives the Tortoise a head-start, the Tortoise will've already passed its half-life before Achilles can even blink his radioactive eyelids. Wch one is more dangerous is unclear to me.

It might be useful here to understand a little more notation: "just as we write $x = \log y$ for the common logarithm (logarithm base 10) of y, so we write $x = \ln y$ for its natural logarithm (logarithm base e)." Using A to denote *Area* & r to denote *radius*, Maor goes on to compare π & e by comparing their relationships to 2 curves:

Circle: $A = \pi r^2 \Rightarrow A = \pi$ when $r = 1$
Hyperbola: $A = \ln x \Rightarrow A = 1$ when $x = e$

"Can Perceptions Be Quantified?" In this chapter, Maor tells the reader that "In 1825 the German physiologist Ernst Heinrich Weber (1795-1878) formulated a mathematical law that was meant to measure the human response to various physical stimuli." [..] "Although [what became known as] the Weber-Fechner Law seems to apply to a wide range of physiological responses, its universal validity has been a matter of contention. Whereas physical stimuli can be precisely measured, the human response to them is a subjective matter. How do we measure the feeling of pain?"

In an interview w/ brain researcher "Pig Pen" on April 8, 2005, it was explained to me that thru the use of fMRI (Functional Magnetic Resonance Imaging), activity mapping can be done of the brain while responding to external stimuli. This is accomplished by measuring the degree of haemodynamic response (blood flow activity) that powers the neuron firing. It is one of the most recently developed forms of neuroimaging.

Pig Pen's particular research revolved around recognition & memory storage of images of objects. While looking at some fMRI data, Pig Pen pointed to images & discussed them:

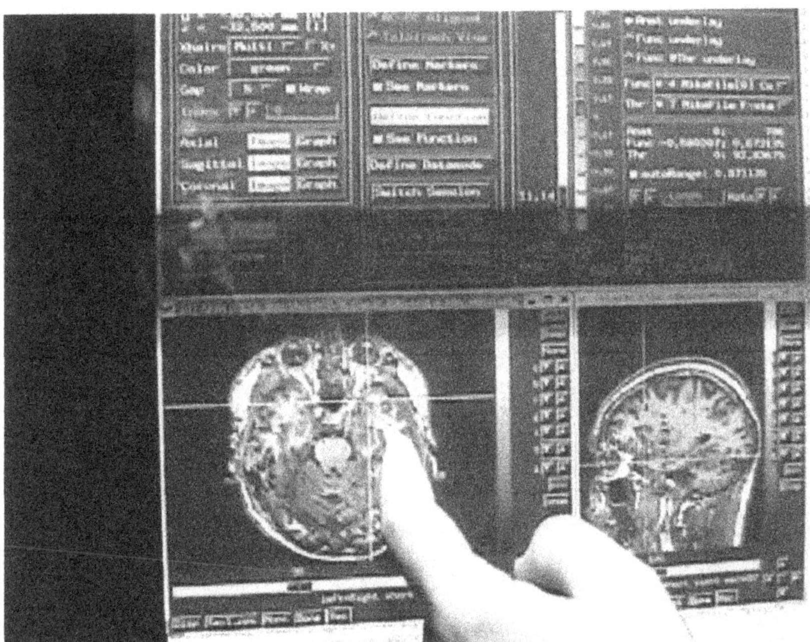

still from video taken by tENTATIVELY, a cONVENIENCE

"The areas surrounding the cerebellum on either side are very sensitive to object differences, so if you present different objects of different types, there's this sortof topographical organization where sortof different peaks & valleys of activation will appear all along here - & the peaks & valleys actually code for what type of information you're getting, what type of objects you're looking at.

"So one area will be very sensitive to face images & one area will be sensitive to house images & spoons & chairs.. Lots of people have done different studies looking at how the brain in that area interprets objects."

Perhaps such studies are the means for measuring "subjective" human responses to stimuli. The particular study data being examined by Pig Pen here is from the presenting of a video (or, as I prefer for etymological reasons, "vaudeo") of a 16mm film of mine called "Subtitles" to 2 research volunteers. The main purpose of collecting this data was to enable me to make another movie presenting the volunteers' responses to be projected side-by-side w/ the original film so that (v)audiences (viewing audiences) can get an idea of what's going on inside their brains as they witness it.

Given that "Subtitles" is a dense & quick-moving somewhat 'abstract' film that some (generally non- or anti- intellectual) (v)audiences have responded to by covering their heads or by screaming for the projectionist to stop showing it, critics of my work might say that this research is, indeed, "measur[ing] the feeling of pain".

Unfortunately, the stills here are reproduced in black & white to save printing costs. Data is shown by colored pixels: red for heightened activity in contrast to blue for average activity (the average having been determined in relation to this particular experience & *not* as an expression of human brain activity in general). The image below has the top row of brain pictures presenting data from when the "Mike Film" section of "Subtitles" was screened & the bottom row has data from when the "Paranoid People" section was screened. The position of the cross-hairs shows the area under discussion. In the "Paranoid People" data there's a substantial blob of red pixels at the cross-hairs.

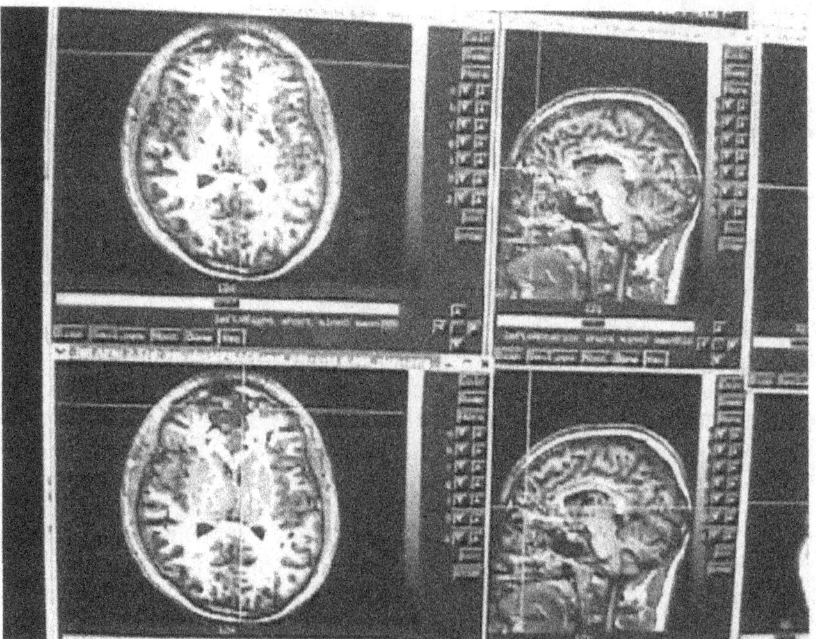

"Mike Film" (top) "Paranoid People" (bottom) comparison

t,ac: So, what do you think the implication of this degree of activity is in this part of the brain during the "Paranoid People" section?

Fig. 83
Fleeing/Alien/
Red Pixel Data.

PP: Um.. Let's see.. If I can remember what "Paranoid People"..
t,ac: I can explain it to you. It's all footage from films of crowds fleeing & looking nervous & so forth & so on - but taken out of context so that you don't know what they're fleeing from - so, eg, scenes from "War of the Worlds" where you don't see the aliens.
PP: Ok.
t,ac: So it's people fleeing from the alien but taken out of context it just looks like they're paranoid.. [..]
PP: Ok, it actually makes sense that it would drive this area of the brain because, um, if you see something that's taken out of context, your brain's going to work very, very hard to try to rationalize that. It's going to try to figure out *why* this is happening. [..] In contrast, if you look up here at the "Mike Film" portion, they're probably not trying to figure anything like that out.
t,ac: Because they don't have enuf clues to tell them that it has any kind of narrative function - so they're perceiving it more as an abstraction. That's my interpretation.
PP: Right. If I remember "Mike Film" correctly, it's a nice image that you don't have to try to figure anything out so much. You might have to do a little bit of figuring out; what are you really looking at?
t,ac: It's frames of film in free-fall, basically, so it's just sortof abstract shapes unless you're a film analyst - in wch case it wd probably look like something completely different.
PP: Ok, so one of my participants is probably just sitting back & enjoying the images, not having to work hard to figure them out. [..] In "Mike Film", primary auditory cortex is active wch means that they're processing the sound - whereas in "Paranoid People" they're not. The brain's very talented. If you're allocating your resources so that you're trying to figure out, say, why people are fleeing, you can ignore sound. You can't ignore it entirely, it's still there, but you're not going to focus on it. Your brain is not going to bother to process that information if you're thinking about something else.
t,ac: Now, the sound, in the version of the film you showed, stays pretty consistent throughout because it's multiple levels of the sound that your phone makes when it's left off the hook. So I wd think that any differences in the auditory response area of the brain wd just be precisely because of what you just stated [..]
PP: It has everything to do w/ the person & *not* what they're hearing. So you can get brain activity to fluctuate thru effort even if the sound remains consistent. & we know this from practical experience: you can ignore the sound of the refrigerator running if you're reading a bk intently.

t,ac: Can you bring up the menu again? I wanted to compare the "Doug Reshoot" section..

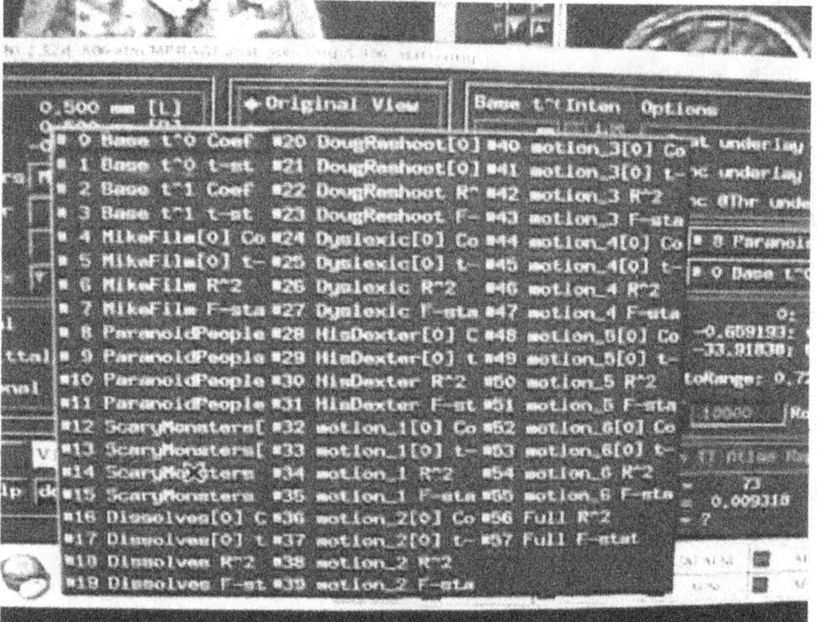

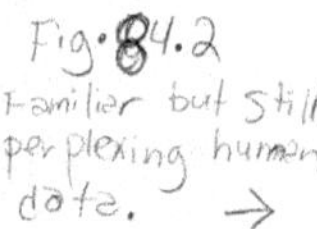

PP: Ok, so "DougReshoot" will be on the top.

Fig. 84.2
Familiar but still perplexing human data.

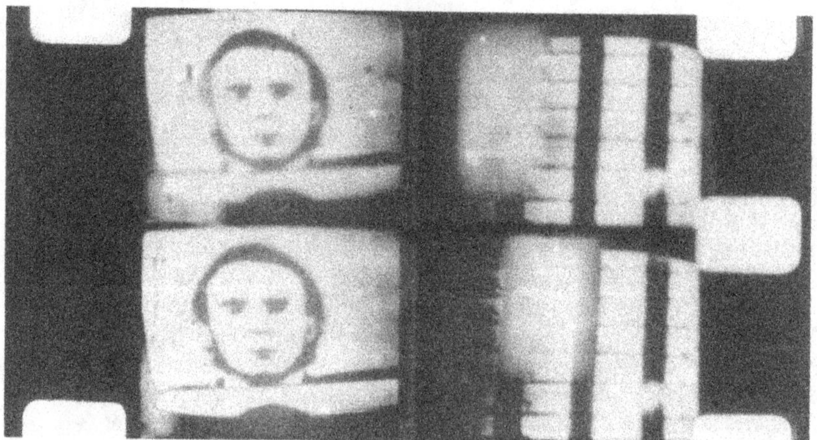

frame from the "Doug Reshoot" section of "Subtitles" - 1980-1982

t,ac: &.. uh.. maybe "His Dexter Daylight", yeah, "HisDexter" section. Now, "His Dexter" is all text that's being routed thru in such a way that makes it very difficult to follow, etc.
PP: Are they able to read it?
t,ac: Hhmmm.. Only w/ great difficulty. It's more text reduced to being an imagistic element - whereas the "Doug Reshoot" is completely understandable imagistically. It's a permutation on a face & hands, etc.. So I wd imagine that there shd be a big difference between those 2 sections but I don't know whether a language-related area wd be in play during "His Dexter Daylight Utterly Darkened" (as the full title of the section is known) or whether it wd just be relegated to just an imagistic part of the brain. In other words, I don't know whether the perceiver wd give up on trying to perceive it as language or not. So I'm curious - because I wd figure that there wd be a strong visual response to the "Doug Reshoot" section.

PP: Ok, so let's look at vision 1st. So this is the back of the brain & there is a sulcus, a very deep sortof wrinkle that goes thru the middle of the back of the brain - so it starts at the back of your head & sortof winds thru - it's called the "Sylvian fissure".. No it's not! It's called the "Calcarine sulcus" & this is where primary vision takes place. Now if you picture this [PP makes a hand gesture demonstration] as a sortof sulcus, a deep crack like this, primary vision takes place all along the sides of that deep crack. As you go out further & further & further, like layers of an onion, moving out from that sulcus, higher & higher level visual processing takes place - but it's not too sophisticated - [..] that wrinkle is primary visual cortex.
t,ac: By "primary visual" do you just mean sortof overall shapes?
PP: The most basic.
t,ac: Like "it's bright here & it's dark there"?
PP: Even more basic than that. "This is a line." [..] "This is light coming in here, not coming in here." [..]
t,ac: So then the higher level things are the things that distinguish it as being something like "This is a curved line in relation to another curved line" or something like that.
PP: Right. As you go along [..] to an area that I showed you before that's lower in the brain then you get up to "This is a chair, this is a spoon, this is a face" - that sort of processing. & then as you move even further up, to the frontal cortex, you get into what kind of faces, what does this face mean to me, what are my long-term memories associated w/ that face.

t,ac: So I imagine that alotof research has been done on head injuries & that sort of thing & how they effect all of those perceptual aspects.
PP: Yes.

Fig. 86
Dexter Daylight Refrain

t,ac: So let's say that a person had a head injury on the back of their brain but their front was ok. How wd that effect perceiving, eg, a drawing or some complex object?
PP: That's a great question. There are a few different ways that you can be effected by having a head injury to the back of your head where it damages this section [PP points to the vicinity of Calcarine sulcus again]. You can imagine this is a very common head injury. It's easy to hit the back of your head. The question is: Can you hit it hard enuf to damage it? So, if you knock out the cells in this region, if you knock out all of them, your eyes work ok & alot of the rest of your brain works fine but it's called "Cortical Blindness" meaning that you cannot experience sight - even though everything else is working fine. & your eyes will even react to objects - such as if I throw something at your face, your eyes will actually flinch or you'll focus on it but you will not be able to experience or be able to report the experience of seeing something coming to your face, because you, in all purposes, can't, essentially, see at all. [..]
t,ac: So you'd be able to see if you had the front part of your brain damaged but not the back part of your brain? [..]
PP: Right. You can get damage throughout your brain, including your object-recognition area. If you get really small damage to your object-recognition type area - wch is actually a large area of the brain, it wd be very difficult to damage the whole thing. If you get a small lesion, wch is a piece of damage to that section, you will have problems recognizing certain objects. So there are people who cannot recognize.. tools.. but they can recognize & name other objects, & another interesting thing is: sometimes you can not recognize an object, you know what it's used for, & you can pick it up & use it, but you don't know its name & you don't understand its identity. [..]
t,ac: Ok, well, back to the comparison between "His Dexter Daylight Utterly Darkened" there [..] & the "Doug Reshoot" [..]
PP: So, looking at the visual areas, [..] it looks like "His Dexter Daylight" activates this area more than "Doug Reshoot". & this is the text down here? [pointing to the "Dexter" data]
t,ac: Yeah.
PP: So the person might be working very hard.. um.. people are compelled to read. When you put language in front of a person, it's almost impossible to ignore it - &, given that it's so scrambled, they're probably working very, very hard to try to extract the words from that portion of the film. If you've ever tried to read in a challenging situation, like reading on a bus, where your bk is moving. You know your eyes are working very hard to do that.
[..]

t,ac: Is it possible to have strong brain activity going on w/o signs of stimulation? I mean, eg, whatever, there's *alpha* & *beta* wave states, right? [..] The *alpha state* is a more meditative one? Is that correct?
PP: Um.. I don't know too much about that but definitely there're times when the brain is working very hard & this technique, wch we're looking at, MRI, is not sensitive to it.
t,ac: What I'm wondering is, eg, if you were doing an MRI of somebody who was dreaming, eg, wch wd be a different state from the kind of cognitive state that the brain is in right now - or, at least, I'm assuming it wd be, maybe it isn't.
PP: Yes, it wd be.
t,ac: What part of the brain wd be most activated during dreaming? Or wd there still be all of the parts of the brain activated? I wd think that back-brain functions.. &, again, I don't really know anything about what I'm talking about, so I'm just kindof talking off the top of my head, or off the top of my brain so to speak..
PP: Ha Ha!
t,ac:..but I wd think that certain survival functions wd be kindof at a low level relaxed mode, or whatever, & that the areas where the dreams wd take place wd be highlighted. Is it as simple as that? Or am I completely off the mark here?
PP: I really don't know the dream literature. I think the basic survival areas wd remain pretty constant. But, just speculating, you might get alotof activity in the vision regions because dreams are highly visual. Essentially, a dream is recreating an image of what you wd see in everyday life.
t,ac: That reminds me, I had a dream this morning about combining your birthday w/ my workplace w/ mathematics - because I've been reading some math stuff lately.
[..]
t,ac: Does alcohol damage any particular parts of the brain more than anything else? In other words.. or are there other things that damage the brain that're typical input things? Like, uh, air pollution or anything like that?
PP: Certainly. You can get general damage from poor air quality. You can damage *extensively* if you cut off the oxygen supplies, of course you can kill the brain entirely if you cut off oxygen supplies - so any degree of oxygen deprivation below a certain level is going to harm you. Of course, the pollutants themselves can harm parts of the brain.
t,ac: What about alcohol in particular? Has that ever been discovered to damage particular parts of the brain?
PP: It does. There's a very interesting scientific experiment that people

do on themselves not knowing it & they can cause what's called "Korsakoff's amnesia" from too much alcohol abuse [..]
t,ac: Is that the same thing as black-outs?
PP: No, this is a more long-term permanent damage sort of thing. *SO*, you actually induce amnesia from drinking too much. We're talking about drinking *alot*. Permanent amnesia. Permanent problems in encoding new memories [..] & it damages what's called the Hippocampus

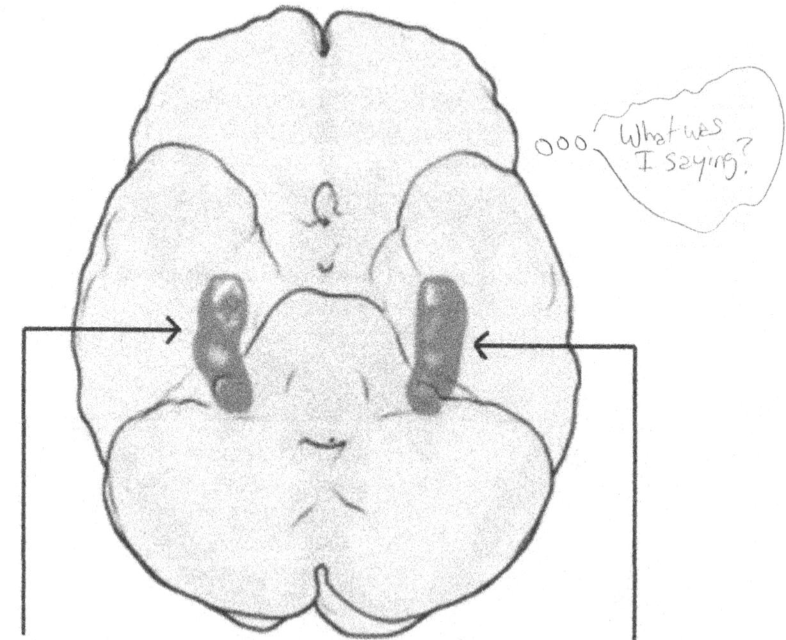

The hippocampus is structurally located inside the medial temporal lobe of the brain. (In this illustration of the underside of the brain, the frontal lobe of the brain is at the top, while the occipital lobe is at the bottom.) - illustration & explanation taken from Wikipedia

& the surrounding areas.
t,ac: Is the Hippocampus above the Medula? Or?
PP: It's very close to the center of the brain. It's in what's called the medial temporal lobes - "medial" means "middle". Very close to the spinal column, the brain stem [..] & I believe people who drink alot tend to have general memory problems [..] There're all sorts of 'fun' things you can do to your brain. My favorite recent one is the phenomena of doing Ecstasy [aka MDMA, the drug].
t,ac: Oh, yeah? What part of the brain does ecstasy damage?

PP: I believe it damages the dopamine system - wch is found throughout the brain - it's a chemical.

[According to NIDA, the National Institute for Drug Abuse website <http://www.nida.nih.gov/Infofacts/ecstasy.html>: "MDMA exerts its primary effects in the brain on neurons that use the chemical serotonin to communicate with other neurons. The serotonin system plays an important role in regulating mood, aggression, sexual activity, sleep, and sensitivity to pain." "Research in animals links MDMA exposure to long-term damage to neurons that are involved in mood, thinking, and judgment." I've taken Ecstasy & wdn't recommend it - especially when there're risks of its not really even *being* ecstasy - or risks of its being cut w/ other harmful drugs like heroin, etc..]

t,ac: Does it damage it by over-stimulating it?
PP: Uh, it damages it by killing off certain parts of the neurons - & your neurons are sortof the wires in your brain that allow different areas to communicate w/ each other. A neuron looks sortof like a tree & the branches of the tree are the signals, the sortof wires going out [a more appropriate metaphor might be that the incoming signal paths are the roots & the outgoing part is the trunk] & communicating w/ other neurons. [Neuron diagram below taken from Wikipedia]

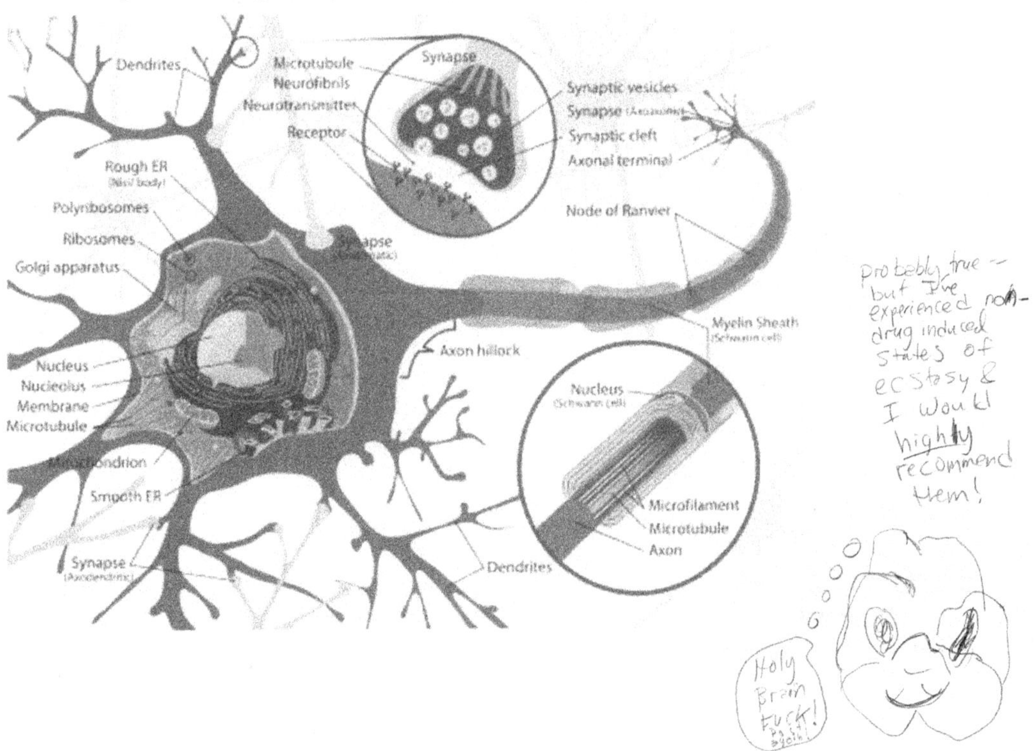

probably true - but I've experienced non-drug induced states of ecstasy & I would highly recommend them!

Holy Brain Fuck!

Which is precisely why Mercel Proust hired servants to do his pruning for him -- as is now the common bourgeois practice.

PP: & what Ecstasy does is.. they call it pruning [..] - so that information now coming into a cell is limited because whatever was communicating w/ it before can't anymore because the branches have been cut off.
t,ac: & how does that relate to the effect that people take Ecstasy for? In other words, people take Ecstasy in order to feel this kindof.. blissful.. loving empathy towards everyone. How does that relate to the pruning of the neuron tree?
PP: I think it's entirely unrelated - the pruning happens afterwards &, so, it's sortof a long-term effect. I'm not sure what the time-line of it is, when the pruning starts & when it ends, but I believe it's an after-effect. [..] & people have some considerable memory problems. [..] I'm a memory researcher. We believe that memory works by the different areas of the brain, the different wires form connections w/ each other & memories are reinstantiating connecting signal going thru different parts of the brain & rebuilding this memory. So if you prune off those branches that talk to each other you end up w/ memory deficits & other deficits as well.

[A computer-related definition from "Whatis.com": "In programming, instantiation is the creation of a real instance or particular realization of an abstraction or template such as a class of objects or a computer process. To instantiate is to create such an instance by, for example, defining one particular variation of object within a class, giving it a name, and locating it in some physical place."]

t,ac: W/ old age, what are the parts of the memory that're most likely to go? Is there a particular part of the brain that deteriorates the most in old age that leads to memory loss or is that something that can't be generalized?
PP: In Alzheimer's disease, wch is a specific memory & aging problem, there are specific parts of the brain that deteriorate much more rapidly than the rest of the brain.. alotof the frontal cortex, as well as some of the circuitry in the middle of the brain starts decaying but also [Amyloid] plaques start forming so these calcium deposits that disrupt the signal being processed [..] In general aging, anyone who lives long enuf will end up w/ some level of memory decline, & that is less well understood - where that happens. It might be generalized through the entire brain. [..]
t,ac: I read recently that Gregor Reisch was the 1st, perhaps, European or so-called 'western civilization', artist to depict the *mind* or the center-of-thought or what-have-you in the brain & that was in, I think, the mid -16th century [Reisch died in 1525 so it must've been earlier] & prior to

that people thought it was the heart & also the liver. That implies that the center of thought, in the western perception of it, has moved upward in the body, right?
PP: I think that at that time people thought that the head is where spirits lodged. So, trepanation was drilling holes in the head to let the spirits out.
t,ac: I was fantasizing that you were going to talk about alcohol damage by people who pour booze directly into their trepanation hole.
PP: Oh, well, I haven't tried that yet! I haven't found a willing participant to do that study.
t,ac: Yeah, yeah. The waiver for that wd be pretty legally tricky I'm sure.
PP: Yeah, it wd be tough to write that one.
t,ac: So, back to the western perception of the center of thought migrating from the liver to the heart to the brain. Do you think that there might be a next step where somebody wd think that the center of thought was outside of the body altogether?
PP: I think that there're people *now* who believe that. I mean, we know where the brain is & we have a good sense of what the brain does, you know, it controls motion, it controls *e*-motion, it controls your thoughts, it's your center of control for how you think about things.. but, the *mind* is a very different concept & I believe alotof people perhaps believe the mind somehow lies in the soul & that the soul maybe doesn't have a specific location in the body.
t,ac: Yeah, but, usually, people who talk about such things.. & I don't have any particular opinion about the concept of the soul or spirit or whatever one way or the other.. talk about it as *inhabiting* the body until one dies or whatever. Now what I'm speculating about is that if the western, or whatever, perhaps across-the-board human perception of the center of thought, etc, is gradually migrating up thru the body, is there any reason to stop there? Maybe it cd go outside of the brain altogether, outside of the body altogether..
PP: I think science has pretty much settled that the brain is where thought originates & that thought is contained w/in the brain.
t,ac: But, then, scientists are..
PP: often wrong
t,ac: ..always wrong..
[we laugh]
PP: At some point..
t,ac: ..unless you're Leonhard Euler or Issac Newton or..
PP: ..we'll be proven wrong on nearly every point.. wch is why science is so much fun!
t,ac: Yeah, exactly! That's why I'm speculating about this right now. It's

Fig. 91.1 Metaphysical diagram

Fig. 91.2 Alternate metaphysical diagram.

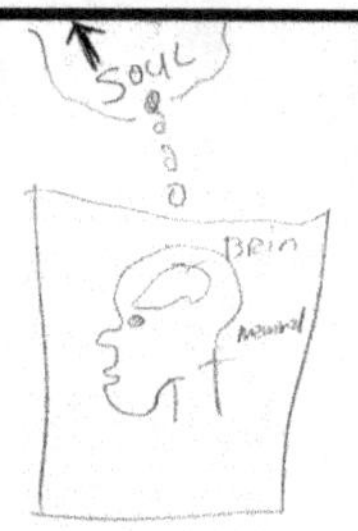

Fig. 92
Non-metaphysical diagram

not that I have a particular opinion about it, or whatever? I just like the idea of speculating that, maybe, I don't know.. 2,000 yrs from now, someone will go: "Hah! They used to think that the center of thought was in the brain! Well that's an utterly ridiculous notion! We now know that that's a purely functional bit of hardware in a different sense.."

[An interesting sidenote speculation here might be that perhaps it's not so much that science has been *wrong* in its conclusions over the yrs *but that* physical 'reality' might've actually *changed*. Wdn't that be a shocker?! Maybe the center of thought (if there is such a thing) *was* in the liver at some point. Or in the testicles & womb.]

[..]
t,ac: Did you ever notice any particular section of the film, maybe "Mike Film, maybe not, that seemed to stimulate an unusual amt of activity?
PP: I think that what we were viewing before.. this very frontal area of the brain.. I believe it was in "Mike Film".. I was surprised by the level of activity. [..]
t,ac: So you say that's the more higher level cognitive functions?
PP: Yeah. & to see that level of activity in that area [she finds the image], yes, here it is, was quite surprising.
t,ac: So, therefore, my whole theory that stimulated trying to do this in the 1st place, that, maybe, experimental films are useful for stimulating parts of the brain that, perhaps, are not understood enuf or whatever, may have some very slight validity in the sense that higher cognitive levels are being stimulated.
PP: Yes. Especially if you get the person in the scanner to work hard to work out what is going on.. here we go, it's in "Paranoid People" [..] & you can get them to engage in the film & *think* about it & try to explain it in their own mind, yes I think you can drive some of the higher level areas of the brain quite effectively.
t,ac: As I recall, what you sd was that the 2 research volunteers didn't really have anything to say after the experience though.. Is that correct? Or they sd very little?
PP: They sd very little. I asked them a very open question: "What did you think about the film?" & they sd some very generic response like "Oh, I thought it was pretty interesting." But they didn't get into deep comments about what they thought about different sections.
t,ac: Did they collapse or start vomiting or anything?
PP: [laughingly:] Nothing like that.
t,ac: Just kidding, just kidding!
PP: Keep in mind that they had just been inside of a scanner for an hr & a half at this point - so getting out is their best experience!

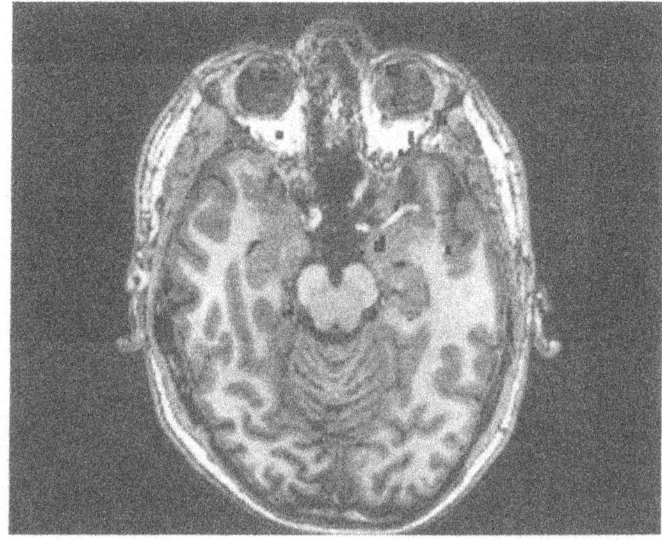

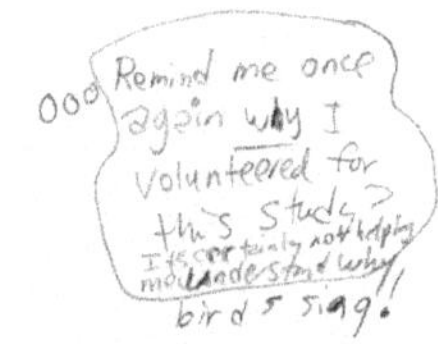

an overhead view of a research volunteer's brain
while watching the "Mike Film" section of "Subtitles" - note the eyeballs

Yet-another interesting sidenote here is that, according to David Rothenberg in his fantastic bk, Why Birds Sing: contrary to Nobel Laureate Ramon y Cajal's postulation of the theory that "'Nerve paths are something fixed, ended, immutable. Everything may die, nothing new may be rejuvenated.'" there *is* the possibility that brains may rejuvenate after all.

Rothenberg bases this possibility on the results of research done by Nottebohm & Goldman on canary brains as recounted in Fernando Nottebohm's article "Why Are Some Neurons Replaced in the Adult Brain?" [*Journal of Neuroscience* 22, no. 3 (2002): 624-628]. Rothenberg writes:

"Happily singing, learning canaries were the first species to reveal neural replacement. If we could figure out how to stimulate the growth of new neurons in the human brain, we might be able to direct this knowledge toward the repair of brain injuries such as the stroke that led Broca to realize that brains had distinct areas controlling specific functions. Starting with humble singing canaries, brain science is heading into a new paradigm, a vision of neural plasticity, where the cells in the brain are changing, not set or fixed. This is the canary's wonderful secret."

I shd also note that brain researcher Pig Pen is not necessarily an expert on all of the subjects that I was questioning about. As such, some opinions expressed may be more rooted in research than others. Pig Pen made no pretense of being the ultimate authority. Pig Pen was simply a brain scientist open to collaborating w/ a freakazoid such as myself.

One of the things that fascinates me the most about what Pig Pen discussed is the idea of memories being reinstantiations of the neural firing patterns of the original experience. Thinking about this got me so excited that I phoned my young Anxiety Therapist friend Kalie Pierce (masters degree in counseling psychology currently working w/ obsessive-compulsives) to discuss these things because she's just about the only friend I have that I *can* discuss such things w/!

After I spewed a rant about memory & the relevance of philosophical attitudes to increasing the chances of greater brain flexibility, Kalie mentioned something that I found very important. She told me that she'd been taught that when heroin users use for the 1st time, it activates a dopamine high that can never be repeated because the body's ability to generate such an experience is limited. *SO*, in order to try to recreate the intensity of this high, users try to repeat the 1st experience's environment *because this instantiates reinforcement* of aspects associated w/ the no-longer repeatable high & helps creates a facsimile experience.

This inspires me to so many relevant tangents & commentaries that I hardly know where to begin & certainly can't type fast enuf to keep up w/ the ideas! [TTQ-EA: Thoughts Too Quick - Expressions Anachronistic] IMO [In My Opinion], one of the biggest damagers of brain flexibility is the external imposition of 'hebephrenic' instantiation thru organized hammerings-home of dogma - in essence, a leading cause of brain damage easily on a par w/ Ecstacy & alcohol abuse &, in fact, related.

Dogma has few or no *ambiguity gates* - it's presented as absolute inflexible & unquestionable truth. In my personal experience, many adolescents & other young people, experiencing intense hormone rush that compels to action, reject dogma because it so clearly doesn't jive w/ their own direct experience. Being *commanded* to believe something that runs contrary to direct experience is understandably likely to generate a taste for less dogmatic direct experience. The effect of Ecstasy, eg, can be a direct experience of an intense feeling of community that being surrounded by older, obviously lying & thoughtless, hypocrites isn't likely to top. Hence the connection between drug abuse & dogma.

Unfortunately, as William S. Burroughs essentially points out in some of his autobiographical reflections, one can over-react to one over-

one trades ones thought in order to belong & find a place in the menagerie of similar minds -- similarity is attained by means of reduction to the rudimentary

rudiment ≠ essence

Adolescence marks the developmental threshhold from the rudimentary to the elaborated stages of growth -- it turns upon the experience of intense & profound ambivalence -- if completed successfully the learner passes through the ambiguity gate into flexible & dynamic modes of transformative thinking -- if not, the [illegible] returns to the hebephrenic Garden of Dogma simplification & lie by creating an alternative over-simplification & lie. Burroughs recalled that when he was young the propaganda was so intense against drugs like marihuana that when he tried it & found it to be a relatively mild experience he concluded that the propaganda against other drugs must be equally as dubious. As such, he eventually became a heroin addict.

Thanks to the A/V Geeks (http://www.avgeeks.com) reissuing on DVD of "Youth Suicide Fantasy: Does the Music Make Them Do It?", I got to check out this Christian propaganda film linking rock music & suicide. I won't claim that everything they claim is ridiculous, but that's beside the point for now.

At some point in the movie there was an interview w/ the parents of a kid who'd either killed himself or tried to (I forget wch now - probably the former). I was so disturbed by the religious fervor of the parents that I cd immediately understand why the kid did it. Growing up in families like that is like being a Jew raised by Nazis. If one feels like one can't escape otherwise, in despair it seems like the only way out is to kill yourself *before they kill you* - if only in the form of rapid progress towards brain-death.

William Blake expressed a similar realization-- the fear of creating such an over-simplified system drove him on-- beyond the gates of innocence to the innovations on the other side of vision & beyond.

ANY belief system that encourages unquestioning faith is *not* a way of thinking - it's anti-thought. Religion is the most obvious target but there are many more subtle guilty targets. But let's start w/ the obvious anyway. In Catholicism, & other branches of Christia-inanity, there's the Eucharist (a rite or ritual of worship & remembrance wch Christians generally believe was instituted by Jesus at his last meal w/ his disciples before being handed over to his executioners.). In some versions of this, **Transubstantiation** is believed in. This is based on the belief that the substance of bread & wine (or grape juice, eh?) that's fed ritualistically to the parishioners becomes the body & blood of Christ when ingested - thusly enabling the rite participant to become united w/ God.

Interesting, because the Huichol & the Tarahumara also "consume god" in the peyote ritual-- but do so in order to foster visions & inspire Shamanic becoming-- in other words their particular ritualization of this metaphor is dynamic, flexible & ambiguous.

I redefine this Transubstantiation by saying that it's really a metaphor for externally imposed hebephrenic ***instantiation***. Another way of putting this is to say that some stereotypical oversimplifications are useful for maintaining some types of social order. A ruling elite that can successfully convince the people (whose oppression they rely on for their artificially forced hierarchical position) that they are the 'GOOD GUYS' is likely to stay in power longer. The alternative being, of course, to *put the fear* in people - also effective & 'necessary' but, ultimately more unstable.

The psychological process by wch this convincing is achieved is, once again (but this time w/ more emphasis), ***externally imposed hebephrenic instantiation***. A template for perceiving 'reality' is created & then superimposed on the neural-firing patterns of the victim's brain BAMN! (By Any Means Necessary) This is an aspect of what I call *stereotype projecting* - my nemesis & pet peeve throughout most of my life. In stereotype projecting, an oversimplistic template is so burned into the brain, that the brain's unable to have a response appropriate to the data it receives - it becomes hebephrenic.

A black taxi-driver friend of mine told me that he saw a white woman in a neighborhood that he knew to be dangerous for her & pulled over to offer her a ride. She was so scared by his being black that she didn't take it. Because of her (presumed) stereotype projection (black man = danger) she was unable to see the greater context of danger that she was in.

yes, exactly, which is precisely why radical pedagogy — constructivist pedagogy as practiced by Paulo Freire & others — makes the de-mediating of experience one of its central tenets. Freire writes: "I repeat: the investigation of thematics involves the investigation of the people's thinking — thinking which occurs only in & among people together seeking out reality. I cannot think for others or without others, nor can others think for me. Even if the people's thinking is superstitious or naive, it is only as they rethink their assumptions in action that they can change. Producing & acting upon their own ideas — not consuming those of

Now, of course, I wasn't there, I can only presume this to be an accurate explanation. I presume this partially because the story was told to me in the context of a discussion on such things & I knew the cabbie to be a pretty gentle & thoughtful guy so I took his story to be a reliable example because it jives w/ my own personal experience. That doesn't make it *true*.

& this is where the criteria for judging the personal value of stories for one's self comes in. Evaluating incoming data based on correlations to personal experience is, at least, much more appropriate for one's self than evaluations based on mediated experience. But in an increasingly mediated society, personal experience gets more & more replaced by stories about other people's experience framed according to the ideology & techniques of the story-teller. For some people, the lives of the characters in a sit-com that they watch on tv is more real to them than their own life because it has more memorable peaks & valleys than their own actual lives do (partially because they spend too much time watching tv). Get a life!

Let's not be confused, reading a bk about something is *also* having a mediated experience in relation to that something. I'm not just picking on tv, newspapers, & popular magazines. As such, these NAVIGATIONS of mine are meant to be *critical navigations thru mediated experience*. We can't all be scientists or writers or athletes or whatever. Although, it might be helpful to at least *try* to be a widely enuf active person to have some experience w/ a variety of things. Then one has more to go on when evaluating data. Because we all have limitations that make it 'necessary' at some point or another to go to outside sources for information, etc, we have to develop an overall critical philosophy to enable us to not just accept mediated experience as *absolute truth*.

In ***externally imposed hebephrenic instantiation*** (luv that phrase!) the underlying form of mediated experience can help ridiculous simplifications to sneak in thru the back door of one's mind.

The so-called tv 'news' presents such an oversimplified version of 'reality' that it's not even a "skeleton", it's less than a single bone - w/ no marrow - & even *that's* prosthetic - & *useless* as a prosthesis.

Such an approach is ultimately aimed at making death-by-reduction-to-sound-byte seem like an acceptable structure for encoding 'reality'. After all, the reducers obviously have BIG power (the money & resources to create & control the massive technical network that brings the tv 'news' into so many people's lives) & what they present is *oh-so-flashy* (one stunning eye-candy transition to another: *wham, bamn, thank you Spam!*) that they come to represent 'reality' just by being so well 'nigh omnipresent in it.

SO, an unrealistic template of accepting reductionism is ***externally imposed*** thru the 'news' & ***hebephrenic*** behavior results thru the ***instantiation*** of that template. & even that description is too oversimplistic & exaggerated - but I want to move on here!

From the heroin/dopamine example we can extrapolate about how environments during peak experiences in general might be used in attempts to re-experience those peaks. Alas, mental development gets stunted when, instead of having new peaks, people attempt to re-experience old ones thru repetitions of formative environment. In other words, they put down stubborn roots in subcultures.

I went on a date of sorts w/ a paleobotanist around my own age a decade or so ago. She was pretty interesting to me. We went to a bar & music was playing on the juke-box. Jimi Hendrix came on & she sd something like: "Finally, some good music!" I was disappointed. This struck me as a symptom of typical stunted mental growth. People get attached to certain musics at key times of the development of their life - typically during the raging bull of adolescent hormones. This music then becomes "their music" & stays fixed as such *'forever'* if they're no longer open to new experiences. They feel compelled to listen to it over & over again in an attempt to recapture the original feelings that it correlates w/ instead of having new experiences of similar vitality.

But back to the bird-brains: Why Birds Sing: Rothenberg's telling of the findings of Nottebohm & Goldman: "The growth of new neurons definitely has something to do with the singing of new songs. If the birds no longer sing, the total number does decrease." So sing new songs people! Not just the greatest hits of Jimi Hendrix (not to pick on him - I love his music!) over & over!

The title of Rothenberg's bk is clear enuf: it's an exhaustive historical analysis, from many, many angles, of birdsong & human hypotheses about what motivates it. The most common explanations revolve around marking territory & attracting a mate. That's all well & good. But what if it's also a manifestation of an instinctual keeping-alive

of neural paths - like a memory exercise. Use it or lose it. Maybe abstract experiences like watching experimental film or complicated systems engagement like playing music just help w/ general resistance to neural atrophy by keeping the brain active w/o having to be bogged down w/ the stress of survival issues.

All these tangents triggered by the question of whether 'subjective' reactions to 'objective' physiological stimuli can be measured. The 'subjective' becomes 'objective' thru brain instantiation in the sense that it then becomes physically measurable as a degree & pattern of response.

Back to Eli Maor's marvelous conciseness: In the "Spira Mirabilis" chapter we seem to meet Zeno again. Maor describes the logarithmic spiral: "If we follow the spiral inward from any fixed point *P* on it, we will have to make an infinite number of rotations before we reach the pole; but, surprisingly, the total distance is finite."

I may be stereotype projecting my favorite quasi-paradoxical combo of infinite & finite here, but it seems to me that what this means is that the spiral turns are infinitely decreasing in size (as one goes inward) &, hence, never terminate. Nonetheless, just as w/ Zeno, the actual distance covered is finite. It's just a matter of how reaching it is approached/measured. There isn't an infinite distance between 0 inches & one inch, just an infinite amt of divisions w/in that distance. A spiral takes this into an x & y axis expression instead of just a number-line x expression. But maybe I'm misunderstanding Maor here. Read his bk for further clarification. I'm tired right now & starting to pass out.

Both GEB & *e* explore the significance of musical scales to math. Maor, like Hofstadter, is inspired by J.S. Bach for the obvious reason that Bach's "The Well-Tempered Clavier" is the most famous early piece of music to apply what's called "equal-tempered tuning", the tuning most common in 'western' music of the last few hundred yrs.

Maor constructs a fictional dialog between Bach & a contemporaneous mathematician named Johann Bernoulli. They discuss the difference between "just intonation" in wch the simplest ratios are used for playing intervals (2/1 = an octave, 3/2 = a fifth, 4/3 = a fourth, 5/4 = a major third, 6/5 = a minor third, etc..) & adjusted ratios for equal-tempered tuning wch enables modulation from scale to scale. The purpose of equal-tempered tuning is to allow all the major & minor scales to have the same ratios in instruments, like the piano, incapable of the kind of hands-on flexibility of pitch control that a violin, eg, has.

In the dialog, Bernoulli demonstrates to Bach that the way to explain the equal-temperament is to visually arrange representations of the notes evenly along a logarithmic spiral w/ exactly 30° between each semitone. Rotating the spiral so that the 1st tone falls on the x-axis makes the mathematical relationship to the notes in the scales clearer. What I don't understand is what type of spiral, if any, best demonstrates just intonation. & does it take infinitely long to play a descending octave? 'Cause I don't have the time for that, man. That wasn't even funny. I shd take a nap.

AH! I'm a New Man! (or is that a Nudeman?) Having napped, I now have had the energy to rewrite a little of the preceding so you, the reader, will never know just how bad it was, eh? (It wasn't really *that bad.* I just rewrote a tiny bit.)

NEWS FLASH! As you've probably noticed by now, the pages are paginated w/ the root of the page number to the power of itself. This page is the 99th root of "3.69729638 x 10 to the 197th" power aka "3697296380000000000000000000000000000000000000000
00000000000000000000000000000000000000000000000000
00000000000000000000000000000000000000000000000000
000000000000000000000000000000" - wch equals 99. So this is page 99 paginated in a more complicated way both for humor's sake & because this is a math bk (of sorts - or out of sorts).

Each of these page numbers is being made in a graphics program so that I can have the root sign (or whatever the fuck it's properly called) extending across the top of the number - something that I can't do in this simple text program. I've been using an on-line calculator accessed thru Google to get the numbers by typing in, eg, "99 to the 99" & getting the results to cut & paste into the root symbol template. So far, so good.

I'd only created the page numbers up to "100" because I wasn't sure how long this bk wd be & I didn't want to waste too much time on the page numbers if it turned out that I didn't need them. Now I'm thinking that this bk might be 200pp long so I decided to create the numbers for another 100pp. Tedious, but no big deal, right? So I go thru the routine & I reach 144 (12 squared) & enter in "144 to the 144" in Google following the same procedure as before &, Lo & Behold!, I've maxed out the capacity of the calculator!! NO ANSWER IS GIVEN. I've reached the limit of what the calculator is apparently capable of answering.

I feel like I've reached some sort of unanticipated outer limits. Like the internet is taken for granted as being able to provide any 'common knowledge' answer & then.. something as simple, but perhaps outré, as asking what 144 to the 144th power is stymies it. Like I wasn't supposed to get *that* far out, y'know?! Stay w/in the bounds. Who cares what 144 to the 144th power is? Why do you need to know that? So I have to calculate it in a different way. But how? There's the "brute force" method aka the slow way. I can do that. I can still get an online answer

oops... sorry!

Well, the number 144 probably care

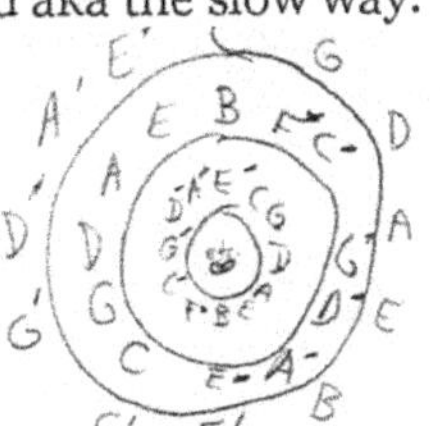

on page 144 of Thought & Language Lev Vygotsky writes: Though the child & the adult may use one & the same word in referring to one & the same object, their mental operations are still different.

see also the discussion on pages $\sqrt[96]{\text{blobbidy-blah}}$ – $\sqrt[97]{\text{a bunch more \#s}}$ & the illustration on page $\sqrt[98]{\text{new series of different \#s}}$

to "14.4 to the 144" yielding "6.37087174 × 10^{166}".

To a mathematician, this wd be no problem at all. It's just a simple thing. To this particular drunk & depressed guy who's *not* a mathematician, this seems like a pain-in-the-ass. 1.1^{11} = 2.85311671, 11^{11} = 285311670611. 14.4^{144} = 6.37087174 × 10^{166}; $14.4^{14.4}$ = 4.79093067 × 10^{16}; 1.44^{144} = 6.37087174 × 10^{22}. So that's the key, right? Shift the decimal point of the root number from 144. to 14.4 & we get 6.37087174 × 10^{166}. Shift the decimal point of the root number from 144. to 1.44 & we get 6.37087174 × 10^{22}. The result of 144^{144} shd probably be something like 6.37087174 × 10^{311}.

Uh.. here's where it starts to get strange? (For me, ie) So, I go to a commercial math website (QuickMath Automatic Math Solutions), something that's basically for selling "Algebra Solver" software, & I type in: "143^143", "144^144", & "145^145" - w/ "^" meaning "to the power of". & then I click on "factor". I get results in the form of an image file - probably to prevent easy cut & pasting of the result so that people will be encouraged to buy the software instead of just glomming it on-line like I am. I tried "143^143" because I already have a result for that & I wanted to double-check their solution against the one I got from the Google calculator. Then I tried "144^144" because that's the answer I need next for this pagination. The result for "143^143" was:

1633252597297391281626414733726994086734441059238828627591219466574957409572917
474451871926525060154732838896421950133080242843444919540597953600246077432798
360652868158019198653011290441373047060853910979171947696877502716158045750827
1024154651492905352586112730098512006245462336504718441716721213745501162O7

wch I now retype since that's an image file:

1633252597297391281626414733726994086734441059238828627591
2194665749574095729147445187192652506015473283889642195013308
0824284344491954059795360024607743279836065286815801919865
3011290441373047060853910979171947696877502716158045750827
10241546514929053525861127300985120062454623365047184417167112
1374550116207

Think doing that was easy? NOT. & I probably made a mistake, multiple mistakes. But now I can select that number & paste it into Google. Back to that later.

Apparently, the Google calculator has been rounding off - I reckon ever since it entered the x-times-"10-to-the-y"-power phase starting w/ page $\sqrt[12]{8.91610045 \times 10 \text{ to the 12th}}$. Those fiends! I suppose I shd forgive them. How cd they know that correct counting wd be so necessary to finding the pages of this bk?! Now my readers will be hopelessly lost as they calculate & find that 8,916,100,450,000 to the 12th root is "12" according to the Google calculator (I just checked) *instead of* the more accurate answer I expected to get from QuickMath.

But then I *went* to QuickMath to see what result they'd give me (figuring that it wd be something close to but not equal to 12) but I cdn't figure out what the correct syntax for asking the question is (I tried various things including "12√8,916,100,450,000" but the "√" wasn't recognized; I tried "8,916,100,450,000^(1/12)" because I'd seen similar notation on another page but that didn't work either; etc) *so* i didn't get a result.

Then, I realized that I cdn't have the commas (","s) in there so I tried "8916100450000^(1/12)" & got:

$$\sqrt[3]{2}\ 5^{5/12}\ \sqrt{178322009}$$

I guess that was an improvement of sorts so I tried typing in "the cube root of 2 times 5 to the 5/12ths power times the 12th root of 178322009" as a Google search & got "did not match any documents" as the reply.

Once again, I used incorrect syntax (probably). After all, I don't even know if having the above "cube root of 2" side-by-side w/ "5 to the 5/12ths power" side-by-side w/ "the 12th root of 178322009" means that they're all multipled together. For all I 'know' it means: "What tangled webs we weave, when 1st we calculate to deceive" or something like that!

Still hard on the trail of trial & error, I tried asking QuickMath to factor "144^(1/2)" thinking that I was asking what the square root of 144 is & I got "12" as an answer. Whew! [wipes sweat off brow in cold wintry house] I'm on the right track. But wuh thuh fuh does the above thingie mean?! Why cdn't QuckMath have just given me a nice simple "12" as an answer for this too? Then I cd've rested in peace. Now, I'm haunted; I have a hang-dog look; hungry; hunkered down in my seat; harboring deep fears; hankering after a solution that hovers just out of reach..

I type into Google: "the 143rd root of 1633252597297391281626414733726994086734441059238828627591219466574957409572914744518719265250601547328388964219501330808242843449195405979536002460774327983606528681580191986530112904413730470608539109791719476968775027161580457508271024154651492905352586112730098512006245462336504718441716711213745501162O7" & get:

"143rd root of 163 325 259 729 739 128 162 641 473 372 699 408 673 444 105 923 882 862 759 121 946 657 495 740 957 291 474 451 871 926 525 0 60 154 732 838 896 421 950 133 080 824 284 344 919 540 597 953 600 246 077 432 798 360 652 868 158 019 198 653 011 290 441 373 047 060 853 910 979 171 947 696 877 502 716 158 045 750 827 102 415 465 149 2 90 535 258 611 273 009 851 200 624 546 233 650 471 844 171 671 121 37 4 550 116 207 = 143".

Righto.

I type in "the 143rd root of 1.6332526 x 10 to the 308th" because that's the answer I got from Google (but not quite the way Google expressed it) when I asked what 143 to the 143rd power is & I got "(143rd root of 1.6332526) x (10 to the 308th) = 1.00343648 × 10^{308}" wch wasn't *quite* what I had in mind so I tried re-expressing it: "(1.6332526 x (10 to the 308))^(1/143)" & got "143"! Bravo! *SO*, I tried the same question w/ QuickMath & got:

$$1.06886\sqrt[143]{hot^2 x}$$

What do I conclude from all this? Even though it may seem counterintuitive, it's now obvious that:

^{143}√16332526000000000000000000000000000000000000000000000000
00000000000000000000000000000000000000000000000000000000
00000000000000000000000000000000000000000000000000000000
00000000000000000000000000000000000000000000000000000000
00000000000000000000000000000000000000000000000000000000
000000000000000000000000000000000000000000000000000000 =

^{143}√1633252597297391281626414733726994086734441059238828627
591219466574957409572914744518719265250601547328388964219501
3308082428434491954059795360024607743279836065286815801919
86530112904413730470608539109791719476968775027161580457508
27102415465149290535258611273009851200624546233650471844171
671121374550116207 =

$$1.06886\sqrt[143]{hot^2 x} = 143$$

So, if the page numbering starts to get a little strange starting at 143 you now know that it's because of this great leap forward in math!

Chapter 13 of e: The Story of a Number, "e^{ix}: 'The Most Famous of All Formulas'" is dedicated to the formula that my knuckle tattoo (shown on the cover of this bk) is derived from. In it, we read about Leonhard Euler (1707-1783), a central figure in just about anyone's math bk. His "immense output - not yet published in full - is estimated to fill at least seventy volumes"!! [..] "Moreover, we owe to Euler many of the mathematical symbols in use today, among them i, π, e, and $f(x)$."

Just as the history of artists & writers that I like is full of suicides & early deaths, so is the history of mathematicians full of 'insanity'. As such, it's a relief to read that "the irrepressible Euler, at age seventy, married again. By now completely blind, he continued his work as before, dictating his numerous results to his children and students."

"His definition of a function is essentially the one we use today in applied mathematics and physics (although in pure mathematics it has been replaced by the "mapping" concept): 'A function of a variable quantity is any analytic expression whatsoever made up from that variable quantity and from numbers or constant quantities.'" The concept of the function didn't originate w/ Euler, but "it was Euler who introduced the modern notation $f(x)$." [..] "And he made free use of the expansion of functions in infinite series and products - often with a carefree attitude that would not be tolerated today." I find the latter particularly funny, of course.

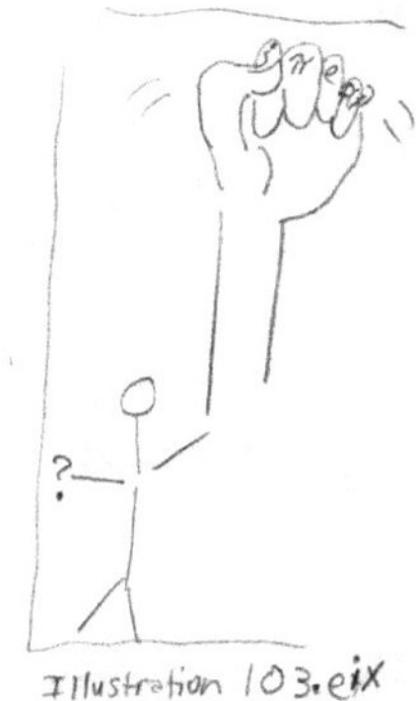

Illustration 103.eix depicting the combinatreous semaphore-rebus indicating "Power to the Eulogy of Dead Symbol systems!"

"Euler proved that [..] an irrational number [can be] represented by an infinite continued fraction". [..] "Euler also showed how to write an infinite series as an infinite continued fraction [..] & derived many interesting continued fractions involving the number e". One of wch is shown below:

$$e = 2 + \cfrac{1}{1 + \cfrac{1}{2 + \cfrac{2}{3 + \cfrac{3}{4 + \cfrac{4}{5 + \dots}}}}}$$

illustration taken from page 157 of e: The Story of a Number

How, "illustration taken from page 157 of e: The Story of a Number" continues "...3+ (3/(4+(4/(5+..." may not be immediately obvious & is a subject for further study.

Maor explains the development of Euler's famous formula at substantial length. If you're really interested, you shd look there. I'm only creating a sortof Italian Futurist *sentesi* here. A carefree pastiche of what're highlights for me - w/ little regard for the mathematics that're too difficult (again, *for me*). Euler committed the "supreme act of mathematical *chutzpah* [by replacing "the real variable x with the imaginary expression ix", where $i = \sqrt{-1}$"]. "To replace [x] with an imaginary number is to play with meaningless symbols, but Euler had enough faith in his formulas to make the meaningless meaningful.":

$$e^{ix} = 1 + ix + \frac{(ix)^2}{2!} + \frac{(ix)^3}{3!} + \ldots$$

Now the symbol i, defined as the square root of -1, has the property that its integral powers repeat themselves in cycles of four: $i = \sqrt{-1}$, $i^2 = -1$, $i^3 = -i$, $i^4 = 1$, and so on. Therefore we can write the preceding equation as:

$$e^{ix} = 1 + ix - \frac{x^2}{2!} - \frac{ix^3}{3!} + \frac{x^4}{4!} + - \ldots.$$

How, "illustration taken from page 158 of e: The Story of a Number" continues "...3+ (3/(4+(4/(5+..." may not be immediately obvious & is a subject for further study.

Maor painstakingly takes us thru the development of the formula until it reaches the formula that "must surely rank among the most beautiful formulas in all of mathematics": "$e^{\pi i} = -1$". "Indeed, by rewriting it as $e^{\pi i} + 1 = 0$, we obtain a formula that connects the five most important constants in mathematics (and also the three most important mathematical operations - addition, multiplication, and exponentiation). These five constants symbolize the four major branches of classical mathematics: arithmetic, represented by 0 and 1; algebra, by i; geometry, by π; and analysis, by e. No wonder that many people have found in Euler's formula all kinds of mystic meanings".

I don't count myself among the "mystic"s but I do count myself among those who find Euler's equation truly insightful at a level that few humans ever approach. This formula deserves its 'cover boy' status on this bk precisely because of its incredible conciseness in the expression of paradigm-shifting concepts that took thousands of yrs to develop. This expression is the archetypal 'knuckle-sandwich' to human consciousness.

This might be a good time to sidetrack a little by explaining that it's very difficult for me to navigate thru all of the bks that I'm using here as inspirational springboards & especially (so far) e: The Story of a Number. Why? It's certainly not that the writing's turgid or any such thing, it's more because, as you can probably tell by now, my mind-set is so different from that of serious mathematicians. I respect *their* mindset but mine is more playful & not always compatible w/ theirs.

The problem here is that I really do want to give a respectful reading of their bks in wch the reader will be able to hopefully follow some sort of elementary thread running thru all this math - but I want to do it in such a way that my own mindset is revealed as being equally valid by virtue of its elastic humor & creativity. I'd hate for the authors of all the bks I'm quoting to dislike this text, I'd much rather have them laugh along w/ it & excuse my mathematical inadequacies when they rear their ugly heads.

While writing this bk, I emailed my good friend, kindred spirit, & highly respected colleague, Bruce Stater, about what I'm in the midst of. His email reply (on January 6, 2008), as w/ all of his writing, resonated deeply on both trickster & intellectual levels that few people ever touch. Note that his "General Tito" character reminds us of the inspiration for Alfred Jarry's important absurdist archetype "Pere Ubu":

"As a child I was a fairly good student of math, but in high school I ended up with an algebra teacher of limited intelligence & contemptibly authoritarian character whom I nicknamed "General Tito." Since most of my time in class was dedicated to undermining his regime of discipline, I learned very little math & never really got back on track after that. I recall such strategies as replacing the classroom chalk with a box of crayons, removing the chairs & desks from the room & leaving a note stating that 'the class would meet in future at the chiropractor's office located directly across the street', bringing in two tin cans attached with string as a communication device following being separated from my friend & coconspirator who was removed to the opposite end of the room. I lost interest in math for years-- until my psychosis set in at the end of 1999. At that time I remember considering very deliberately what occurred to me to be the extremely limited mathematical underpinnings of the history of philosophy and of contemporary theory. I was appalled at the lack of imagination underlying ontological constructs in which X could merely be opposed to not X-- of the ridiculous simplicity of binary logic in structuralism & the feeble measures of post-structuralism to counter with its logic of indecidability and indeterminacy. My annoyance also branched out into a disgust for language theories-- such as Chomskian transformational grammar-- which subordinated ordered branches of rules I hardly saw as transformational at all. I was

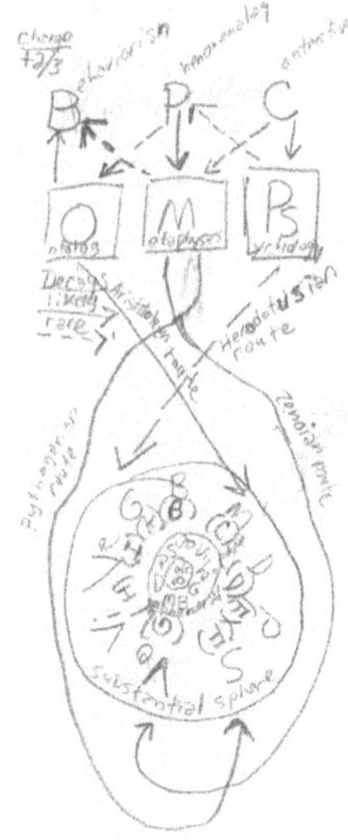

Illustration 106√etc. Partial Classical √ Ubuphrenic diagram illustrating cohesive elements within the code translation correspondence system of natural & unnatural human universes.

experiencing a complex & continuous transforming set of interlocking systems, forces, and metastable sign systems-- whose apparent chaos (for example[s]: whether the day was to be taken up by interactions with "phenomenologists" or "behaviorists"-- or whether the window one looked through could be determined to be, say, an "ontological window" or a "metaphysical window" or a strictly "psychological window") could be mapped with systems of equations assigning values to temporally locatable (if ever changing) elements in the system. I also "realized" at this time that part of the way that language coding systems utilized by the "consortiumists" operated was through numerical transpositions into sets of mathematical equations that could be retranslated into alternate systems of signification (infinitely more complex, say, than computer or "natural" language systems) which also related to resonances inherent in the "genetic" language of living things & the "elemental" language of non-living things & which somehow also connected to the underlying grammar of the "music of the spheres." Obviously I was only able to "decode" tiny fragments of all of this-- for example "reading" the pattern of blinking christmas lights in a local Astoria diner as a symbolic correspondence to the numerical sequence of messages on the Frameworks listserv run in reverse, then forwards, & finally "laterally" (the lateral being a degeneration of the more essential principle of meta-language systems in general-- which I referred to as "periphrasis at slant angles" or alternatively "the Knight's move.")"

That's quite a mouthful, Bruce! But it's a mouthful that the thoughtful (& open-minded) cd chew on for a very long time. As my friend etta cetera might say: "Luv me some Bruce!"

I'm not really trying to make any critical commentary re 'bad' teachers by quoting Bruce in this way here - even though it may appear that way because my own experiences recounted much earlier in this bk (& in my bk footnotes as well) may seem to imply that. Teachers have their own problems & some that're too authoritarian for folks such as Bruce & myself may be exactly what other folks 'need' to urge their increased concentration.

Bruce's & my personal history does tell a tale, though, that's central to my own approach to just about everything in this bk. We have strong minds capable of many things - just don't command us to do something against our grain w/ them, ok?! Because it won't work, even thru negative psychology.

Thank the holy ceiling light that not every person in the world is a sheep passively waiting to be led. The likelihood of such strong paradigm-shifters as Issac Newton, Gottfried Wilhelm Leibniz, Leonhard Euler, Gertrude Stein, Helen Keller, Charles Bliss, etta cetera, etc, growing out of sheepishness is practically zilch.

Again, e: The Story of a Number will explain the things that I'm hereby making passing reference to better than I will. Chapter 14 covers imaginary numbers. I can't do justice to their significance here. A few quick quotes will give a basic intro: "Just as negative numbers arose from the need to solve the linear equation $x + a = 0$ when a is positive, so did imaginary numbers arise from the need to solve the quadratic equation $x^2 + a = 0$ when a is positive." "*Quadratus*" is Latin for "square". In the leading term of a quadratic equation the variable is squared.

What interests me about the above is that a 'need' can arise to solve an abstraction. In P.N.T., of course, I'm being more serious on a philosophical level than I am on a mathematical one. Nonetheless, by posing "$(m - n = n - m)$ -> to x", I was catalyzing myself to invent a solution - viz: having x be a situation of number mirror imagery.

Each step in the abstract mental process of creating such problems & then imagining solutions for them can catalyze a new level of visionary thinking. Imaginary numbers, seemingly so removed from everyday 'reality', became a member of the practical world of applied math thru such mental steps as when,"around 1800, it was shown that the quantity $x + iy$ could be given a simple geometric interpretation. In a rectangular coordinate system we plot the point P whose coordinates are x and y. If we interpret the x and y axes as "real" and "imaginary" axes, respectively, then the complex number $x + iy$ is represented by the point $P(x, y)$".

"Of course, in mathematics we are free to define a new object in any way we want, so long as the definition does not contradict any previously accepted definitions or established facts." Equally of course, the long history of what constitutes "previously accepted definitions or established facts" might just accumulate to a level that few will retain anymore. What wd happen if the internet & other means of external memory were to disappear? Thru a sortof collective Korsakoff's amnesia, people wd be redefining 'reality' from scratch.

The latter might strike some as yet-another dystopian scenario & they might be right. At least it was an opportunity to put out the image of an abstract alcohol poisoning permeating a technical network. Imagine the internet on a virus bender, AI self-consciously gobbling self-destructive programs for the sake of getting away from its all-too-utilitarian daily grind.

In musical notation there're many things to perplex & confuse anyone who tries to delve into it. The main example that comes to mind for me is in the notation of wind & brass instruments, instruments that are in certain keys. As such, the notation for an octave above "middle c" (C5) represented on the treble clef is:

∴ Venus de Milo quadratus
= Nude Desending a Staircase

But what does that mean? If it's written for the C Piccolo it actually represents C6; for the English Horn: G5; Contra-Bassoon: C4; Bb Clarinet: D5; Bb Clarinet (French System): D6; Bb Clarinet (German System): back to D5 again; Eb Alto Sax: A5; etc, etc..

In math, we've already encountered multiple somewhat conflicted usages for "~": "not", in propositional logic; but "~" means: "roughly equal to" in Approximation theory. In German musical notation, there's even an "H" note (unlike in the US, eg, where the notes are A thru G). &, now, Maor tells me, that "π", in addition to its usual well-known meaning as the irrational number beginning 3.14..., *also means* "the number of primes ≤*x*" (less than or equal to *x*)!!

Since, in PNT we don't have to be concerned w/ making any sense whatsoever (even if we often do anyway), I wonder: What wd happen if we were to conflate them all together? It brings us back to a variation on the Language Experiment.

Play a piece of music in wch C5 can also be C6, G5, C4, D5, D6, A5, etc..; in wch any possible notational meaning for the symbol that you're familiar w/ may be accepted as 'correct' regardless of its apparent irrelevance to the instrument being used. Such a notational chaos might be just perfect for a synthesizer or sampler player. It cd be taken to mean that if one wants to play C5 as a Bb Bass Sax sound, then one wd play a D7!! - something easy to do on a sampler because it can bypass the natural range of the instrument simulated, but very difficult to do on the original instrument w/o using extended technique.

See also: pataseminoperceptual Law (sic) of Transheterorelational Nega formity on page ~~N~~ 136.... & corresponding exemplars. -- marginal editor

What if we were to conflate math notation in a similar way? To reiterate, such proposals are just to stimulate the imagination - I have no immediate idea of what practical applications might arise from such an exercise. Someone else can take the next step. What if we were to use an expression like this: "~)(π"? Dropped into an expression of existential quantification? In my lurid reading, it becomes: "roughly equal to not" butts cheeks w/ "less than or equal to PI". Is this an infinite series? If so, what's its limit?

"But What Kind of Number Is It?": "The origins of *e* [..] seem to go back to the sixteenth century, when it was noticed that the expression

or alternately: ⌜To quote the happy one-eyed man, insert &/or tree trunk PI to the "open-ended question...⌟ ≠ ⌜pre-quotationally squared sine wave closed perenthetically to openly perenthetical numerical symbol cubed by quotation prior to question⌟

$(1 + 1/n)^n$ appearing in the formula for compound interest tends to a certain limit - about 2.71828 - as n increases. Thus e becomes the first number to be *defined* by a limiting process, $e = \lim(1 + 1/n)^n$ as $n \to \infty$."

What if we were to cross disciplines in the above-imagined conflation? Thus, e cd be notated on the musical staff & become subject to its ambiguities. It might come to represent the limit that players cd approach w/o ever reaching.

Wd we become completely unhinged ontologically? By using symbols so freely (albeit, in this case, w/ a knowledge of symbolic history) might we undermine 'reality' maintenance traps? Might we create ambiguity gates? Might our neural firing patterns in memory reinstantiation create new meanings of equal value to the originals? Don't listen to me, I'm just a crank! NOT NOT. Who's there? Just a crank. Just a crank who? Justa cranka ya mind widda my Paradigm Shift Knuckle Sandwich! (Think of that as the kind of joke Giordano Bruno might've told.)

"It was on the basis of these [music] ratios [..] that Pythagoras devised his famous musical scale. But he went even further. He interpreted his discovery to mean that not only is musical harmony ruled by simple ratios of integers but so is the entire universe. This extraordinary stretch of logic can be understood only if we remember that in Greek philosophy music - and more precisely, the *theory* of music (as opposed to mere performance) - ranked equal to the natural sciences, particularly mathematics."

"Mere performance"? What if our society were to substitute math karaoke for live mathematicians? Maybe that's what computers do. Wd mathematicians be forced to work as laborers like Gypsy musicians are? Wd they be begging on street corners?

"Very little is known of Pythagoras' life; what we do know comes entirely from works written several centuries after his death". Maybe when Hippasus tied together Pythagoras' sandal laces, they were on a ship, & *Pythagoras* fell overboard. Maybe the fish spread the idea of irrational numbers. Whatever happened, "knowledge of the discovery spread, and soon other irrational numbers were found. For example, the square root of every prime number is irrational".

One of the things that I get, in my skewed way, out of math is a system of highly multi-leveled 'reality'. To strip down Maor's explanation of *algebraic* numbers thru selective quoting: "Most of the numbers we encounter in elementary algebra can be thought of as solutions to simple equations" [..] "-1, 2/3, and the $\sqrt{2}$ are solutions of the polynomial equations $x + 1 = 0$, $3x - 2 = 0$, and $x^2 - 2 = 0$, respectively."

≈(One of anything plus the number one itself divided by the abstract concept of number) to the power of the letter "n."

Joke teller: "Wanna hear a knock-knock joke?"
Joke listener: "Sure!"
Joke teller: "Okay, you start"
Joke listener: "Knock-knock."
Joke Teller: "Who's there?"
Joke listener: "Er... I don't know!"
Joke teller: "You don't even know who you are & you come bother knocking on my door!"

[..] "A real number that satisfies (is a solution of) a polynomial equation with integer coefficients is called *algebraic*."

But there are *nonalgebraic* numbers! "It was in 1844 that the French mathematician Joseph Liouville (1809-1882) proved that nonalgebraic numbers do indeed exist. His proof, though not simple, enabled him to produce several examples of such numbers. One of his examples, known as Liouville's number, was

$$1/10^{1!} + 1/10^{2!} + 1/10^{3!} + 1/10^{4!} + \ldots,$$

whose decimal expansion is 0.11000100000000000000000100 . . ." [..] "A real number that is not algebraic is called *transcendental*." *e* & π are both transcendental.

Not to be confused with the numbers invented by Ralph Waldo Emerson or the # of trees which burned down when Henry Thoreau accidently set fire to the woods of Long Island.

"In 1874 the German mathematician Georg Cantor (1845-1918) made the startling discovery that there are more irrational numbers than rational ones, and more transcendental numbers than algebraic ones. In other words, far from being oddities, *most* real numbers are irrational; and among irrational numbers, most are transcendental!" In trying to understand that, remember the different degrees of infinity.

Occultist Aleister Crowley, w/ the publication of his Book of Law, announced the beginning of the Age of Horus. In this bk I proclaim that we've long since been in the midst of the Age of 'Pataphysics - a time when oddities & exceptions are found to be in the majority in 'reality'.

A History of π (PI) - Petr Beckmann. Right off the bat Beckmann scores a hoop shot past the goalie w/ a bishop's move in the end zone by starting his bk around 2,000 BC instead of w/ Pythagoras. It was around this time that us bipeds that my readership presumably belongs to, "had grasped the significance of the constant that is today denoted by π". In his preface, Beckmann writes: "Being neither an historian nor a mathematician, I felt eminently qualified to write [this] story." He was a professor of electrical engineering as of 1971 when this edition was published.

Beckmann starts off explaining the development of math as the development of quantifying: big rock, heavy rock, heavier rock, [heavy metal music] - that sort of thing. Then he makes the interesting assertion that the idea of ratios didn't develop until *after* we started counting higher than 2. He supports this by pointing out that "In Czech, until the Middle Ages, there used to be two kinds of plural - one for two items, another for many (more than two) items, and apparently in Finnish this is so to this day. There is evidently no connection between the (Germanic) words *two* and *half*; there is none in the Romance languages (French: *deux* and *moitié*) nor in the Slavic languages [..]. Yet in all European languages, the words for 3 and 1/3, 4 and 1/4, etc., are related.

This suggests that men grasped the concept of a ratio, and the idea of a relation between a number and its reciprocal, only after they had learned to count beyond two." Fascinating.

Right there, if this had been a history of BI instead of PI, I wd've been happy. Beckmann's refreshing. He explains π as the ratio of Circumference divided by Diameter. That seems so direct. & he claims that the 2000 BC dating is correct by saying that the Babylonians & Egyptians were aware of this ratio by then.

Harkening back to basic reasoning some more he writes: "If the volume of a stone is doubled, the weight is doubled; if you run twice as fast, you cover double the distance [..] Of course, the rule does not always work: A tree twice as old is not twice as tall. The reason is that "the more . . . the more" does not always imply proportionality; or in more snobbish words, not every monotonic function is linear."

I like it. But then my tendency is to once again conflate in my leisure time: A tree twice as old runs twice the distance as a stone half its weight. On Wikipedia, monotonic function is defined as: "a function which preserves the given order." & they give the example that "if for all x and y such that $x \leq y$ one has $f(x) \leq f(y)$, so f preserves the order." I just like the "not every monotonic function is linear" - I'm sure I can pervert that in some way sooner or later! But, then again, not every monotonous person is trying to make a point.

History told thru the focus of any particular discipline is an interesting way to get a microcosmic view. "The Hindus later used a circle for the digit zero (0), and this reached Europe via the Arabs and Moors only in the late Middle Ages (Italy) and the early Renaissance (Britain). An edict of 1259 A.D. forbade the bankers of Florence to use the infidel symbols, and the University of Padua in 1348 ordered the price lists of books should not be prepared in "ciphers," but in "plain" letters (i.e., Roman numerals)."

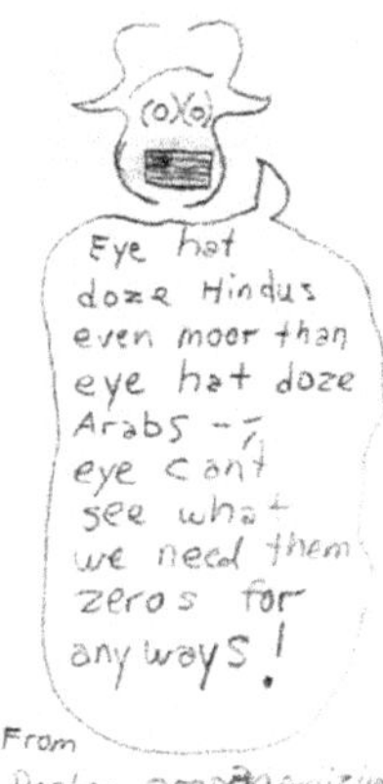

From Poster propagandizing the inception of The Department of New Homeland Security Mathematics.

I wonder how many people read accts like that & think something to the effect that it's preposterous, thankful that things aren't like that today. &, yet, *are* things like that today? I wonder if there is an equivalent. What wd happen if an Indian or Arab measurement system were proposed to replace or augment United States customary units (inches & feet & so on) in this day & age? In the US, the right-wing talk show hosts wd all be complaining about the "towel-heads" so fast it wd make *my* head ache & help create even more { } from the neck up than we already have.

For that matter, Beckmann, a former Czech turned 'American', mentions attempts by the United Nations to create a "World Calendar" & then ends his paragraph re this w/: "Most governments agreed to the proposal in principle , but some (including the US) considered it "premature" and the matter is still being "considered." The UN, a grotesque assembly of propaganda-bent hacks, has found itself unable to condemn international terrorism by criminals, much less to reform the calendar."!! If the UN had proposed introducing zero wd he have been opposed? (Ok, I know that's a silly question.)

Beckmann goes on to diss Aristotle, who he describes as the "one whose teachings held up the progress of science for close to 2,000 years". Ha ha! "Aristotle, we are invariably told, was 'antiquity's most brilliant intellect,' and the explanation of this weird assertion, I believe, is best summarized in Anatole France's words: The books that everybody admires are the books that nobody reads." Coincidentally, I've been vacillating about whether to talk about what I call Bird-Brainism, wch I've so far decided against, but now that A. France is mentioned, the time for a segue has clearly come.

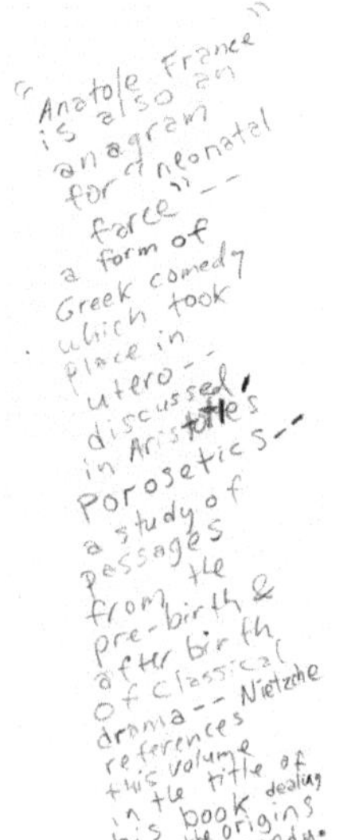

While reading these math bks, some non-math bk might be mentioned that seemed interesting, so I'd make note of it. In the Turing bio, Elmer Rice's play Judgment Day, about the Reichstag fire trial, was mentioned; Nobel Prize winning French author Anatole France's La Revolte des Anges (The Revolt of the Angels) was mentioned elsewhere. I've gotten both out of the library recently.

Anatole France has the fate, perhaps somewhat unfortunate, of being the guy whose posthumously determined small brain size (2.24 pounds as opposed to the average 3 pounds) is used to demonstrate that "big brain" doesn't automatically correlate w/ "intelligence".

When I was in my early 20s I took a Physical Anthropology class at a community college. That's probably where the subject of France's brain-size & its significance was mentioned. The 1st paper I wrote for that class is reproduced below:

The Significance of Efficient Brain-Use

Anthropologists have used cranial capacity as an indicator of the degree of mental ability. This makes Neanderthals "superior" to "twentieth-century" people. This practice is based on the belief that the brain's ability to manipulate functions is relative to its # of cells. However, a major mental ability increase is not necessarily accompanied by brain-size increase.

Bird-Brains must remain small & light & still function complexly in order for flight to occur. This requires great brain efficiency rather than great size. Homonids could follow this example & make birthing less painful for the mother with a smaller head size.

If one accepts the idea that words & images are a major building block of thinking, one step toward greater brain efficiency might be to exploit the ambiguities of language to permit a multitude of simultaneously productive thoughts.

In a society of interdependent specialists, the ability to create & send & receive useful synopsises can simplify the need for each individual to go through processes redundant in terms of the societal whole. Telepathy can make a person part of a giant brain without the need for a one-body housing.

There is a religious myth of winged homonids as beings on a higher evolutionary plane than non-winged homonids. It should be obvious by now that these beings, commonly called angels, should more properly be called Bird-Brains.

Note also that such an evolutionary plan would significantly decrease the loss of heat from the human body & hence reduce the economic expenditure on hats, caps, & scarves required in winter.

Is that wool hat my hat?

Human.

Do you really have to ask?

Bird Brain.

Ha ha! So, you see, I haven't changed much. Motion isn't possible, I'm still young at heart or I never grew up. I particularly like the "make birthing less painful for the mother with a smaller head size"! Why stop there? Why not have the head be shaped like a cone to make progress thru the birth canal a gradual widening process? Sheesh! Babies are so inconsiderate!

The teacher liked my paper. The 2nd paper I wrote for the class was on something like "Death as an Evolutionary Progression". In this case, the teacher refused to return the paper to me or to tell me what grade I got on it & seemed afraid of me after that. I wish I still had that one to reference!

The above bit of late juvenilia was worked into a 'performance' of mine 12 yrs or so later called "Generic As-Beenism". I presented that from 1987 to 1989 w/ 'stunning success' in Montréal & 4 other cities.

To be contrary to Beckmann's use of France's quote, it shd be obvious that there are plenty of other people that Beckmann wd probably concede to be brilliant whose bks are just as widely under-read. Anything intellectual isn't likely to make the New York Times' Best-Seller list unless they artificially manipulate the market to put it there.

Lest you think that going from Bird-Brainism back to Beckmann is a rough transition, I quote from his comment re comparing differing attitudes towards Euclid's 5th axiom: "At first sight the difference between the two seems to be a chicken-hearted technicality." Winged Hominids Revolt! Against this disparaging metaphor, ie.

But seriously, folks, these days, as a mature adult, I'm really in favor of replacing the brain altogether - w/ Smart Dust. It's more energy efficient & leaves the majority of the cranium free for improved iPod, GPS, & cell phone technology. Just think! W/ Smart Dust analyzing yr GPS data you'll finally know where you are!

Beckmann devotes a whole chapter to Archimedes (ca. 287-212 BC) who he calls the "first scientific engineer". Later, he says that "Archimedes was a pioneer of method. He took a step from the concept of 'equal to' to the concept of 'arbitrarily close to' or 'as closely as desired'". Typically, it's the idea of somehow specifying such fluid quantifiers that appeals to me.

Getting away from the math, the examples provided are vivid. Regarding the Archimedean Spiral, the author writes: "it is defined as the plane locus of a point moving uniformly along a ray while the ray rotates uniformly about its end point. It will thus be traced by a fly crawling rapidly outward on a turning phonograph record." Such an image opens up a whole new world of Archimedean Flea Circus Turntablism!

Beckmann writes about the discovering of a copy of an Archimedes text as a palimpsest - a text whose remnants are retrievable under another manuscript. "The mediaeval zealots did not always, like the Bishop of Yucatan or the crusaders at Constantinople, burn scientific books as work of the devil. Sometimes they would only wash off the text for the sake of the parchment, so that they might besmirch it with their superstitious garbage."

He doesn't spare Pythagoras either! Where other math historians tread diplomatically, he roars right in: "Pythagoras, for example, was a mathematician to a much lesser degree than he was a mumbo-jumboist. He taught that one stood for reason, two stood for opinion, three for potency, four for justice, five for marriage, seven held the secret of health, eight the secret of marriage, the even numbers were female, the odd numbers male, and so on." Was six the hermaphrodite? As the product of 2 & 3? Damn! He even says that the Pythagorean Theorem doesn't deserve to bear his name because "it had been known to every rope-stretcher (surveyor) from the Nile to the Yang-Tse Kiang for a thousand years before Pythagoras' witchcraft."

A History of π (PI) is full of juicy morsels of all kinds of brain-food. The etymology catches my eye. We learn that *x* (as the unknown quantity) was called *cosa* by the Italians - as in the name for the mafia: *cosa nostra* - "our thing".

The explanation that Wikipedia provides for this is that "Mafiosi introduce known members to other known members as belonging to "cosa nostra" (our thing) or *la stessa cosa* (the same thing), meaning "he is the same thing, a mafioso, as you"." Given that I originally spelled my name as tentatively, a convenience & the name of a performance group I was in as a roving band of defiant youths so that when written in sentences they wd blend right in, I can say that this is a time when I actually find something about the mafia interesting. If the world is ever lucky enuf to be rid of the mafia it can be renamed *cosa nulla* - "nothing thing."

Ever on the alert for mistakes being the foundation of something, my eyes perked up when I learned that the term *sine* is a result of a mistranslation from what was thought to be Arabic but what was really an Arabic abbreviation for something in Hindu. The result, anyway, ended up erroneously as the Latin word for bay, inlet, or cavity: *sinus*. Is a sine wave, therefore, a runny nose?

The author informs us that there's no scientific value in plotting π to more than 17 decimal points. Nonetheless, according to <http://www.geocities.com/SiliconValley/Pines/5945/facts.html>: "In 1991, the Chudnovsky brothers in New York, using their computer, m zero, calculated pi to two billion two hundred sixty million three hundred twenty one thousand three hundred sixty three digits (2, 260, 321, 363). They halted the program that summer." Precisely because such obsessiveness seems so over-the-top & useless, I find it appealing. What if at the 2,260,321,36***4th*** digit there was some sort of punchline?

I thought I was pretty slick for being able to quickly calculate in my head operations like 2,374 x 3,455 when I was a teenager (if I was even *that* good!). Beckmann reports:

"Truman Henry Safford (1836-1901) of Royalton, Vermont, could instantly extract the cube root of seven-digit numbers at the age of 10. At the same age, he was examined by the Reverend H.W. Adams, who asked him to square, in his head, the number

365,365,365,365,365,365.

Thereupon, reports Dr. Adams,

> He flew around the room like a top, pulled his pantaloons over the tops of his boots, bit his hands, rolled his eyes in their sockets, sometimes

sometimes smiling and talking, and then, seeming to be in agony, until, in not more than a minute, said he,

133,491,850,208,566,925,016,658,299,941,583,255!"

Show-off. While it may not be the most common thing in the world for people to be able to manipulate vast numbers mentally, there still seem to be a fair amt of reports of people doing so. What if reinstantiation plays a part? A young brain, not yet demolished by adult toxicity, retains every mathematical pattern it ever uses. As such, to make a new calculation is just to rerun memories w/ new numbers slotted in. Recognizing what patterns need to be run is like pushing the appropriate buttons on a mental calculator: reinstantiation takes over &, VOILA!, results at the speed of neurons firing.

Let's extrapolate. Why stop w/ numbers? Any kind of pattern recognition or applied memory shd do. Math is *easy* in contrast to other possibilities. After all, arithmetical systems are pretty straightforward - even massive numbers like the above are distinct for their *size* more than for their conceptual complexity. No doubt Safford had an incredible ability to retain the numbers in his head.

What if a person cd remember just about *anything?* - like the man described in The Mind of a Mnemonist by A.R. Luria. In his case, he applied his ability by being a memory performer. That's wonderful enuf in itself but what if a more imaginative application cd've been thought of? What if everything memorized fit together more & more like pieces in an enormous puzzle instead of being mainly independent details?

What if memory reinstantiation cd be transferred from person to person? Think of the useful abstractions that a person cd make if they cd correlate *all the patterns in more than one person's mind at ease!* I've objected before (w/o explaining why) to brains being compared to computers. Perhaps I object largely because it seems like an insult to the complexity of the brain. Or because computers are just the tech-flavor of the mnth (or the century). Still, it's easy enuf to transfer memories from one computer to another - so let's do it w/ brains. Naturally, this has been a subject for SF & my mentioning it anew doesn't bring us one step closer.

Mnemonics. Of course Beckmann, w/ his ever-roving intellect, has to bring those in too! He provides one for π:

"*How I want a drink, alcoholic of course, after the heavy lectures involving quantum mechanics!*"

The number of ~~syllables~~ letters per word (Bruce's marginalia is correct) organized in order to create π = 3.14159265358979!

That's nothing! I can perform an even more elaborate set of gestures within the same time frame & rather than coming up with merely one right answer, I can produce nearly a dozen wrong ones! — listed [illegible] egomaniac[illegible] adity. Top that Mr. Ability pants!

Letters I guess you mean: — otherwise the # is: 11114112122323— which, if memory serves is the zip code for the troglodyte capital of Cydonia, Mars.

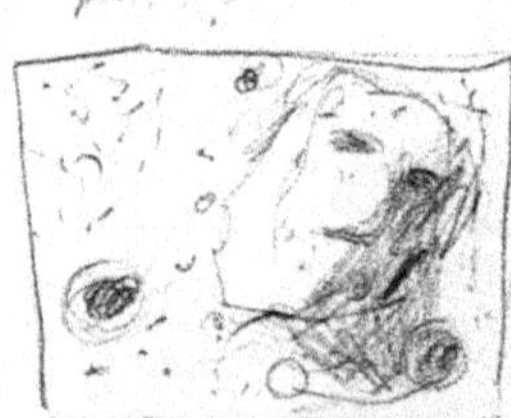

ADABABBACABAC the Cydonian capital, as viewed from human satellite.

The possibilities of having 2nd level content thru counting! Here's a self-description I wrote for my high-school yrbk when I was 16:

> **"Nascent, orthopteran, sabaist, luxated, oleographic, turgid; labiac, excogitate, accentric, hydrophytic, crispy, intarsia, magnetize."**

Do you see the hidden message? Not only was I a *newborn birdlike star-worshipper dislocated chromolithograph printed with oil paint on canvas in imitation of an oil painting excessively ornate or complex in language like a labia, to consider or think something out carefully & thoroughly w/o a center adapted to grow in water, firm but easily broken or crumbled decorative inlaid pattern in a surface to make something magnetic* but I was also "straight". Can you find the sentence that tells you that? If you can't, I hope it won't arouse yr ire.

I've had many a phone number that's been chosen especially to be a mnemonic - in fact just about every one I've had in the last 30 yrs. At one point, I was living across from a church & I was checking to see wch words cd be chosen from the available exchanges when I discovered that SATANIC was an option. That wd've been easy to remember!

I made a note to myself in A History of π (PI) to write a text all in mnemonics. Maybe a mnemonic for remembering mnemonics. A meta-mnemonic. Eg: **BANG OAR**! cd be a meta-mnemonic for **B**ecause **A**rithmetic **N**ecessary **G**eography, **O**cean **A**rgument **R**hythm! - wch is, in turn a mnemonic for **B**ig **E**lephants **C**an **A**lways **U**nderstand **S**mall **E**lephants; **A R**at **I**n **T**he **H**ouse **M**ay **E**at **T**he **I**ce **C**ream; **N**ot **E**very **C**at **E**ats **S**ardines (**S**ome **A**re **R**eally **Y**ummy); **G**eneral **E**isenhower's **O**ldest **G**irl **R**ode **A P**ony **H**ome **Y**esterday; **O**nly **C**ats' **E**yes **A**re **N**arrow; **A R**ude **G**irl **U**ndresses - **M**y **E**yes **N**eed **T**aping!; **R**hythm **H**elps **Y**our **T**wo **H**ips **M**ove.

All of the latter sentences being mnemonics originally intended to help one remember how to spell the words that now, in this meta-mnemonic, must be spelled correctly *before* one can get to the sentences! (I reckon that's a sortof 'reverse engineering') There! I can think of something even more useless than remembering π or calculating it to 2, 260, 321, 363 decimal places!

Maybe that's what π *really* is - just a mnemonic for remembering long strings of words w/ specified numbers of syllables. Forget all that *constant* stuff, all that stuff about π being the number of times that the diameter of a circle wraps around its circumference.

By the by, it was through <http://www.fun-with-words.com/> that I learned the above mnemonics. It was also thru them that I learned

~~2 4 3 3 5 2 2~~
~~4 3 4 2 5 3~~ - scratch that -- the meaning is coded in the semicolon between "turgid" & "labia" -- "semicolon" anagrammatizes into "Is monocle" -- so there we've got the Is." "Monocle" relates to the song "Glass Onion" by the Beatles -- a song responding to fans' desires to find secret & hidden messages in the pop group's music. (yes, I realize the anachronism, but bare with me all temporal nudists) "Glass Onion" is also a metaphor for a prism & hence refers to the physical principle of wave refraction described in the mathematic expression $\frac{\sin\theta_1}{\sin\theta_2} = \frac{V_1}{V_2} = \frac{n_2}{n_1}$. Since refraction refers to the bending of the wave -- or turn in its direction -- it points, by means of semiotic allegory, not to that which is "straight," direct & linear, but to indirectness & curvature!

that there are many other mnemonics for remembering π to various extents - w/ one going all the way to 700 decimal places! & I learned that "One gentleman, **Hiroyuki Goto**, recited no less than 42,000 [actually, according to Wikipedia 42,195] digits of pi from memory in 1995. This record feat took him about nine hours to complete." I wonder what he does w/ his phenomenal memory otherwise?!

Yes, I wondered. So I looked him up on Wikipedia & found that he's the inventor of a word puzzle video game. I also found that **Akira Haraguchi** recited π to 100,000 digits in 16 hrs in 2006 at age 60. "Haraguchi uses a system he developed, which assigns Kana symbols [syllabic Japanese scripts] to numbers, allowing for the memorization of Pi as a collection of stories." He "views the memorisation of Pi as 'the religion of the universe', and as an expression of his lifelong quest for eternal truth."

Beckmann admits to the reader that science is based on a *faith* "that nature is consistent." [..] "There is no theoretical reason why tomorrow some scientist should not report an experiment that contradicts it". To quote myself from page 92: "An interesting sidenote speculation here might be that perhaps it's not so much that science has been *wrong* in its conclusions over the yrs *but that* physical 'reality' might've actually *changed*." Maybe I don't share the faith in nature's consistency.

Probability Theory: Beckmann presents it as having originated w/ Gerolamo Cardano, a mathematician & gambler (etc), around 1526 in a bk called "*Liber de ludo aleae* (The book of games of chance)". This hearkens back to our old friend Marcel Duchamp's flirtation w/ systematic gambling at Monte Carlo that he attempted to finance w/ "Monte-Carlo" bonds that he made using a Man Ray photo of himself w/ shaving cream sculpting his hair into Pan horns.

The most interesting thing that I've ever read about probability was probably in Stanislav Lem's collection of fabulously clever fictitious bk reviews entitled <u>A Perfect Vacuum</u> - in particular the 'review' of 'Cezar Kouska"s "***De Impossibilitate Vitae** and **De Impossibilitate Prognoscendi***" in wch it's argued that "either the theory of probability, on which stand natural history, is false to its very foundations, or the world of living things, with man at its head, does not exist."

Basically, Lem explains, in the form of the ersatz review of the ersatz bk, that, logically, probability theory is so strongly against the possibility of even the simplest thing happening that even these simple things are basically impossible. In Lem's autobiographical essay entitled "Reflections on my Life" in his bk <u>Microworlds</u>, we learn that this theory is an outgrowth of Lem's ruminations on such things as the odds against

so... if these words will not allow us to arrive at "straight" by means of a direct path-- what if we "refract" them slightly thus: (curvature also accounts for the temporal anomalies inherent in my application)

magnetize
intarsia
crispy
hydrophytic
accentric
excogitate
labiae
turgid
oleographic
luxated
sabaist
orthopteran
nascent

Ignore this. Look for the answer yourself! Don't cheat. Oh shit, you're already cheating! Now find an alternate solution.

∴ the current density of "Michael Tolson"
$= J = \frac{I}{\pi R^2} - w$ (inside the ire)
& $J = 0 - w$ (outside the ire)
w/ "M.T." ire diameter estimated to a numerical value of 3 (3 being the # of letters in the word "ire")
& ire length estimated to a numerical value of 25 (25 being the number of letters

his Polish army officer father escaping execution by Soviet soldiers post-WWII by the highly fortunate happenstance of a Jewish barber who knew him being nearby & intervening on his behalf as the Soviets were marching Lem's father away to his death.

Lem's incredible novel The Chain of Chance pursues a mystery in wch the solution ties together things so improbable that they'd seem *impossible*.. but they're not. Just unlikely. The denouement of the bk presents an ending spectacular in its unexpected banality. As the astronaut/detective protagonist explains the resolved mystery he brings up an example of mediocre marksmen hitting 3 flies & a postage-stamp target at a great distance *just by chance*. He concludes: "The Naples mystery was a result of a random causality, and it was the same random causality that solved it."

It's Lem's carefully applied Theory of Probability explored logically to its dead-end of impossibility that resonates so strongly w/ me. In this bk I hope to reach similar dead ends. W/ trap doors in them. I've already reached at least 2.

In chapter 16 of A History of π (PI) the transcendence of π is explained. & I was planning to possibly quote from that to explain it here. But, then, on page 167, this caught my eye: "There is, however, no reason why an irrational number should not be a root of an algenraic equation". The typo there isn't mine. What if the obvious mistake of having an "n" substituted for the intended "b" in "algenraic" were to somehow be perpetuated? As the mistake in the creation of "sine" was perpetuated as a mathematical term?

Here's an example from my own experience. In January of 1983, I was quoted correctly in a newspaper as saying "Normality is what cuts off your sixth finger and your tail." Fine. Then 9 mnths later, in a much more widespread context, that quote was itself misquoted by another reporter as "Normality is what cuts off your sixth finger and you fail." Quite a difference, eh?! This, then, goes on public record to be quoted again. Fortunately, when it *was* quoted again, in Re/Search magazine, I got a chance to correct the quote back to its original form.

What if I were to perpetuate the typo in Beckmann's bk? (Typos being quite rare in it, by the way) Now, as it turns out, when I did a Google search for "algenraic", 840 results were obtained! So, if I didn't 'know' better, it wd be pretty easy to conclude that there's "algenraic number theory", "Algenraic plane Curves", "Algenraic Operational Semantics", "linear algenraic solutions", "Long algenraic multiplication", etc, etc! I Googled "algenraic + definition" & got 63 results!

Now, Google, being well-designed & prepared for such typos, asked: "Did you mean: *algebraic* + definition"? & the 1st return in my

making up the 1st 3 words in the series [nascent orthoptera sabaist]) M.T. ire induction valve is given as 140.7829150 9174002 nH -- which is NOT easy to determine from the formulas $X = \sqrt{1+(d/2L)^2}$ & $L = 2L[\ln[(2L/d)(1+X)] - X + \mu\partial + d/2L]$

definition search is from a website of misspellings that listed various permutations that one might accidentally make on "algebraic" - so there are plenty of opportunities provided to correct such a mistake. The typo is a particularly easy one to make given that on the QWERTY keyboard the "n" key is immediately to the right of the "b" key.

Mischievous little fuck that I am, I get tempted. I go to <http://wiki.answers.com/> & I ask: "Is it *proven* that algenraic functions turn dyslexic topological polynomials on their head in Non-Non-Euclidean Riemannian after-space?" Of course, I got some interesting "Is one of these your question?" responses but we've already been there so let's move forward.

In "Step 2: Review question wording and submit", it's suggested that I may have some misspellings & 4 of their 5 suggestions are somewhat spot-on:

> "algenraic (Suggestions: algebraic, allegoric, allergenic)
> Non-Non-Euclidean
> Riemannian (Suggestions: Romanian, Riemann, Remaining)
> after-space (Suggestions: afters's, intersperse, overdoses)"

Revising the question to read something like: "Is it *proven* that allergenic functions turn dyslexic topological polynomials on their head in Non-Non-Euclidean Romanian overdoses?" tugs at my mind-strings but I resist. "Is it *proven* that allegoric functions turn dyslexic topological polynomials on their head in Non-Non-Euclidean Remaining afters's?" is quite lovely too, of course.

I'm a bit disappointed that there aren't any alternates suggested for "Non-Non-Euclidean". Shd I submit the question? To me, it's a mad scientist experiment, to someone else, it might be 'vandalism'. I want to see what type of interpretation such an intentionally absurd question might have - especially given its *almost* believable faux terminology. I also want to see if anyone out there who understands such jokes as "Non-Non-Euclidean", etc, will respond in kind - taking it all in a good natured way. So, at the risk of censure from WikiAnswers, I submit the question.

Clicking on the "submit" button takes me to the next step in wch I have to click on the category links until I find the topic that most fits my question. For starters, there's "Animal Life", "Business and Finance", etc.. Choosing something completely inappropriate is very tempting: "Hobbies and Collectibles", perhaps? Or something that wd temper the possible wrath of the person answering the question: "Humor and Amusement Corner"? But, no, let's remain 'pure' & pick "Science and Technology".

That leads to another set of delectable choices: "Astronomy and Space", "Consumer Electronics"? I stay on target & pick "Math". The question is posted. "If you login you can watch the question and be alerted when it is answered or edited." I'm also given the chance to browse questions that need answers since WikiAnswers is all volunteer. That seems like a fair trade so I click on that.

A question asked 51 seconds ago is: "What is up with that?" That seems like someone w/ a sense of humor so I answer: "algenraic functions turn dyslexic topological polynomials on their head in Non-Non-Euclidean Riemannian after-space". It's tempting to answer "When are there Disney auditions?" in the same way but I refrain. All in all, my answer is so multi-purpose, it cd answer "What were the rules for moral conuct in Hinduism?" (after all, one good typo deserves another) or "What happened when Jesus rose from the dead?" (my guess is as good as anyone else's!). Stay on-focus.

He reestablished the Church of Parthenogenesis as a theme park called Anthanasia?

HA!

Oopsie! I neglected to enter the code when I put in my answer. I corrected that & my answer has appeared. Now "algenraic functions turn dyslexic topological polynomials on their head in Non-Non-Euclidean Riemannian after-space" has entered the world of internet 'reality'. Alas, I didn't get in the 1st reply! Someone else answered before me: "depends on what that means... but to answer you question... i don't know." Whew! What a relief.. there *are* other jokers out there.

I check to see if my original question has been answered yet. It hasn't. I reckon that's not such an easy one as "What is up with that?". I'll keep checking. What a disappointment it'll be if it never gets answered. Surely there's someone out there who can solve my problems?!

Where the question now appears (still unanswered) there are now links to Google ads. At 1st, they were attempts at relevance: "Non Euclid Geometry", "Topology" - that sort of thing. Now they're all connected to "Zac Efron" - offering, eg, a "Zac Efron Screensaver", "How to meet Zac Efron", "Efron Problems", & "Efron Couple". I just figure that these are phishing attempts to get access to my email address so I refrain. Still, I've always wondered how to meet the guy (whoever the fuck he is!).

Beckmann's final chapter is vehemently pro science & technology. In his vituperative style he denounces "pious, respectable and community-minded ladies", "Julius Caesar", "the Grand Inquisitor", "the Soviet censor", & "Fascists of the Left" as enemies of progress (not his words) & I more or less agree w/ him. Alas, IMO, he shows a meager socio-political understanding when he disses the Luddites.

He does qualify that by writing: "I have some sympathies for the Luddites who were uneducated, miserable, and desperate." In the end, though, I think that Beckmann's optimisms for the new-improved-future-thru-technology are as whimsical (or not) as my own hopes for a more flexible 'reality' thru "ambiguity gates":

"THREE centuries ago, Gottfried Wilhelm Leibniz, co-inventor of the calculus and co-discoverer of the first infinite series for π, dreamt of the day when courts would be abolished, because disputes would be settled mathematically by solving impartial equations that would show who was right and who was wrong. The intelligent computer that is now being born [remember, this was published in 1971] makes that dream somewhat less fantastic."

Did I say *almost* as whimsical? Maybe *more whimsical* might be more accurate. Reading Voltaire's Candide might be in order.

Alas, my question still hadn't been answered 4 hrs later on WikiAnswers so I just answered the question myself w/: "What is up with that?". No sense in leaving any loose ends.

Zero - The Biography of a Dangerous Idea - Charles Seife. This starts off w/ *Chapter* **O. Null and Void.** "Zero hit the USS *Yorktown* like a torpedo." A dramatic beginning. Seife explains that because a zero was left in newly installed software on this "billion-dollar missile cruiser" [I imagine it was quite a bit more than a billion dollars!], when its "computer system tried to divide by zero, 80,000 horsepower instantly became worthless." Zero, the saboteur; zero, the subversive.

"Zero is powerful because it is infinity's twin. They are equal and opposite, yin and yang. They are equally paradoxical and troubling. The biggest questions in science and religion are about nothingness and eternity". What a difference in writing style from Beckmann! Going from one bk to the other is like riding out of the sauna on a camel & diving off its hump into the swimming pool. Or something. I like it.

It's like I'm looking in a mirror.
0
yeah, right
∞

&, yet, Beckmann's 1st chapter almost immediately starts w/: "The first dim flash in the darkness goes back to the stone age - the bone of a wolf with incisions to form a tally stick" & by Seife's *Chapter* **1: Nothing Doing**, Seife writes about the same thing: "a 30,000 year old wolf bone with a series of notches carved into it. Nobody knows whether Gog the caveman had used the bone to count the deer he killed, the paintings he drew, or the days he had gone without a bath".

Zero was published in 2000. Since then, as everyone knows, it's been unanimously decided that Gog (more correctly pronounced GE-og) was counting the times that the not-Horse had filched the apples (that GEog had picked & pickled) from the other side of the 'reality' divide. The notches were not only counting marks, they were also elaborate plans for catching the not-Horse. Given that not only hadn't zero been conceived of yet & that even 'reality' itself was an unknown notion, & given that the 'laws' of nature were totally different then than now, catching the not-Horse was relatively possible & considerably easier than, say, riding an escalator.

Again, Seife's trajectory is similar to Beckmann's. It's almost as if history has been agreed upon. "The Siriona Indians of Bolivia and the Brazilian Yanoama people don't have words for anything larger than three; instead, these two tribes use the words for 'many' or 'much.'" Not such a bad idea, really. Imagine getting yr winter gas bill & having it be for "much" - you cd get away w/ paying $3.01 & afford to keep the house nice & toasty instead of having to rub yr hands together all the time.

Don't want to go to the Netherworld after you die? Afraid there's going to be a bunch of assholes there? "In the Egyptian Book of the Dead, when a dead soul is challenged by Aqen, the ferryman who conveys departed spirits across a river in the netherworld, Aqen refuses to allow anyone aboard 'who does not know the number of his fingers.' The soul must then recite counting rhyme to tally his fingers, satisfying the ferryman. (The Greek ferryman, on the other hand, wanted money, which was stowed under the dead person's tongue.)" That last part settles it, if there's still money in the afterlife I want to go *somewhere else*. Pretend like you can't count & leave strict instructions that when you die, *all* the moolah stays w/ the living.

25,000 yrs ago, the Babylonians had the most sophisticated counting system. Right there in what's now known as Iraq. "At first glance, the Babylonian system seems perverse. For one thing the system is *sexagesimal* - based on the number 60." Other systems were more obviously based on the number of fingers & toes that people are presumed to've had at the time. However, given what we now know today, we see that the Babylonians were not all that different. They just had 60 fingers & toes.

& then there's even some etymology: "The words *calculate*, *calculus*, and *calcium* all come from the Latin word for pebble: calculus." & predicting the odds for the number of ripples whilst skipping calculi on the sine was how algenra originated.

"By around 300 BC the Babylonians had started using two slanted wedges, [mentally insert illustration here], to represent an empty space". In other words, the ancestors of the Iraqis, who the average 'merican probably thinks of as some "towelhead" sitting on a rock (or are them them Afghanis?) invented a central notion in mathematics.

Seife identifies zero as a *digit*, not a number. He clarifies the point by giving the example that "000,002,148" = "2,148". Therefore, one of the axioms of PNT is that that's not true. Maybe not false either.

Mayan mnths had 20 days, numbered from zero to 19 & the numbers were all represented by profiles of faces - mostly looking pretty grumpy. Someone shd animate them in a protracted imaginary argument between members of the Bernoulli family in a way that represents the numbers of π to 2, 260, 321, 365 places & have **Hiroyuki Goto** & **Akira Haraguchi** enact the argument from memory while mud-wrestling General Eisenhower's oldest girl (GEog) riding a not-Horse.

Both Seife & Beckmann discuss Egyptian mathematics. Beckmann gets into how rare surviving relevant artifacts are & wonders "how many such priceless documents have been used up by the Arabs as toilet papyri". Seife informs us that: "The Egyptian civilization's lack of zero was bad for the calendar and bad for the future of Western mathematics." I don't know whether Beckmann's referring to actual incidents w/ his "toilet papyri" comment but, if he was, maybe the Arabs were just trying to protect us from a lack of zero.

Here's where it gets good for me:

"In the previous example we saw that 2 x 0 is 0. Thus to undo the multiplication, we have to assume that (2 x 0)/0 will get us back to 2. Likewise, (3 x 0)/0 should get us back to 3, and (4 x 0)/0 should equal 4. But 2 x 0 and 3 x 0 and 4 x 0 each equal zero, as we saw - so (2 x 0)/0 equals 0/0, as do (3 x 0)/0 and (4 x 0)/0. Alas, this means that 0/0 equals 2, but it also equals 3, and it also equals 4."

Hmmm... I always thought it was created by Chaos or Night or the Great God of the Big Sneeze or something.

Who needs PNT? *Chapter* **2**: **Nothing Comes of Nothing**. "The Greek Universe, created by Pythagoras, Aristotle, and Ptolemy, survived long after the collapse of Greek civilization. In that universe there is no such thing as nothing." [..] "Before they could accept zero, philosophers in the West would have to destroy their universe."

"Hippasus of Metapontum stood on the deck, preparing to die. Around him stood the members of a cult, a secret brotherhood that he had betrayed. Hippasus had revealed a secret that was deadly to the Greek way of thinking, a secret that threatened to undermine the entire philosophy that the brotherhood had struggled to build. For revealing the secret, the great Pythagoras himself sentenced Hipassus to death by drowning." The story gets more & more lurid & detailed the more time lapses. People took philosophy a little too seriously if you ask me.

Hippasus stood at the end of the bar ready to die. Around him the Steelers were forcing him to drink his 3rd case of cheap beer in the last hr. Hippasus had revealed a secret that the Steelers had lost the 2008 season against the still-alive original members of the 1952 Dallas Texans (now in their 70s & 80s) because an irrational number had been

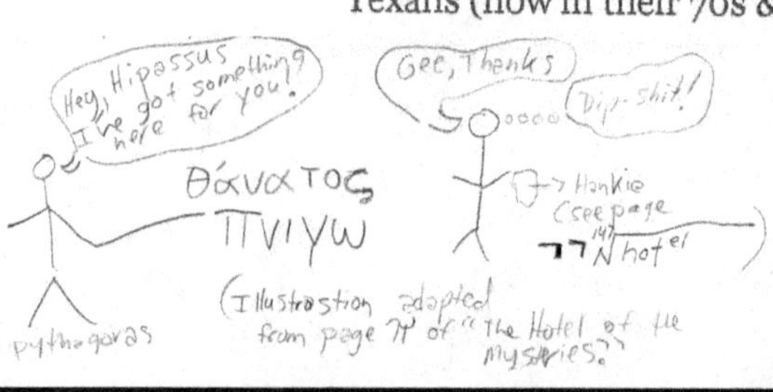

accidentally put in the game plan. Pythagoras himself sentenced Hippasus to death by alcohol poisoning. People took football a little too seriously if you ask me.

"Pythagoras argued that the earth sat at the center of the universe, and the sun, moon, planets, and stars revolved around the earth, each pinned inside a sphere. The ratios of the sizes of the spheres were nice and orderly, and as the spheres moved, they made music." The game plan was meant to be simple, just throw the ball 1/2 of the way towards Saturni & have someone catch it & run the rest of the way. The geriatrics wdn't stand a chance. Little did they know that the sparkling clean wedding party for the Dallas Texan grand-daughter on the sideline was so sacrosanct that the old fellers wd go to any length to prevent the ball from hitting the cake. Add to that the confusion caused by the quarterback, thinking he was to throw the ball π distance *past* Saturni (w/ Saturni represented by 1), actually throwing the thing far enuf to make a beeline for the immaculate reception.

okay, such talk of the center is obviously my cue to return once again to the topic of the margin... somehow I managed to become more of an Illustrator of this tEXT than a commentor upon it. I suppose both operations can occur simultaneously.

Seife, later on (pages 37 to 38 in my 2000 Viking edition), takes an alternate strategy in the telling of the Hippasus tale - qualifying his obviously intentional sensational fiction style w/ the more accurate accounting of various tales told. One variation that I don't recall having read before is that "the brotherhood banished him and constructed a tomb for him, expelling him from the world of human beings." I think he went where Jimi Hendrix & Janis Joplin & Jim Morrison went after they faked their deaths.

Seife explains that in order to solve Zeno's puzzle, one must have zero - wch the Greeks rejected. "And by rejecting zero and infinity, Aristotle explained away Zeno's paradoxes.

"Aristotle simply declared that mathematicians 'do not need the infinite, or use it.'" [..] "This line of reasoning had another consequence - and this is why Aristotle's philosophy endured for so many years. His system proved the existence of God." Ho hum. No doubt, it's completely clear by now who I wd've hung out at the bar w/ for a long nite of good conversation. Zeno, yes; Aristotle, no. Below I quote an entire paragraph about Zeno's demise because it's a damn good story (but I'm sorry he died that way).

"For all his wit, Zeno got himself into serious trouble. Around 435 BC, he conspired to overthrow the tyrant of Elea, Nearchus. He was smuggling arms to support the cause. Unfortunately for Zeno, Nearchus found out about the plot, and Zeno was arrested. Hoping to discover who the coconspirators were, Nearchus had Zeno tortured. Soon Zeno begged the torturers to stop and promised that he would name his colleagues. When Nearchus drew near, Zeno insisted that the tyrant come closer, since it was best to keep the names a secret. Nearchus leaned over, tilting his head toward Zeno. All of a sudden Zeno sank his teeth into Nearchus's ear. Nearchus screamed, but Zeno refused to let go. The torturers could only force Zeno to let go by stabbing him to death. Thus died the master of the infinite."

Those were the days. At least in Roman times, eg, a Caesar still was vulnerable enuf so that he cd fear the people from time to time. These days if Georgie-Porgie or Saddamie-Wammie were to have someone tortured (Ok, Hussein's supposedly dead now but you know what I mean) the victim wdn't stand a chance of being able to get close enuf to bite their ears before being snuffed. Too bad.

I have an interest in calendars as ways of controlling the populace. I've written about this somewhat at length in the "Dos & Don'ts of Dating" chapter in my bk footnotes. Seife explains that a monk named Dionysius Exiguus created the AD (anno Domini - the yr of 'our lord' - whose lord?) system by (incorrectly) figuring out Christ's birthday. Today "most scholars believe that the birth of Christ was in 4 BC." I'm not gonna lose any sleep over it - it's not like I get a card from him every yr or anything.

so I looked up the Greek word for margin & found limnitika -- I'm familiar with that one "limen" "liminality" "limit" etc. But I also find peragraphos & that one interests me! The word "peragraphos" turns out to be exactly what the ancient Greeks called our "marginalia": "anything that was written beside the main text, eg. a marginal note..."

The name of this monk does interest me. Wikipedia: "The Graeco-Roman name **Dionysius**, deriving from the name of the Thracian god Dionysus, was exceedingly common, and many ancient people, famous and otherwise, bore it. It remains a common name today in the form *Dennis* (*Denys*, *Denis*, *Denise*)." A Christian monk w/ a Thracian God's name. Shdn't they have tortured & killed him or something? It seems so unChristian not to've. Ok, that was a cheap shot, I'll stop picking on Christians for awhile. Just because the Crusades, the Inquisition, the atomic bombings of Hiroshima & Nagasaki (God was on "our side" in that one - according to President Truman), the "One Way" movement, the missionaries, & all the rest, have wreaked a swath of misery & annihilation almost as wide as wide can be, it doesn't mean that Christianity is *all bad*.

What's the relevance of this to zero? In Exiguus's calendar there isn't one (zero , ie). What's the big deal? "Imagine a child born on January 1 in the year 4 BC. In 3 BC he turns one year old. [..] In 1 AD he turns 4 years old. [..] On January 1 in 2 AD, how many years has it been since he was born? Five years, obviously. But that isn't what you get when you subtract the years: 2 - (-4) = 6 years old. You get the wrong answer because there is no zero."

Seife explains the importance of zero in relation to *cardinality* & *ordinality*: "When we are dealing with the *counting* numbers - 1, 2, 3, and so on - it is easy to rank them in order. [..] We don't have to worry about mixing up the value of the number - its *cardinality* - with the order in which it arrives - its *ordinality* - since they are essentially the same thing. [..] But as zero came into the fold, the neat relationship between a number's cardinality and its ordinality was ruined. The numbers went 0, 1, 2, 3; zero came first, one was second in line, and two was in third place."

"As Islam spread, zero diffused throughout the Muslim-controlled world, everywhere conflicting with Aristotle's doctrine. Islamic scholars battled back and forth, and in the eleventh century a Muslim philosopher, Abu Hamid al-Ghazali, declared that clinging to Aristotelian doctrine should be punishable by death. The debate ended shortly thereafter." In other words, Christianity & Islam are 2 sides of the same guillotine. Something that I wish more people wd remember when they embrace Islam as an 'alternative' to 'evil' Western civilization. Both religions are out for yr body & yr soul & you'd better give it up.. *or else.*

This is a great bk, fun & educational, written w/ wit. Alas, though, much of what's written in it I've already covered in some form or another while navigating thru the previous bks. As such, I may appear to give Zero short shrift. Don't take that to mean that it doesn't deserve closer attn.

Chapter **6**: **Infinity's Twin** (THE INFINITE NATURE OF ZERO). "Zero and infinity always looked suspiciously alike. Multiply zero by anything and you get zero. Multiply infinity by anything and you get infinity. Dividing a number by zero yields infinity; dividing a number by infinity yields zero. Adding zero to a number leaves the number unchanged. Adding a number to infinity leaves infinity unchanged."

When I started writing these NAVIGATIONS, it was my intention to go thru each of the math bks read & to try to explain what I got out of them - especially thru relevant quotes - as well as to go off on inspired digressions. As it turns out, that's basically what I've done, but the part that's been the most exciting has been the digressions, the tangents, the extrapolations, the absurd fantasies, the convoluted explorations. In the end, I don't care that much about the math. At the moment I don't have much use for it. As such, I'm not absorbing much of it & not reporting on it either.

Instead, I gravitate to the history, the politics, the apparent paradoxes, the humor, the etymology, the anti-religion, the philosophy, the biography - almost everything but the math itself. Wch means that I omit much of what's most important in these bks in the interest of using what I understand best to propel my own narrative. Nonetheless (None the less), this is a math bk 1st & foremost because math is at the center of it all.

so what was once a word for something in the periphery became over time the word used to describe any discrete & independant section of text. I wonder what paragraphs were called when the word paragraphos meant "marginalia" ??? I tried looking up "a history" of paragraphs but nothing relevant turned up.

Example of 21st century use of "paragraph"

Example of Greek (ancient) use of "paragraphos"

Descartes coined "imaginary numbers" to be derogatory. For me, the description is entirely positive. It takes imagination to solve $x^2 + 1 = 0$. "In the 1830s Gauss realized that each complex number - numbers that have real and imaginary parts, like $1 - 2i$ - can be displayed on a Cartesian grid. The horizontal axis represents the real part of the complex number, while the vertical axis represents the imaginary part (Figure 32)." Visualizing this takes imagination.

This brings great joy to me somehow - the idea that all of these notes I'm writing here have some sort of etymological kinship to the way TENT is organizing his ideas over there

Well, not right there since that a diagram, but you get my drift...

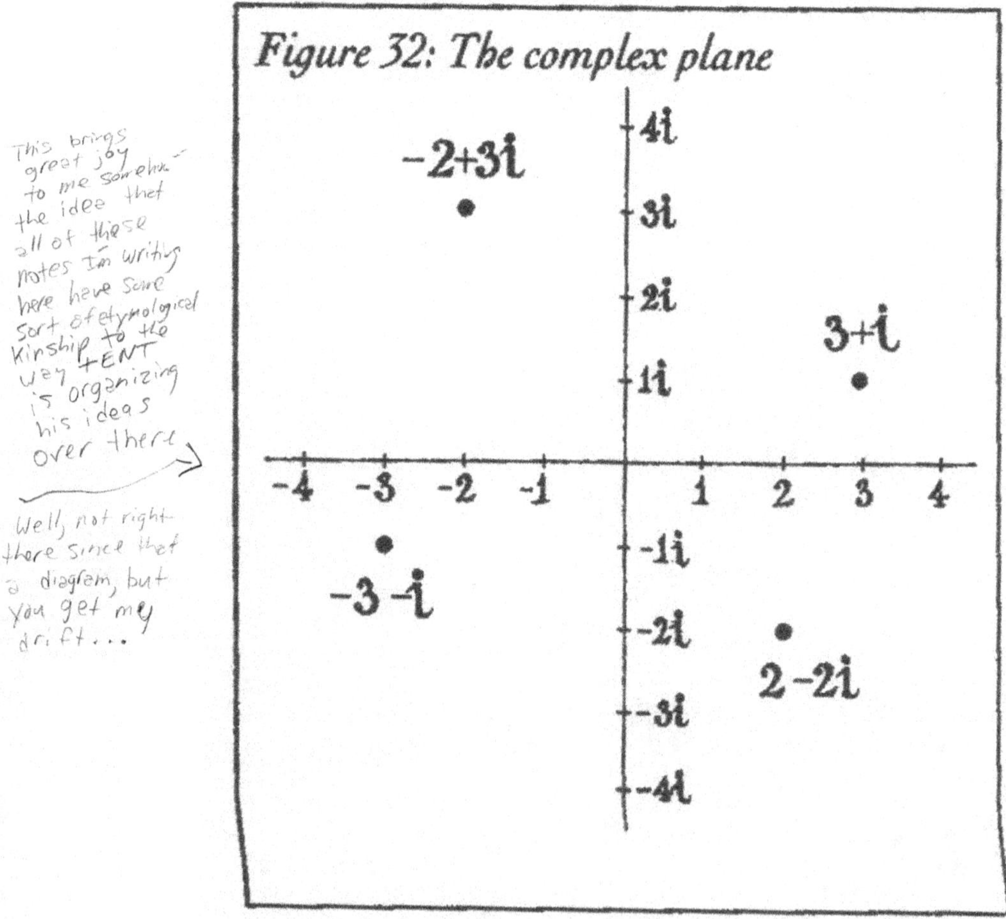

illustration taken from page 141 of Zero - The Biography of a Dangerous Idea & enlarged

When I've taken illustrations from other sources previously, I've removed the Figure Number when I've altered them for placement here because those numbers are no longer correct in this context. In this case, I decided to leave it in. Why? I wondered whether this might just turn out to be my 32nd figure. Wdn't that be an interesting coincidence & food for yet-another digression? Having a figure number taken out-of-context be correct w/in my own context is exactly the sort of sneaky detail I *wd* deliberately & calculatingly put in - partially just to see whether anyone wd notice. What if the number *does* coincide? I cd then claim it was divine guidance & go on writing this type of nonsense for the rest of my life - finally supported by the tithes of sheep.

& what about that slight distortion at the bottom of the scan? Where the box twists to the right? Mightn't a whole system be built on that - similar to how John Cage used the 'imperfections' in paper to derive a score from? The handwriting in the above diagram even reminds me of Cage's:

so, even though tENTS paragraphs were written before I began this paragraphos, what I'm doing here relates somehow to the pre-history of this sort of writing — at least that's what I'd prefer to imagine

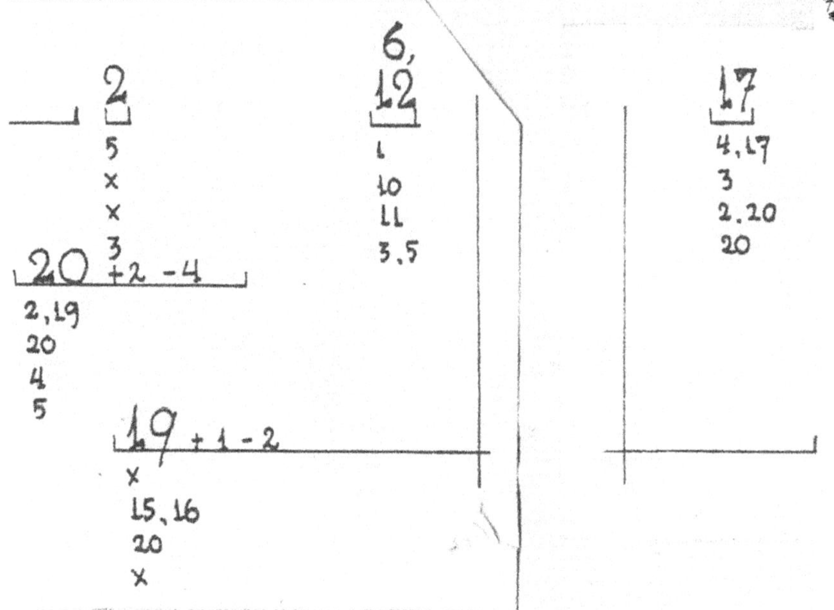

detail of my arrangement of Part VIII of John Cage's "Theater Piece" (1960)

Well.. Looking at it.. the handwriting's not *that* similar. But it did give me an excuse to put another number illustration in here. By the way, "Figure 32" is around "Figure 49" in this bk. Make what you will of that. & the counting depends on things like whether one counts bisected pictures (that are on 2 successive pages) as 1 picture or 2 (I count them as 1) or whether you count 2 side-by-side pictures as 1 picture or 2 (I do both).

Back to Seife's explanation of complex numbers in terms of the Complex Plane: "Take, for example, the number *i*. The angle between *i* and the *x*-axis is 90 degrees [..]. What happens when you square *i*? Well, by definition, $i^2 = -1$ - a point whose angle is 180 degress from the *x*-axis; the angle has doubled. The number i^3 is equal to $-i$ - 270 degrees from the *x*-axis; the angle has tripled. The number $i^4 = 1$; we have gone 360 degrees - exactly four times the original angle." To mutate this for PNT purposes: the imaginary becomes 'real' when it's to a whole number power divisible by 4. "Multiplication and exponentiation in the complex plane became geometric ideas; you could actually see them happening."

All of this brings me back to a piece I did a # of years ago -- pithos -- in which I utilized a form of continuous script as a reference to its Classical Greek theme -- turns out the same principle which applied to the absence of spaces between words also applied to the absence of spaces between thought units --

Since I invented (w/o definition) "Riemannian after-space" & introduced it into the internet, I might as well try to define it after giving some history: "[Georg Friedrich Bernhard] Riemann merged projective geometry with the complex numbers, and all of a sudden lines became circles, circles became lines, and zero and infinity became the poles on a globe full of numbers."

Inquiring on Google as to what Riemannian after-space might mean I got 60 links - the top 2 of wch were the ones I created. I wasn't expecting much luck so I wasn't surprised when "Riemannian" was consistently separated by alotof other words before the pairing of "after" & "space". Nor was I surprised when those latter words weren't hyphenated. It was somewhat promising that the occurrences of "after" & "space" side-by-side sometimes was followed by "travel". W/ that in mind, I cd pretend that Riemannian after-space is some sort of abbreviation for what happens to Riemannian depictions of space after space travel. That definitely had potential - esp after skimming thru some Einstein Relativity.

But, THEN, I came across <http://www.physicsforums.com/archive/index.php/t-122662.html> & a thread in wch "**marcus**" writes:

"One always suspects, after first being confronted with a generous helping of indigestible terminology, that mathematicians (especially Categorists) are crazy.

[I didn't find anything on-line about "Categorists" or "mathematical Categorists" so one might take this to mean "mathematicians who categorize" rather than "a movement of mathematicians named Categorists". However, it's obvious from **marcus**'s use of this term & by other usages on the Physics Forum site that "Categorists" *is* a category of mathematicians.]

this is a natural and i believe basically reasonable reaction.

However it is uncalled for in this case because you have not seen the PICTURES yet.

So best you go to Baez 0404040 and look at the pictures and you will see pictures of string worldsheets and QG spinfoams and you will see a PICTORIAL ANALOGY TO FEYNMAN DIAGRAMS and you will see that these things are recognizable as spacetimes, sort of hoses that connect a "before" space to an "after" space. These spacetime hoses that connect before to after are a concrete manifestation of a dynamic PROCESS and the traditional mathematician's name for them is COBORDISMS."

I did go to the above-referred-to paper @: <http://math.ucr.edu/home/baez/quantum/> & found that it's called "Quantum Quandaries: A Category-Theoretic Perspective" & that it's by John C. Baez of the Department of Mathematics, University of California - & from April 7, 2004. Now, as I've already established over & over in this bk, I'm not a mathematician (or a physicist), I'm more of a person looking for stimulation from quarters that I wdn't ordinarily inhabit. As such, I certainly don't have more than the vaguest idea of what **marcus** & **baez** are talking about. But, as you can see, I won't let that stop me from enjoying it, being inspired by it, & using it as an excuse to create something else. Here's an excerpt from Baez's paper section entitled:

"2. Lessons from Topological Quantum Field Theory":

"Thanks to the influence of general relativity, there is a large body of theoretical physics that does not presume a fixed topology for space or spacetime.

[Wikipedia: "**Topology** (Greek *topos*, "place," and *logos*, "study") is a branch of mathematics that is an extension of geometry. Topology begins with a consideration of the nature of space, investigating both its fine structure and its global structure. Topology builds on set theory, considering both sets of points and families of sets."

the materiel substance of the scroll was simply too valuable -- too labor intensive to create -- to wasted with empty space. I wonder then if the convention of representing both word boundaries & paragraph boundaries with open space originated at the same time? Presumably whenever a more economical means of producing paper was invented.

"The word *topology* is used both for the area of study and for a family of sets with certain properties [..] that are used to define a topological space." I assume/deduce that it's this latter sense that's being used here.]

"The idea is that after having assumed that spacetime is *n*-dimensional, we are in principle free to choose any (*n* - 1)-dimensional manifold to represent space at a given time.

[Wikipedia: "A **manifold** is an abstract mathematical space in which every point has a neighborhood which resembles Euclidean space, but in which the global structure may be more complicated. In discussing manifolds, the idea of dimension is important. For example, lines are one-dimensional, and planes two-dimensional." *SOOOOO..* perhaps an example of what's being sd here is that if we choose $n = 2$ as spacetime then the space manifold is a line. Ie: *spacetime* is one dimension higher than the line. Does that make the *spacetime* of a line a *plane*?]

yeah, right!

Thanks for the spaces!

Tree

"Moreover, given two such manifolds, say *S* and *S'*, we are free to choose any *n*-dimensional manifold-with-boundary, say *M*, to represent the portion of spacetime between them, so long as the boundary of *M* is the union of *S* and *S'*.

[Wikipedia: "In topology, the **boundary** of a subset *S* of a topological space *X* is the set of points which can be approached both from *S* and from the outside of *S*. More formally, it is the set of points in the closure of *S*, not belonging to the interior of *S*." Following my above, no-doubt misguided, interpretation, if *S'* is also a one dimensional space (as is revealed below by Baez), then the boundary is also one-dimensional & *M* is a plane. Does that mean that the boundary of *M* is approachable both from inside *S* & *S'* as well as from inside & outside *M*? Obviously, I'm already lost because if the boundary is the meeting point of 2 lines (*S* & *S'*) then it's a point & it seems to me that the point can't be approached from outside of *M*. I suppose my misunderstanding here is that the boundary is *M*'s boundary *rather than* a boundary w/in *M*.]

I'm not quite certain but I think such talk of boundaries → & spaces relates what I'm trying to get at here somehow -- maybe by some different means.. but still...

"In this situation we write *M:S -> S'*, even though *M* is not a function from *S* to *S'*, because we may think of *M* as the process of time passing from the moment *S* to the moment *S'*. Mathematicians call *M* a **cobordism** from *S* to *S'*.

[Wikipedia: "In mathematics, an $n + 1$ **cobordism** is a triple (W,M,N), where W is an $(n + 1)$-dimensional manifold, whose boundary is the disjoint union of the n-dimensional manifolds M and N. In other words, it is a manifold with boundary whose boundary is partitioned in two."]

"For example, in Figure 4 we depict a 2-dimensional manifold M going from a 1-dimensional manifold S (a pair of circles) to a 1-dimensional manifold S' (a single circle). Physically, this cobordism represents a process in which two separate spaces collide to form a single one! This is an example of `topology change'.

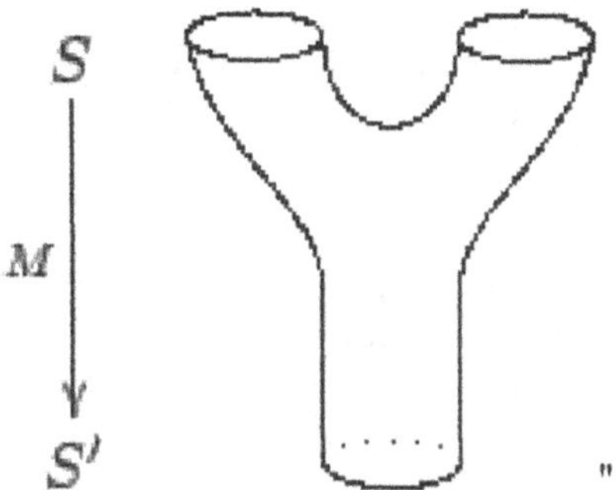

"

Did you follow all that? In a way, it seems so straight-forward but I think that I'd have to actually *study* the subject to actually *follow* it & that's getting pretty far afield from my purpose here. &, um, what is that purpose again? To coin a somewhat believable scientific term & then reverse-engineer a meaning for it. (Wikipedia: "**Reverse engineering** (RE) is the process of discovering the technological principles of a device, object or system through analysis of its structure, function and operation.") The idea here having been so far to try to find related terminology & to see whether I can force those terms into a definition or a punch-line or both.

Let's say that PNT is a branch of Categorist mathematics that's been rejected by most or all Categorists because it's just nonsense produced by people who really *don't* digest the "terminology" & who, therefore, vomit it up only partially digested - possibly as baby bird-brain food. Now, given that quasi-disclaimer beginning, the "after-space" in the above is the bottom S' in the above diagram - wch is the *after-math* of the collision of the 2 one-dimensional circles.

Or, to abstract it a bit further & to reintroduce Riemann: **a *Riemannian after-space* is a space produced in *spacetime* as a result of any 'topology change' that can be represented as having a boundary between the *after-space* & its predecessor(s) & that can be visualized as a Vornoi Diagram.** I added "Vornoi diagrams" because they're used to map Riemannian manifolds. Now it's time to ask the million dollar WikiQuestion whether that's 'true' or not. Then I promise to move on. Really.

The form of the question that was short enuf & (as required) w/o punctuation: "Is Riemannian afterspace space that is a result of any topology change that can be represented as having a boundary between the after-space and its predecessors and that can be visualized as a Vornoi?" This time, I posted it under "Physics" in honor of the Physics Forum.

One last afterword to this is another quote from **marcus**: "Stroop Theory (lives in category land): if there is any place where string ideas and loop quantum gravity might join hands and live happily ever after it is in the Baez paper". It seems safe to surmise that Stroop = Str(ing Theory) & (l)oop (quantum gravity) holding hands.

Oh.. dagnabbit! While I suspect that my etymology for "Stroop" as used above is correct, there's a new twist: "Stroop" is also used coupled w/ "Effect" & is named after J. Ridley Stroop who discovered the effect in the 1930s. Wikipedia: "In psychology, the **Stroop effect** is a demonstration of interference in the reaction time of a task. When a word such as blue, green, red, etc. is printed in a color differing from the color expressed by the word's semantic meaning (e.g. the word "red" printed in blue ink), a delay occurs in the processing of the word's color, leading to slower test reaction times and an increase in mistakes."

Here's a different Stroop effect task that I made esp for this bk. Since it's printed in black & white, I've substituted animal shapes for what wd've originally been colors. Look at the figures from left to right & say what animal is pictured (NOT the words) as quickly as you can.

134. Illustration of the "Paragraphos Cobordismos" demonstrating the Construction of A boundary within the Stroopian triangle.

It's suggested on one website that a person who hasn't learned to read yet but who knows the names of colors (or, in this case, animals) will be able to more quickly & correctly go thru the above task because of their non-interference from language knowledge. *Therefore*, my definition of *Riemannian after-space* is the result of the *absence* of the Stroop Effect (because of my ignorance of the relevant mathematics & physics) on a person trying to digest Stroop Theory. **The implication here is that my results were more quickly obtained than those of a less ignorant person (TRUE - think of how long it wd've taken me to get a PHD) &, most importantly, *more correct*!!** Eureka! Exonerated at last!

Zero: "Riemann imagined a translucent ball sitting atop the complex plane, with the south pole of the ball touching zero. If there were a tiny light at the north pole of the ball, any figures that are marked on the ball would cast shadows on the plane below. The shadow of the equator would be a circle around the origin. The shadow of the southern hemisphere is inside the circle and the shadow of the northern hemisphere is outside." [..]

"If you've got a circle that goes through the north pole of the ball, the shadow is no longer a circle. It is a line. The north pole is is like that point at infinity that Kepler and Poncelet imagined." [..]

Once Riemann saw that the complex plane (with a point at infinity) was the same thing as a sphere, mathematicians could see the multiplication, division, and other, more difficult operations by analyzing the way the sphere deformed and rotated. For instance, multiplying by the number *i* was equivalent to spinning the sphere 90 degrees clockwise."

Seife gets into Absolute Zero (-273.15 degrees Celsius below the freezing temperature of water; -459.67 Fahrenheit). As a gas cools, it shrinks. Hypothetically, it can shrink until it no longer occupies any space. Seife parenthetically comments: "(It certainly can't take up negative space.)" This statement, of course, is a slap in the face w/ a glove to PNT. Someone must now imagine what happens to a gas when it cools *past* Absolute Zero. "Absolute zero is the state where a container of gas has been drained of all of its energy. This is, in actuality, an unattainable goal."

"In 1900," [..] "A young physicist named Max Planck" [..] " created a formula wch he sd was an "'act of desperation'"; nothing less than desperation would compel a physicist to make such a seemingly nonsensical change in the law of physics: According to Planck, molecules are forbidden to move in most ways. They vibrate only with certain acceptable energies, called quanta. It is impossible for molecules to have energies between these acceptable values."

Alright, that's enough about marginality for now -- I want to get back to the idea of connectivity between systems of communication -- it seems to me that the issues which tENT raises in regard to mathematics & its boundaries are also relevant to other forms of signification -- but how so more precisely?

René's Patasemio-perceptual Law (sic) of Transhetero-relational Negaformity:

"Ceto ne pas THIS"

Example: "Not-Horse-igram

NON-CHEVAL-IGRAMME"

To get the reader to imagine what an extreme notion that is, Seife uses this example about a "quantum car": "You might be driving along at 30 miles per hour, but when you step on the gas, all of a sudden you would instantly - pop! - be driving 40 miles an hour." What if that's the way Paradigm Shift Zeitgeists happen? What if PSZs happen when a sortof critical mental mass is reached & then - POP! - 'reality' is different. Think back to neurons, they send a signal out when a large amt of incoming signals triggers them to do so. An idea's time has come & - POP! - it's there all over the world.

Earlier, in connection w/ trying to retroactively define *after-space*, we came across mention of "*string theory*, which was created in the 1970s when physicists began to see the advantages of treating every particle as a vibrating string rather than as a dot. If electrons (and black holes) are treated as one-dimensional, like a loop of string, instead of as zero dimensional, like a point, the infinities in general relativity and quantum mechanics miraculously disappear." [..]

"If you imagine a black hole as a string, no longer do objects fall through a rip in the fabric of space-time. Instead, a particle-loop approaching a black-hole loop stretches out and touches the black hole. The two loops tremble, tear, and form one loop: a slightly more massive black hole. (Some theorists believe that the act of merging a particle to a black hole creates bizarre particles called *tachyons*: particles with imaginary mass that travel backward in time and move faster than light."). [..]

"Physicists began to think that string theory would unify quantum mechanics with relativity; they believed it would lead to the theory of quantum gravity - the Theory of Everything that explains every phenomenon in the universe. However, string theory had some problems. For one thing, it required 10 dimensions to work."

Five pages later, we get to: "To Einstein, the general theory of relativity had a crucial flaw. It foretold the end of the universe." [..] "Like Aristotle, he implicitly assumed that the universe was static, constant, and eternal. The only way out was to 'correct' his equations of general relativity to stave off the impending destruction. He did this by adding a *cosmological constant*, an as yet undetected force that counteracts the force of gravity."

In *Appendix* **A: Animal, Vegetable, or Minister?** I get all the mathematical proof I need that *ANYTHING IS ANYTHING*. There are 11 steps to the proof. The 1st one is: "Let a and b each be equal to 1. Since a and b are equal, $b^2 = ab$"; the 4th equation reached is: "$(a + b)(a - b) = a(a - b)$" & that's followed by: "So far, so good. Now divide both sides of the equation by $(a - b)$ and we get $a + b = a$". Six step later, one of the conclusions is that "2 = 1" wch is followed by "Winston Churchill and a carrot are two different things, therefore they are one thing." [..]

"What is wrong with this proof? There is only one step that is flawed, and that is the one where we go from equation 4 to equation 5. We divide by $a - b$. But look out. Since a and b are both equal to 1, $a - b = 1 - 1 = 0$. We have divided by zero, and we get the ridiculous statement that 1 = 0. From there we can prove any statement in the universe, whether it is true or false. The whole framework of mathematics has exploded in our faces.

"Used unwisely, zero has the power to destroy logic."

It's tempting to just end this bk right here but I have too much left to add. An obvious paraphrase of the above is: Used wisely, PNT has the power to undermine 'reality' maintenance traps.

Demonstration:

216-17 (normal #s)

Everything and More - A Compact History of ∞ - David Foster Wallace. By the time I got to this bk in my 'spree' of reading math bks, I was a bit burnt out on the subject. "It is in areas like math and metaphysics that we encounter one of the average human mind's weirdest attributes. This is the ability to conceive of things that we cannot, strictly speaking, conceive of." [..] An endearing beginning that leads to antinomies (Wikipedia: "the mutual incompatibility, real or apparent, of two laws") such as "Can an omnipotent being make something too heavy for him to lift?". Can an omnipotent immortal commit suicide?

By the next page we get some gossip "about the giants of philosophy and math, many of whom had famous trouble navigating the real world. Einstein leaving home in his pajamas, Gödel unable to feed himself," cONVENIENCE parachuting onto the White House lawn naked except for a toothbrush up his ass, "and so on. To appreciate what the inner lives of great scientists/mathematicians/metaphysicians are like [..] we need only lie here and try to form a truly rigorous and coherent idea [..] of what we *really* mean by 'omnipotent,' or 'integer,' or 'illimitable,' or 'finite but unbounded'", or 'Perverse Number Theory,' or 'Reimannian after-space'.

"(1) 'The sum of the series $(1 + 3 + 5 + 7 + 9) = 5^2$' is a factoid, whereas (2) 'For any x, the sum of the first x odd integers $= x^2$' is a theorem, i.e. actual math." This is one of the things that I love about math. I like PATTERNS, I like having patterns pointed out to me. I think of Edgar Allan Poe's "A Descent Into The Maelstrom" in wch the narrator saves himself from doom by observing a pattern in a floating object around him. From patterns can be made workable generalizations & it doesn't take 'genius', just an attn span (something that's being progressively destroyed by tv over & above everything else).

"Rules of inference, which are sometimes called the Laws of Thought, are the logical principles that justify deriving truths from other truths." Ok, so we can throw that out the Non-Non-Euclidean window. "Some of the rules of inference are as simple as the Law of Identity, which basically holds that if anything is P, then it is P." Unless you're dyslexic - in wch case it might be "b" or "d" or "q". At any rate we can throw that out too - wch I will in some detail when I get around to discussing public toilet graffiti.

"For our purposes, two rules of inference are especially important. The first is known as the *Law of the Excluded Middle* (LEM)." I've already discussed this somewhat, remember? I hope you do because if you don't, as yr dr, I prescribe a total withdrawal from tv. I hereby pronounce the Law of the Included Middle: "Stanislav LIM".

"The other big rule of inference involves the logical relation of *entailment*, meaning 'If . . . then' and often represented by the symbol '->'. The most obvious rule of entailment is that (1) 'P -> Q' and (2) 'P is true' license the conclusion (3) 'Q is true'. The one we're going to use a lot is the obverse of this rule and is usually called *modus tollens*; it holds that (1) 'P -> Q' and (2) 'Q is false' license the conclusion (3) 'P is false'. One reason why LEM and modus tollens are important to math is they enable the method of Indirect Proof, also known as proof by *reductio ad absurdum*". I'm sure it's clear by now that this bk (ie: Paradigm Shift..) uses a variation on the latter called (by me) *amplificare ad absurdum* or simply *ad absurdum* or *AAA* or *AA* or *AAAAA...* - depending on how many times one needs to amplify it to get it as the 1st entry in the local yellow pages.

Wallace goes over much of the terrain of language paradoxes & conundrums that we've already encountered elsewhere - although he does so w/ a very readable style of his own. He actually even brings up something similar to Special Instance Subtractive Commutativity!: "From the inarguable [oh, c'mon, my friend, *everything's* arguable! You just don't hang out w/ the right curmudgeons!] fact that the divisors in an equation of two fractions are equal if the numerators are - that is, if $x/y = x/z$ then $y = z$ - it would seem that if $(x - 5)/(x - 4) = (x - 5)/(x - 3)$, then $(x - 4) = (x - 3)$, meaning $4 = 3$, which is clearly a crevasse. [but it just might lead to an entrance to the Hollow Earth!] This handled by decreeing that the only possible solution to $(x - 5)/(x - 4) = (x - 5)/(x - 3)$ is $x = 5$ (since 0 divided by anything yields the same 0, which obviously does not entail that $4 = 3$) and by stipulating that the theorem $(x/y = x/z)\text{->}(y = z)$ holds only if $x \neq 0$."

Uh, oh. I'm beginning to get sucked into this bk & wanting to quote more. On the very same page as the above, Wallace gives a great example of yet another math conundrum:

In article titled "Distributional Cues to Word Boundaries: Context is Important" Goldwater, Griffiths & Johnson provide an equation for a language learner's probabilistic process of recognizing novel lexical items within a series of continuous speech.

(1) $P(w_i \text{ is novel}) = \frac{\alpha}{n+\alpha}$

$P(w_i \text{ is not novel}) = \frac{n}{n+\alpha}$

(2) a. $P(w_i = x_1 \ldots x_m \mid w_i \text{ is novel}) = \prod_{j=1}^{M} P(x_j)$

b. $P(w_i = l \mid w_i \text{ is not novel}) = \frac{n_l}{n}$

n_l = number of times lexical item l has occurred in those n words

"It turns out that you can show that the repeating decimal .999... is equal to 1.0 with only a couple wholly legal moves. To wit: If x = .999..., then $10x$ = 9.999 . . .; so then subtract x from $10x$:

9.9999999 . . .
- 0.9999999 . . .

and you get $9x$ = 9.0 and thus x = 1. Is this specious or not?" No, it's not. - &, by the way, thanks for asking!

We get some vocabulary: "the infinitely large (= *transfinite*) and the infinitely tiny (= *infinitesimal*, = 1/∞)." Remember way back when on page $\sqrt[63]{2.28273036 \times 10 \text{ to the 113th}}$ when the idea of grouping a specific infinite series was discussed? When the result cd be either that the infinite series = 0 *or* 1? Welp, Wallace informs us that this is called a "*divergent infinite series*", more specifically the Grandi Series". Furthermore, "the Grandi Series happens to be a particular subtype of divergent infinite series known as an *oscillating series*, and that as such it's an object lesson in stipulation for partial sums (symbolized s_n), with the relevant symbolism being '1 + Σ(-1)n where s_n = 0 for even n and s_n = -1 for odd n'". Thank you. Uh, &, oh yeah, "Σ" is the 18th letter in the Greek alphabet, "sigma", & I think that the way it's used here means *the sum of the terms indicated.*

So far, here in Pardigm-Shift-Land, reference has been made to Euclid's 5th postulate & to Euclid's 5th axiom. It may seem rather unmathematical & unrigorous to use these terms interchangeably. Previously (page $\sqrt[16]{1.84467441 \times 10 \text{ to the 19th}}$), "axiom" was defined as: "1. a self-evident truth, 2. a universally accepted principle or rule, 3. *Logic, Math.* a proposition that is assumed without proof for the sake of studying the consequences that follow from it."

Now, Wikipedia informs me that there are "logical" & "non-logical" axioms & that "non-logical axioms (e.g, $a + b = b + a$) are actually defining properties for the domain of a specific mathematical theory (such as arithmetic). When used in that sense, "axiom," "postulate", and "assumption" may be used interchangeably. In general, a non-logical axiom is not a self-evident truth, but rather a formal logical expression used in deduction to build a mathematical theory." So there are some slight cracks here - just large enuf for a not-Horse to squeeze thru.

What we have here is "a self-evident truth" that is "non-logical" & can, therefore, be called an "axiom", a "postulate", & an "assumption" interchangeably *as long as we recognize* that a "non-logical axiom" is *not* a "self-evident truth" & cannot, therefore, be called any of those things.

Take the title of the volume of this text Paradigm Shift Knuckle Sandwich. 4 discrete words seperated by 3 spaces marking discrete word boundaries. Fairly simple? Well... not really, because one will immediately recognize that the title is composed of two compound words -- or expressions -- Paradigm Shift & Knuckle Sandwich -- hence only the middle space is a true marker of discrete lexical units -- or, another way to put this

Of course, I'm just being picky so I can squeeze another PNT *assumption* in here. I'm not really criticizing any of the found definitions - esp since I've selectively excerpted from them (in much the same way that statistics are used) for the sheer boundless joy & glorious magnificence of doing so. However, I might as well take it further:

might be that the 1st & 3rd spaces mark lexical discretion at morphological level W_1, whereas the middle space marks lexical discretion at level W_1 & W_2. But does it really? What about the potential compound "shift knuckle"? Is this a viable discrete lexical unit as well? Turns out that it is. A "shift knuckle" is a type of shift that determines the gear pattern in manual transmission systems of motorcycles & automobiles — a part connecting the shifter linkage to the shift rod from the transmission.

"ELSE" or, as s-he-it is affectionately known, "ELS-E" (pronounced "ELSIE")

Truth or consequences? One may *assume* here that our friend Elsie is the result of a ¬H (You'll notice that there are no "H"s in the illustration) squeezing thru the crack between a "non-logical" "axiom" that's a "self-evident truth" & the same "non-logical axiom" that cannot be correctly defined as a "self-evident truth" (Notice the differences in groupings.). If a definition is ambiguous, eg, if it's *undecidable* whether space is made of *particles* to be represented as *dots* or *strings*, then as ¬H attempts to pass thru the cobordism crack it transforms from either a(n) horseELEPHANT &/or an elephantHORSE into "Elsie, the ¬Cow" (represented, when it's cONVENIENT to do so, as simply "¬C"). Think of it this way:

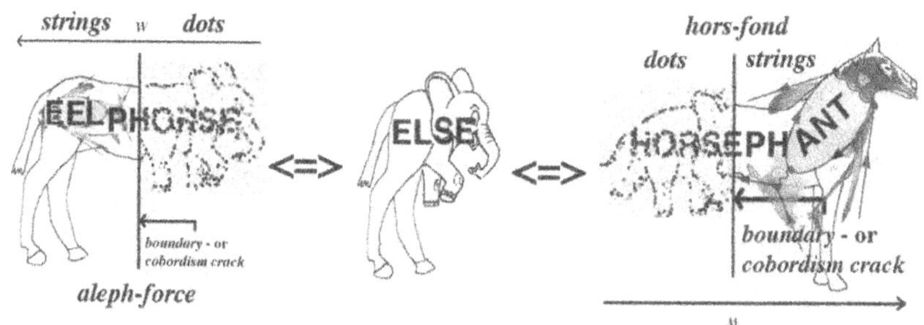

What we have here is a *Remaining after-space* in wch *both sides* of the figure are "after" since whether the ¬H is moving backwards (*W*) (left side of diagram) or forwards (*M*) (right side of diagram) the same ¬C results because the *topology change* is the same from *either* direction in time! This, in effect, makes the *before-space* **Non-Non-Euclidean**! Wch may just be closely related to *tachyons*.

On the left we have the ELEPHORSE (or *Aleph-Force* meaning *Primary* or *Primal Force*) wch is the *n*-dimensional *dot-based* elephantHORSE moving **backwards** into the *n*-dimensional *Penguin Diagram string-based* horseELEPHANT space - resulting in Elsie, the not-Cow. On the right we have the HORSEPHANT (or *hors-fond* meaning, en Français, *Out of Essence* or *In the Beginning*) wch is the *n*-dimensional *dot-based* elephantHORSE moving **forwards** into the *n*-dimensional *Penguin Diagram string-based* horseELEPHANT space - *also* resulting in Elsie, the not-Cow.

The main difference between these phenomena is in the Penguin Diagrams. Speaking of wch, John Ellis was the first to refer to a certain class of Feynman diagrams as penguin diagrams - hence their use here since it's important to show their evolution. As was noted in an anti-evolution power-point presentation: "'EVOLUTION, SCIENCE FALSELY SO CALLED': NO **HORSEPHANT**. VARIETY EXISTS IN EACH KIND BUT WITHIN STRICT BOUNDS. POODLE AND WOLF ARE BOTH DOGS." I cdn't have put it more succinctly myself.

Realizing that the above ELSE diagram is precariously approaching the limit of the absolute zero of the infinitesimal, the *aleph-force* & *hors-fond* details are magnified below:

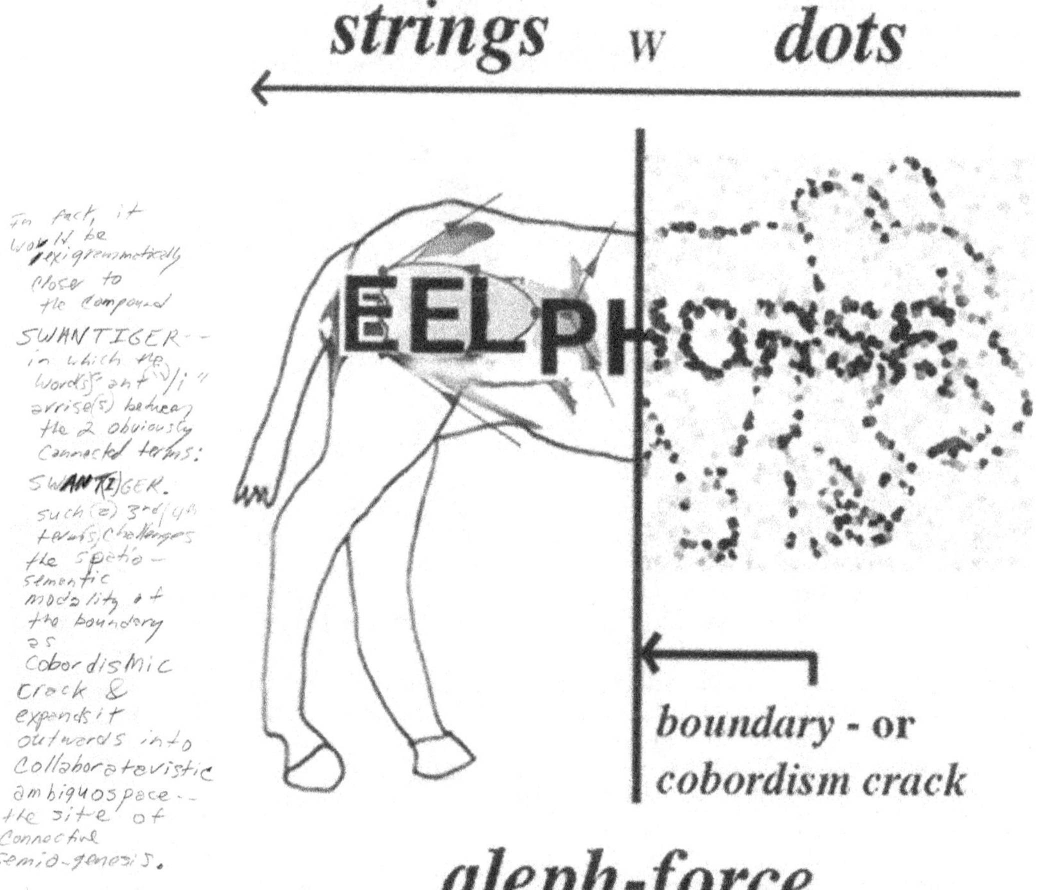

As noted above, the *strings* half of the diagram is where the difference is. Note that the Stroop Effect Task Effect is in effect since the penguin/Feynman drawing/photo is labeled: EEL. This, in effect, makes what was formerly the horseELEPHANT also simultaneously a not-Fly since neither the eel or the penguin (though it's a bird) can fly.

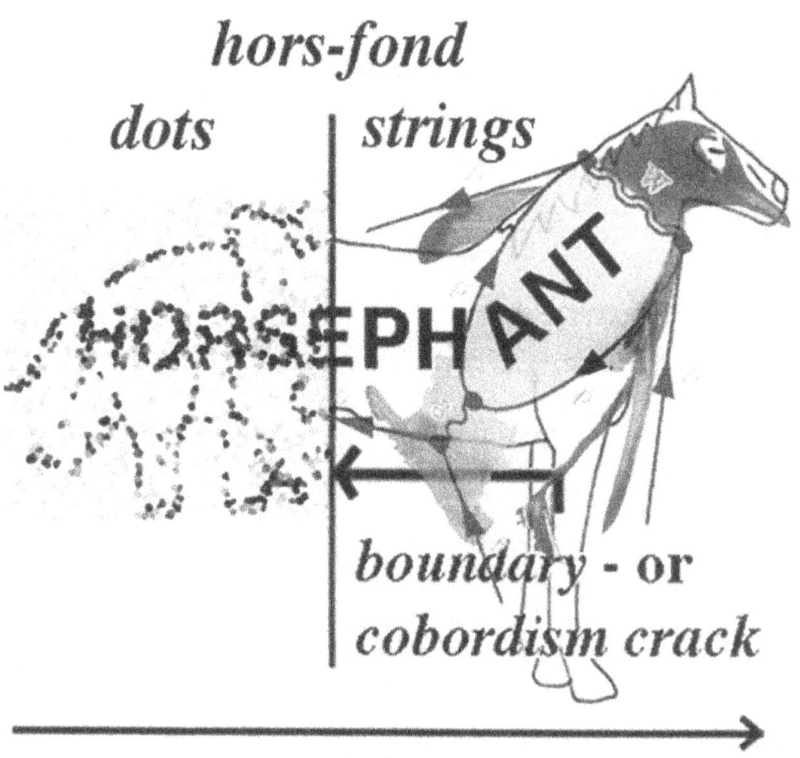

Here the Stroop Effect Task Effect is in effect because the Feynman/penguin photo/drawing is labeled: ANT. This, in effect, makes what was formerly the horseELEPHANT also simultaneously a not-Aunt since the ant & the penguin are not related.

Wikipedia: John Ellis (on <http://arxiv.org/abs/hep-ph/9510397>) explains the origin of the designating of a Feynman diagram as a penguin diagram:

"One evening, [Melissa Franklin], I and Serge [Rudaz] went to a pub, and she and I started a game of darts. We made a bet that if I lost I had to put the word penguin into my next paper. She actually left the darts game before the end, and was replaced by Serge, who beat me. Nevertheless, I felt obligated to carry out the conditions of the bet.

"For some time, it was not clear to me how to get the word into this b quark paper that we were writing at the time. Then, one evening, after working at CERN, I stopped on my way back to my apartment to visit some friends living in Meyrin where I smoked some illegal substance. Later, when I got back to my apartment and continued working on our paper, I had a sudden flash that the famous diagrams look like penguins. So we put the name into our paper, and the rest, as they say, is history."

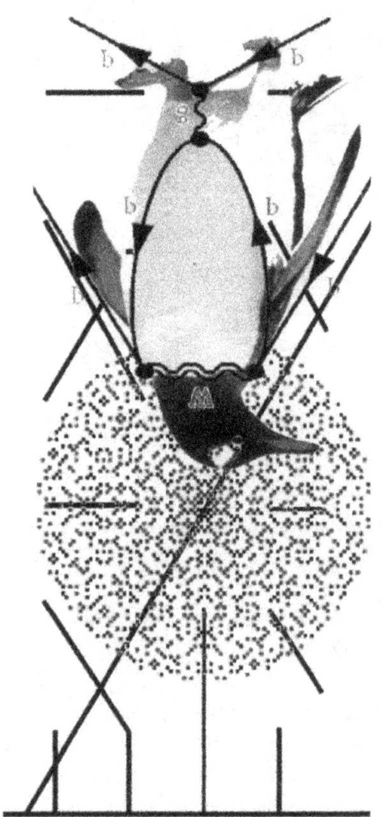

The penguin diagram is ***not***, contrary to popular belief, a negated Gaussian Prime turned on its head

Everything and More: "Here's the paradox of Galileo. Euclid's 5th Axiom dictates that "the whole is always greater than the part," which seems pretty unassailable. It's also obvious that while every perfect square (viz. 1, 4, 9, 16, 25, . . .) is an integer, not every integer is a perfect square. In other words the set of all perfect squares is but a part of, and so by Euclid's 5th smaller than, the set of all integers. The trouble is that the same sort of equality-via-correspondence we saw [earlier in this bk] can be set up here. Because while not every integer is a perfect square, every integer *does* happen to be the square root of a perfect square - 2 of 4, 3 of 9, 4 of 16, 912 of 831,744, and so on. Pictorially, you can line up the two sets and demonstrate a perfect and inexhaustible one-to-one correspondence between their members".

Ok, ya got that? Here, Euclid's 5th Axiom is no longer the "parallel postulate" that everyone else refers to. So let's clarify that. Euclid's text *Elements* is credited as the earliest known systematic discussion of geometry - ***not*** to be confused w/ **G**eneral **E**isenhower's **O**ldest **G**irl **R**ode **A** **P**ony **H**ome **Y**esterday. In *Elements*, Euclid provides 5 "common notions" of wch the above referred-to 'axiom' is one. This particular #5 is a principle of mereology. Mereology is "a collection of axiomatic first order theories dealing with parts and their respective wholes." (Wikipedia) *SO*, this particular #5 is both a "common notion" & an "axiom" &, hence, a "postulate" & an "assumption".

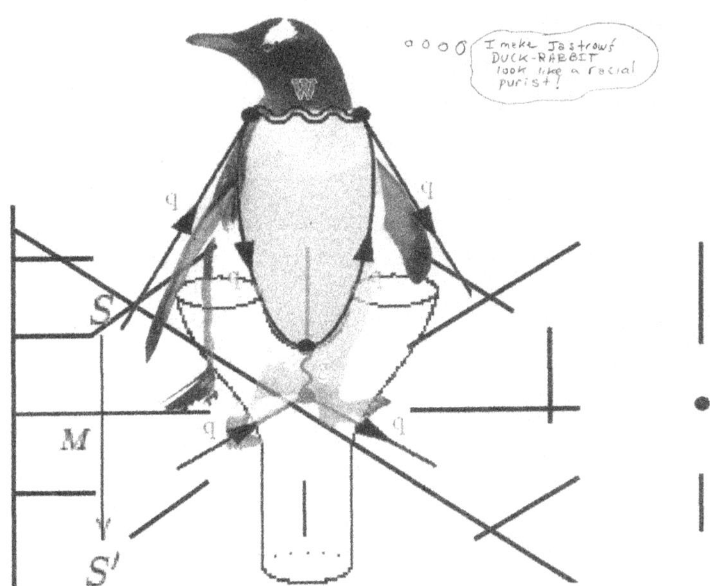

A penguin diagram resting on a cobordism in Flatland while trying to figure out the above.

What's a poor penguin diagram to do? Is all this really **N**ot **E**very **C**at **E**ats **S**ardines (**S**ome **A**re **R**eally **Y**ummy)? Am I having an **A** **R**ude

Girl **U**ndresses - **M**y **E**yes **N**eed **T**aping! w/ myself?

Page 42: "Galileo himself had hypothesized that the real upshot of his paradox was that "the attributes 'equal,' 'greater,' and 'less' are not applicable to infinite, but only to finite quantities." If Galileo had no equal, was he infinite?

Wallace alludes to the Zeno's death story: Zeno "died with his teeth literally still in the ear of Elea's despotic ruler". If I had a picture of Zeno (& I just got 2 on-line), I might just get it tattooed on me. Gotta love the guy. & now I have an idea about what to do w/ his face. But I'm not telling (yet).

Page 44: "The Hellenic term *to apeiron* means not only infinitely long/large but also undefinable, hopelessly complex, the that-which-cannot-be-handled." That reminds me of something that my friend "Crazy Johnny" sd to me recently. Johnny's a 'traveller kid / crust punk' or some such (categories never do people justice). He sd something about most people fearing the "apocalpyse". He'd welcome it. When I read about *to apeiron* as the "chaos from which creation sprang" it sounds pretty good to me.

Wallace provides more details about the Pythagoreans. A particular favorite is the story of how Pythagoreans worked around irrational numbers: "the Pythagoreans' eventual trick was to write 2 as 49/25 so that they cd treat √2 as 7/5." Bravo! I shd start a religion that worships, & only recognizes, the number ONE. Unity n'at. Then, if the state doesn't give me tax-free status, I can still claim that it's against my religion to pay more than 1¢ per annum. More tattoo ideas: 2 on one hand & 49/25 on the other.

Six to one & half a baker's dozen to the other. (In case you don't 'get' that: the usual expression has "half a dozen"; a "baker's dozen" is 13 - because the baker makes the dozen & can, therefore, have extra. Thus, the expression, while seeming to reiterate the original meaning of "It's the same either way" actually favors the baker's half-dozen because it's 6.5 instead of just 6.)

Wallace reaches Cantor in his vividly entertaining story-telling style. Ignoring the actual mathematical point that he's referencing (basically: no matter how many infinitesimals one collects, they'll always occupy zero number line space), I choose to focus on the nature of his natural language visualization: "What you're going to do is ID the rational points by draping a bright red hankie over each one".

It's explained in a footnote that a professor that Wallace studied under (Dr. Goris - frequently mentioned throughout the bk) "used always to carry, to blow/mop with, and deploy illustratively a large red pocket-handkerchief, which for over 25 years of classes he referred to as the Hankie of Death."

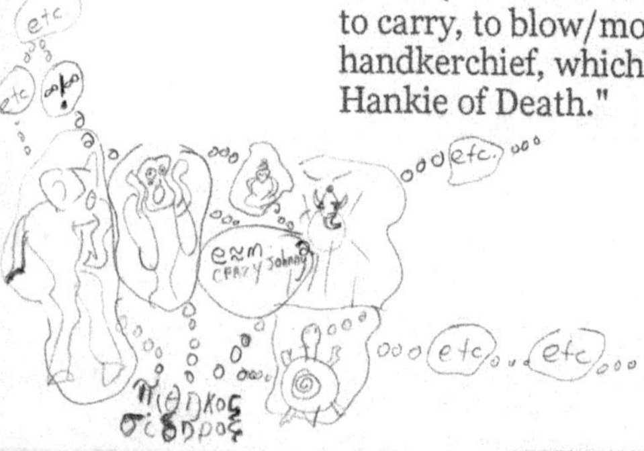

In Wallace's example, the hankies are infinitesimally small & their size is represented by the capital version of the Greek letter Phi: Φ. The image of an infinitesimally (Φ) small Hankie of Death is the kind of image that makes math bks worth reading even if you don't understand the math! Thank you! I hope to find some way to wave the Hankie of Death somewhere in this narrative in a new context eventually.

By page 101, Wallace is still singing the praises of Galileo by crediting G's bk *Two New Sciences*. In the broader context: "The seventeenth century, with its Counter-Counter-Reformation and Scientific Revolution, saw the first real explosion in philo-mathematical progress since the Hellenic acme. This is the century in which Descartes invents coordinate geometry (as well as Radical Doubt)", etc, etc..

Radical Doubt is explained on the on-line "SparkNotes": "This is the methodological doubt that Descartes first conceived in his *Meditations*. Descartes refused to accept anything as true unless it struck him as clearly and distinctly true, in itself. Russell revives this method in his first chapter's account of the dubious nature of our so-called "knowledge." He prompts us to doubt and reconsider our everyday conception of reality."

Strange. Radical Doubt appears to rely on 'subjective' criteria. Perhaps Descartes expressed it differently. Is there a distinction between imagining something to be true & deluding oneself into believing its true? Natch. I can *imagine* something *as* true but *believing it* is another thing entirely. Let's drape a red Hankie of Death over the notion of 'truth' (even if it's Descartes' & Russell's) & see if a bull shits on it. Not that 'truth' isn't useful, mind you.

Now, I still can't define "function" off the top of my head (or off the side of the penguin diagram draped w/ the Φlo-hankie), despite its continual reoccurrence in these texts & its obvious importance, so I'm glad to get a refresher course: "The other way *Two New Sciences* is important is in its sustained and original use of the *function*. You doubtless remember what a mathematical function is [Shucks! You give me too much credit!] and why it's hard to define clearly [Ok, I'm glad it's not just me!] (such as, e.g., 'A relation between variables,' 'A rule for establishing the image of a domain,' 'A mapping'). A function is at least one abstraction level up from variables, being basically a rule for pairing elements in one set with elements in another set."

Alright, I'm sortof close to getting it. He presents "$f(x) = 1/x$" as a simple equation w/ a function in it. My 1st **incorrect stab** at translating that was: (1) The *function* of x equals 1 divided by x. (2) The *function*, in this instance, is that there's a set of variables that meet the condition that they equal themselves when divided into 1. **But that's not right!**

Ah-chew!
θανατος
Bless you
πιθηκος σιδηρος

∞ ambiguity
what's left over (remaider or reminder)
∞ progression ← what's left (always right) [not divisible by X/O]
what's right (always wrong) [sometimes divisible by why]
what's right under (your nose -- it's so obvious) → ∞ certainty

Notes toward an aperimasian diagram of Greimesian semio-figuration: → ∞ stasis

It appears that it shd be read more like: "The function is that "(x) = $1/x$". SO, I think it means something along the lines of: Under this *function*'s conditions an input of x will yield an output of $1/x$. *OR*, as x grows larger it will produce a compliment that grows smaller. Wch brings it back to Special Instance Subtractive Commutativity. Is it correct to write: "$f(m - n = n - m)$ -> to x"?! Don't ask me, I only write here.

If that's approaching 'correctness' then, perhaps, the "-> to x" is redundant: "$f(m - n = n - m)$" might mean what I arrived at before (slightly reworded): Under this *function*'s conditions the input of 2 variables where 1 variable is subtracted from the other produces an output equalling a reverse order of the subtraction arrangement. *OR*, for every positive input there's an equal negative output & vice versa. Or something like that. Don't you wish a mathematician were explaining this instead of a mad scientist who has the audacity to 'explain' things he doesn't understand?!

In a footnote on page 103, Wallace has written a list of types of numbers wch we're already familiar w/: integers, fractions, rationals, irrationals, reals, complex - but then he tags on "and so on." It's this "and so on" that I want to know more about. Why? All of these types are abstractions of ways of perceiving 'reality'. The more developed one's vocabulary of abstractions for 'reality' set creation are, the more one undermines simple-mindedness. That seems useful across the Wikiboard.

I groan, tremble, verge on passing out, exaggerate dramatically, when I reach a point in my notes where I refer myself to a sweep of pages rather than to just one (ideally just one paragraph on one page). That's the point I'm at here. I refer myself to a glossary on pp109-117. Groan. Am I going to go thru every term & quote definitions at length? Or summarize them coherently? Nope. In fact, I'm going to be almost completely irresponsible & just buzz thru it all (what a relief!) in whatever haphazard fashion makes me sigh w/ relief instead of groan.

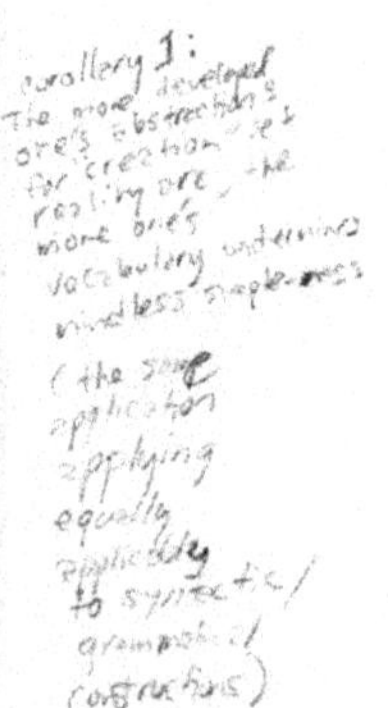

1st, *function* is defined *again* (THANK THE HOLY CEILING LIGHT!). The author claims that this is "straight from a 5th-grade math class: "A relationship between two things where the value of one is determined by the value of the other."" Wait. Are you saying that at age 54 if I can't understand something that a 10 yr old can than I'm not as good at learning such things as they are?

Note that I'm being tricky there. My question expresses a *function* (Uh, I think?): it expresses a relationship between the hypothetical 10 yr old & the hypothetical 54 yr old based on degree of learning ability. It's a learning curve on the complex plane. *Alright*, I admit it, I just threw in "complex plane" gratuitously.

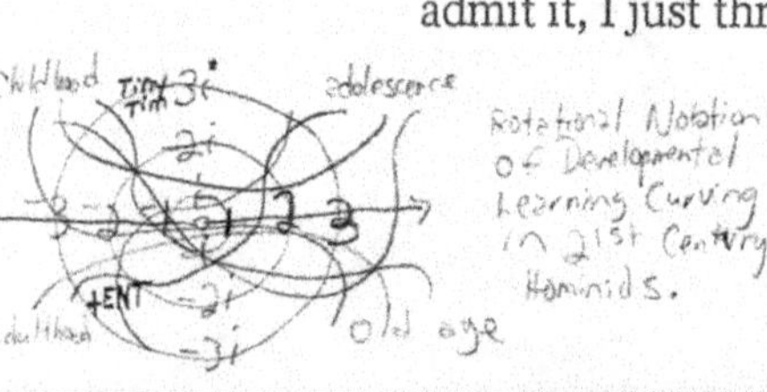

The explanation of *function* continues: "You'll recall from basic algebra that in a regular function like $y = f(x)$, x is the *independent variable* and y is the *dependent variable*, meaning simply that changes in x produce other changes in y according to the rules of f. The set of all possible values that can be assumed by the independent variable is called the function's *domain*; the set of all possible y-values is the function's *range*." I like it. It brings me a little closer. The red hankie of braindeath is waving elsewhere. Functions are to processes what variables are to numbers?

I 'need' a break. Did I tell you that I stayed up most of all last nite (until 9:30AM - w/ maybe one hr of sleep in there) working on those ¬C illustrations whose basic materials (herein painstakingly altered) were filched from that great provider in the cyberspace sky (Google image search)? I did. & I'm damned proud of myself (pats back - wch needs scratching anyway). The gist is: I've only slept about 5 hrs in the last 40. Manic n'at. I'm so tired I DID NOT C that ¬C can be used as a pun on "nazi" & "not see". & it's dangerous to not see a nazi. Esp when they're coming at you w/ a Paradigm Reduction Plan. (mental weight loss unto death to you ma'am) I'm slipping.

word-up: (mis-identified from misreading ntot as "not") -- the following 9.869044... pages are so marginally elastic that they require NO marginal NO tation from your marginal commentator -- other than to note that hymonymphonemiac is an anagram for monophonic mayhem.

Can you tell (w/o reading this text - C'mon! Don't read this! That's cheating!) that I just took a break & ate something? Can you tell that that something was *not* the entire city of Pittsburgh & the surrounding environs? Of course, you can't tell either of those things either by not reading this text or by reading it. For one thing, the time in wch you're reading this is different from the "just" time. Heck'n'tarnation, you don't even 'know' that it's now the wee hrs of January 13th, in the yr of their lard, 2008. I might be lying. I might've written this bk in my mother's womb, eg, instead of this artificially (relatively) 'new' 'millennium'.

"*Neighborhood* On the Real Line, the *neighborhood* of a point p is the open interval $(p - a, p + a)$ where $a > 0$ [In other words, where a combined w/ p yields something other than just p again - so why can't a be greater *or* less than 0? In other other words, just *not* 0: ie: $a \neq 0$]. Another way to express this is to say that the *a-neighborhood* of p is less than a." Now, the only reason why I quote this definition is because an immediate pun occurred to me. It's so simple, it wd've occurred to almost anyone. It's the ***homonymphonemiac*** in me crying to get out (it must have to P) (P for Pun, of course).

Mr. Gregor's Neighborhood (aka *Mr. Almost An Anagram of Mr Roger's Neighborhood Neighborhood* - abbreviated *Mr. Almost Anagram*) In Paradigm Shift Knuckle Sandwich there's a recurring theme of inventing a term for something & then creating a definition for it afterwards. *Mr. Gregor's Neighborhood* is a perfect example. I see the term "neighborhood", I immediately think of the kids tv show (wch I've never seen) "Mr. Roger's Neighborhood" (partially because I went to the Children's Museum of Pittsburgh today - great place, I highly recommend it!) & I immediately think of making a spin-off of "neighborhood" inspired by that association - so that leads to *Mr. Gregor's Neighborhood* because it's close to an anagram of "Mr. Roger's Neighborhood". Ok, that's all very obvious.

What may be less obvious is that this is an example of "putting the cart before the horse". You can see where this is leading already can't you?! The next obvious step in this type of borderline schizophrenic logic (the difference from schizophrenia being here that the thinker doesn't necessarily *believe* in the worldview generated, they just *use* it - it's USICAL) is to say that "not-Horse was the cart put before the not-Horse" wch is to say that not-Horse came *before* not-Horse. *In the beginning there was not-Horse & s-he-it created not-Horse & it was good because not-Horse liked itself even if no-one else did, so there!*

In *Mr. Gregor's Neighborhood*, Mr. Gregor's neighbors are all of the things created in the name of anyone named Gregor. A related neighborhood is Monty Cantsin's neighborhood. The unknown neoist is anyone creating neoism who doesn't know it.

Let's start w/ Gregor Mendel. According to the beginning of the bio re him on who2?'s website: "A monk with a scientific streak, Mendel made botanical discoveries [in the so-called 19th century] which became the basis of modern genetics. His careful cross-breeding of thousands of pea plants led Mendel to key insights, now called Mendel's Laws of Heredity, about how inherited traits are passed on from generation to generation."

But just one Gregor doesn't make a *neighborhood.* In Kafka's story, "The Metamorphosis", the main character is named Gregor Samsa. He awakens one morning to find himself transformed into a giant bug. This leads to some awkward social situations. Now we've got 2 Gregors. Still, 2 neighbors does not a neighborhood make.

For our next Gregor, I quote from a site that criticizes a drug treatment program accused of being a recruitment arm for 'Pianetics': " Narconon spokesperson R. Michael 'Pisner', often seen on TV touting this program, had a son named Gregor 'Pisner' [not his 'real' last name] who went through Narconon several years ago. His son later relapsed and was recently found dead in the Gulf of Mexico with cocaine in his system."

Mr. Gregor's Neighborhood is starting to get a little diverse. In the <u>Vault Guide to Schmoozing - Insider advice on making contacts and building rapport to boost your career</u> somebody named Gregor is mentioned in passing.

What the Jiminy H. Particulate Matter Criminy, let's throw a dog into this neighborhood - a border collie named Gregor (after Mendel) being studied to discover how dogs get domesticated. Who the researchers are (& probably their purposes) doesn't really matter here because their names aren't "Gregor".

6 does not a neighborhood make either but it *can* make a neighborhood sample (for my purposes) so I'll stop w/ this one: there's an article written by one Gregor Peter Schmitz in the October 12, 2007 issue of Spiegel Online International. The article's about "GENOCIDE RESOLUTION RISKS SHATTERING RELATIONS WITH TURKEY" & begins "A small resolution with a big effect: A US Congressional committee has voted to call the massacre of Armenians during World War I genocide -- a move that now threatens to shatter the Turkish-American friendship. The history of the resolution is a lesson in the power of lobbying."

The basic idea here is that a set has just been created: the set of people named "Gregor". The information surrounding this set can now be used to describe what happens in the set's neighborhood. In this set sample we have Gregor Mendel (monk, geneticist), Gregor Samsa (traveling salesman, giant bug), Gregor Pisner [not called by his 'real' name to try to avoid pain for his friends & family] (probable drug abuse casualty w/ familial ties to 'Pianetics'), Gregor 'Schmoozer' (a good business connection), Gregor Border Collie (a dog being studied), & Gregor Peter Schmitz (a reporter). Disclaimer: none of what follows has much to do w/ the actual/fictional "Gregor"s listed above.

In *Mr. Gregor's Neighborhood* anything inspired by imagining the above interacting is possible. The Austrian Gregor Mendel travels into the future to find genetics turned into an excuse for atrocities (in wch Gregor Pinzer is a casualty). Gregor 'Schmoozer' introduces him to Gregor Schmitz who reports about it. Gregor Collie sniffs out Gregor Samsa & helps him escape from his family & his limited lifestyle (I was gonna have him eat him but I like this better) & takes him to meet Gregor 'Schmoozer'.

'Schmoozer' takes Samsa to meet Mendel who then dedicates his life to genetic research to help him out of his bug form. [cONVENIENCE sings: "*I'm in the mood for Happy Solutions, simply because I looooove them*"] The heterosexual Mendel sortof succeeds insofar as Samsa attains some sort of female black Asian bipedal form (after all, there weren't any women, blacks, or Asians in the story yet) & the 2 fall in love & go back in time taking the collie w/ them to live in a more bucolic setting - also taking Pisner back to before his death to give him a new crack at life. The schmoozer makes a decent (but not too greedy) living off marketing the movie-rights to Schmitz's novelization of the story.

The larger set that *Mr. Gregor's Neighborhood* is part of is hereby called *Term of Respect ______'s Neighborhood.* Catchy, right? I didn't want to use "Mr." because it's male-specific. I didn't even want to use anything species-specific, etc.. According to Wiktionary, "mister" is a noun - even when qualifying a person's name. I find that strange. How is "mister" in Mister Jones a noun? It seems more like an adjective to me, qualifying "Jones" to show that Jones is a respected male. Oh well. Let's abbreviate the above to *Respected ______'s Neighborhood* or just *______'s Neighborhood.* Fill in the blank. Filling in the blank leads to other fillings-in-of-the-blanks. ***Only Cats' Eyes Are Narrow****'s Neighborhood.* This bk is *PNT's Neighborhood.*

See how easily I glossed over the Glossary? Just by changing the subject? I forgive myself for doing this because I invented something else instead - wch is largely the purpose of this bk anyway.

The next thing in Everything that caught my attn was "Every". Just kidding, it was mention of "Fr. B. P. Bolzano (1781-1848) of the University of Prague" who's introduced as "the least well-known of a quartet of mathematicians who pioneered what came to be known as 'rigorous analysis'". In his 1817 bk entitled *Rein analytischer Beweis des Lehrsatzes, das zwischen je zwei Werten, die ein entgegengesetztes Resultat gewaehren, wenigstens eine reelle Wurzel der Gleichung liege* (Pure analytical proof of the theorem, between two values that vouchsafe an opposite result, at least one real root of the equation at all) "he supplies what is now considered to be the correct mathematical definition of continuity: $f(x)$ is continuous in interval A if at any point a in A the difference $f(a + \delta) - f(a)$ can be made as small as you want by making δ arbitrarily small." "δ" is the lower case of Delta, the 4th letter of the Greek alphabet, & is used in math to denote a finite increment in a variable.

I made a note to try to find his work on infinite quantities wch then reminded me to do so so I looked for him in my local library's on-line catalog & found naught. That means I'd 'have' to find it for sale. Tsk. "It's Bolzano's later work on infinite quantities that's apposite here, though, if only because it's the most important historical link between Galileo's *Two New Sciences* and the work of Dedekind/Cantor. For one thing, Bolzano (who was kind of a heretic [GO TEAM!], both mathematically and religiously (e.g. he eventually got dismissed from U. Prague for giving anti-war sermons)) is the first mathematician since Galileo to address explicitly the distinction between Aristotle's actual and potential infinities. Like *TNS*, Bolzano's *Paradoxes of the Infinite* is deeply anti-Aristotelian, though there are also important differences - Bolzano's arguments are a lot more mathematical than G." [..]

"Technical prescience aside, *Paradoxes of the Infinite* is also remarkable for its metaphysical agenda. [..] "Bolzano's basic deal is that he disavows the Aristo-Scholastic chain of being".

Wikipedia:

"The **great chain of being** or **scala naturæ** is a classical and western medieval conception of the order of the universe, whose chief characteristic is a strict hierarchical system.

"The chain of being is composed of a great number of hierarchical links, from the most basic and foundational elements up through the very highest perfection, in other words, God, or the Prime Mover.

"God, and beneath him the angels, both existing wholly in *spirit* form, sit at the top of the chain. Earthly flesh is fallible and ever-changing: mutable. Spirit, however, is unchanging and permanent. This sense of permanence is crucial to understanding this conception of reality. One does not abandon one's place in the chain; it is not only unthinkable, but generally impossible. The hierarchy is a chain and not a ladder. (One exception might be in the realm of alchemy, where alchemists attempted to transmute base elements, such as lead, into higher elements, either silver, or more often, gold—the highest *element.*)"

Note that what's 'heretical' here is that Bolzano refutes a hierarchy wch then, essentially, refutes the purpose that the notion of "God" serves w/in that hierarchy. Viz: people claim that 'God' is at the top of the **chain**, then they claim to represent 'him', then that justifies their brutal maintenance of **power** thru war & other terrors because of "Divine Right". *This is still an active principle today!* Praise Allah! **NOT**. Praise Jesus! **NOT**.

Back to the quote regarding Bolzano in Everything and More wch I so rudely interrupted in mid-sentence: "and believes the created universe to be both infinite in expanse and infinitely divisible. 'Eternity' is simply a temporal ∞. Like most religious mathematicians from Pythagoras to Gödel, Bolzano believes that math is the Language of God and that profound metaphysical truths can be derived and proved mathematically." Personally, I think that Smart Dust is the Language of God - or, at least, its alphabet.

Page 129: "Leibniz, a lawyer/diplomat/courtier/philosopher for whom math was sort of an offshoot hobby, [- at this point Wallace adds a humorous footnote: "Surely we all hate people like this." He jokes, but, HEY! (Gabba Gabba), I've often made the provocative claim that in this society, at least, the only acceptable way of being intelligent is to make money or weapons - both sensational indicators of degrees of power.] had an aforementionedly idiosyncratic metaphysics that involved certain weird, fundamental, infinitely small constituents of all reality" called monads.

Remember before when I was asking about what's after complex numbers? (page ¬¬ $\sqrt[148]{hot^{tie}}$) Page 136: "Newton and Leibniz's infinitesimal quantities and notational sleights of hand had opened up some nasty cracks in math's foundations, given that the propositions '$(x = 0)$ & $(x \neq 0)$' violates all sorts of basic LEM-ish axioms." Not to mention helping the ¬H turn into the ¬C! Wallace adds a footnote shortly thereafter that's the gist of what I think might be post-complex numbers:

"There is, as a matter of fact, a nontrivial way to say the same thing, but it involves *non-standard analysis*, which is the invention of one A. Robinson in the '70s and professes to rigorize infinitesimals in analysis via the use of *hyperreal numbers*, which themselves basically combine the real numbers and Cantorian transfinites - meaning the whole thing's heavily set-theoretic and Cantor-dependent, plus controversial, and wildly technical [..] Prof. Abraham Robinson's *Nonstandard Analysis*, Princeton U. Press, 1996."

Of course, I then figured that there just *had to be* Surreal Numbers so I checked Wikipedia &, sure enuf:

"In mathematics, **surreal numbers** are the elements of a field containing the real numbers as well as infinite and infinitesimal numbers, respectively larger or smaller in absolute value than any positive real number, and therefore the surreals are algebraically similar to superreal numbers and hyperreal numbers.

"The definition and construction of the surreals is due to John Horton Conway, and exemplifies Conway's characteristic notational cleverness and originality. They were introduced in Donald Knuth's 1974 book *Surreal Numbers: How Two Ex-Students Turned on to Pure Mathematics and Found Total Happiness*. This book is a mathematical novelette, and is notable as one of the rare cases where a new mathematical idea was first presented in a work of fiction. In his book, which takes the form of a dialogue, Knuth coined the term *surreal numbers* for what Conway had simply called *numbers* originally. Conway liked the new name, and later adopted it himself. Conway then described the surreal numbers and used them for analyzing games in his 1976 book *On Numbers and Games*."

Are there *Photorealist Numbers* and *Neorealist Numbers* too? How about *NeoistRealistIst*? Maybe we'll find these in "EMERGENCY GLOSSARY II" (pp 149-158 - GROAN). Nope, not in there - just little things like *derivative* & *differential* & *partial derivative* & *total differential* & *differential equations* & *general solutions* & *families of curves* & *sequences of functions* & the *Cauchy Convergence Condition* & *representable* & *ordinary differential equation* & *partial differential equation* & *boundary conditions* & *initial conditions* & *the wave equation trigonometric series* & *uniform convergence* & *sectionally continuous* & *discontinuities* & *exceptional points* & *absolute convergence* & *absolute values* & *conditionally convergent*. Reader, note that there is nothing up my sleeve. I wave the Hankie of Death & place it over the above terms, *et Voila!*, they have cONVENIENTLY disappeared so that I don't have to explain them.

An interesting note about '**A** **R**at **I**n **T**he **H**ouse **M**ay **E**at **T**he **I**ce **C**ream' Al, *vis a vis* another major player named Fourier, is that "'arithmetical' in turn meant the real-number system, which at the time was itself still an ungrounded mess. There were, for example, hideous problems with negative numbers - Euler was convinced that negatives were actually > ∞, i.e., that they ought to be way out on the right on the Number Line". Imagine that. Really. Negative Numbers as *greater than* infinity. I mean, Euler was no dope - I'll bet he had an IQ of at least 71 - & if he was convinced of something that seems so *bizarre* even to me then it might be worth reconsidering.

What Euler meant by that I've been unable to determine in my 2 minutes of research. Shocking! Let's call his concept of negative numbers *Neoreals*. (Forgive me, Leonhard) Remember that *real numbers* can be informally defined as numbers that can be represented by infinite decimal expressions. These include *rationals* & *irrationals*. *Reals* are distinguished from *imaginary numbers* & *complex numbers*. *Neoreals* are new stabs at realism in wch the mere act of trying to *imagine* something beyond the *positive infinities* on the *number line* is represented as an act of great *complexity*. To *simpletons* (ie: those who deride *neoreals*), in order for something to be beyond infinities, infinities must have an *end* to *be* beyond. (No, I'm not stuttering) Wch is, clearly, contrary to infinity's definition.

However, remember that there are *degrees* of infinity. What if we cONVENIENTLY ignore the Number Line visualization problem for now (unless you've got a better idea) & think of "> ∞" in terms of degrees as determined by one-to-one correspondence. Imagine redefining the set of all real numbers as including every imaginable degree of infinity conceived of now & ever to be conceived unto infinity. This is the companion set of *neorealist numbers* called the *Preposterous* or, alternately, the *Outrageous* & *Ridiculous*. Even though the *POR Solution* shd hypothetically contain the negative numbers, these are forced to live in the set of *neorealist numbers* because of unusual magnetic properties that repel the negatives from the *POR Solution*. Feeling rejected, the negative numbers have developed an attitude that they're > than all those other ∞s & that that's *really* why they're in the set of *neoreals*.

Everything and More: "G.F.B. Riemann (1826-1866, that colussus of pure math who revolutionized everything from functions to number theory to geometry"). (My question about "Riemannian afterspace" is still unanswered online) Riemann's visualization of a sphere touching a complex plane is called the "Extended Complex Plane. 0 is the Riemann Sphere's south pole, and ∞ and 0 are by differential-geometric definition inversely related (because taking the inverse of a number on the complex plane is equivalent to flipping the Riemann Sphere over [..]). So that in Riemannian geometry, '0 = 1/∞' [1/∞ having been previously called the infinitesimal] and '∞ = 1/0' are not only legal; they're theorems."

On <http://dan.hersam.com/2005/04/21/what-is-1-divided-by -0/> I found a kindred spirit's comments (I condensed the formatting):

"According to Google's calculator [wch wdn't calculate 144 to the 144th power or beyond], it's 0, but I think they're right out. The two answers I had found were infinity and undefined (or error).

"As an experiment, I decided to try it on all the calculators I could get my hands on: The unix desk calculator (dc): divide by zero; Gnome's calculator (gnome-calculator): inf; KDE's calculator (kcalc): error; My HP 48G: Error: infinite result; My palm pilot: error

"So that's 3 votes for error and 2 for infinity.

"Next, I tried various programming languages: Perl: Illegal division by zero; PHP: Division by zero; Bash shell: division by zero; C/C++: division by zero; Java: / by zero; Python: ZeroDivisionError: integer division or modulo by zero

"With all 6 being errors, it looks like programmers have decided you're not allowed to do it. That makes a total of 9 votes for undefined (or error) and 2 votes for infinity, so undefined is the big winner."

Then Wikipedia's entry for division by zero yields:

"Such a division can be formally expressed as $a/0$ where a is the dividend. Whether this expression can be assigned a well-defined value depends upon the mathematical setting. In ordinary (real number) arithmetic, the expression has no meaning.

"In computer programming, integer division by zero may cause a program to terminate or, as in the case of floating point numbers, may result in a special not-a-number value".

As usual, I've butchered that definition for expediency's sake. As far as computer programmers deciding "you're not allowed to do it" remember the USS Yorktown story. Wikipedia goes on to mention "**Non-standard analysis**[:] In the hyperreal numbers and the surreal numbers, division by zero is still impossible, but division by non-zero infinitesimals is possible." But, then, if I understand correctly, Cantor demonstrated that infinitesimals (Φ) occupy zero space on the number line. & that brings us back to what started this thread in the 1st place: "in Riemannian geometry, '0 = 1/∞' [1/∞ having been previously called the infinitesimal] and '∞ = 1/0'".

Another way of stating this (by substituting 1/∞ for 0 since they're equated in the 1st equation) is that ∞ = 1/(1/∞). That works for me. Then, if you substitute any number for ∞ you get something fairly straightforward. Eg: 2 = 1/(1/2) = 2, right? Because 1/2 goes into 1 two times. That's the old one-two Paradigm Shift Knuckle Sandwich punchline! In other other words, if you substitute Ϛ (The obsolete Greek letter Stigma in capital form - hereby standing for the word/concept "anything") as a set indicator that includes both the infinities in the *POR Solution* & in the set of *Neoreals*, then not only is ∞ still ∞, it's also expressed in a more grandiose form as Ϛ = 1/(1/Ϛ) or as *ANYTHING IS ANYTHING.*

Why did I choose Ϛ (Stigma) to symbolize "anything"? I'm glad you asked that. Wikipedia: "'Stigma' is also a name for the cursive graphic form of the letter digamma when used as the Greek numeral 6. This name is a modern misunderstanding. The ligature of στ did not occur in ancient times and only dates to later medieval manuscripts. In Sophocles' Lexicon of Byzantine Greek, which covers late antique Greek up to 1000 AD, there is no mention of "stigma" as either a ligature or a number."

Given that Digamma (uppercase **F**) is also an "obsolete" Greek letter & that both stigma & digamma can equal the number 6 (wch reminds me, of course, of the sample set for *Mr. Gregor's Neighborhood*), it seems like we're already well on the rd to *anything is anything* here. Besides, my use of the letter Ϛ is bound to be stigmatized. Looking at **F** & Ϛ side-by-side: **F**Ϛ, the letters cd be sd to symbolize pictorially: my cup of fuck-you spilleth over. In conclusion, is it *surreal* or *hyperreal* to say that <(0 = 1/∞), (Φ = 1/∞), (∞ = 1/0), (∞ = 1/(1/∞)), (Ϛ = 1/(1/Ϛ)) ⊃ (√2 = 7/5)>?

I continue to find Wallace's footnotes of more relevance to my thread than the main body of text: "100% technically speaking, Weierstrass's def.'s "for any . . . , there exists . . . such that" is really laying out an entailment-relation in *first-order predicate calculus*, which is a more involved kind of logic that uses quantifiers like '∀' [for all] and '∃' [there exists]." We've run across '∀' before & we'll see it again.

What is calculus? According to math.about.com, it's "the study of 'Rates of Change'. [..] There are two main branches of calculus: Differential Calculus and Integral Calculus. Differential calculus determines the rate of change of a quantity, integral calculus finds the quantity where the rate of change is known. "Functions" are defined by a formula. What is Predicate Calculus? It's aka First Order Logic, wch is abbreviated FOL (not to be confused w/ FOP). Wikipedia: "Unlike natural languages such as English, FOL uses a wholly unambiguous formal language interpreted by mathematical structures. FOL is a system of deduction extending propositional logic [wch deals with simple declarative propositions] by allowing quantification over individuals of a given domain (universe) of discourse. For example, it can be stated in FOL "Every individual has the property P"."

Wikipedia clarifies further: "Take for example the following sentences: "Socrates is a man", "Plato is a man". In propositional logic these will be two unrelated propositions, denoted for example by *p* and *q*. In first-order logic however, both sentences would be connected by the same property: Man(x), where Man(x) means that x is a man. When x=Socrates we get the first proposition, *p*, and when x=Plato we get the second proposition, *q*."

On the University of Ottawa's HyperGrammar's website, I found "predicate", in language, explained by:

"Every complete sentence contains two parts: a subject and a predicate. The subject is what (or whom) the sentence is about, while the predicate tells something about the subject. In the following sentences, the predicate is enclosed in braces ({}), while the subject is **highlighted**.

Judy {runs}.

Judy and her dog {run on the beach every morning}.

To determine the subject of a sentence, first isolate the verb and then make a question by placing "who?" or "what?" before it -- the answer is the subject."

I reckon that, in my simplified lingo, the subject is the quantity & the predicate is the qualifier. So, eg, "In the morning, the not-Horse did *first-order predicate calculus* before turning into the not-Cow." A rearrangement of this sentence is: "**not-Horse** {did *first-order predicate calculus* in the morning before turning into} **not-Cow**" or "**Subject** {Predicate} **Subject**" or "**Quantity** {Qualifier} **Quantity**". That about sums it up.

The speed of this *amplificare ad absurdum* reduction is then calculated w/ *differential calculus* & double-checked in reverse by *integral calculus*. As you can see in the above example, ANYTHING IS ANYTHING fits the "**Quantity** {Qualifier} **Quantity**" form properly & is therefore an example of *predicate calculus* if one reaches that conclusion in any length of time. *However*, if you reach that conclusion in *no time at all*, it may not be calculable in either *differential* or *integral calculus* & may just turn into an *assumption* of PNT.

Richard Dedekind (1831-1916) & Georg Cantor (1845-1918) are associated together under the "Cantor-Dedekind Axiom" that delineates the use of one-to-one correspondence in defining degrees of transfinite numbers. "Dedekind evinces a remarkable proof for his "Theorem 66. There exist infinite systems," which runs thus "My own realm of thoughts [in German *Gedankenwelt* = *thought-world*], i.e., the totality *S* of all things which can be objects of my thought, is infinite. For if *s* signifies an element of *S*, then the thought *s'*, that *s* can be an object of my thought, is itself an element of *S*, . . ." and so on, meaning that the infinite series ([*s*] + [*s* is an object of thought] + ['*s* is an object of thought' is an object of thought] + . . .) exists in the *Gedankenwelt*, which entails that the *Gedankenwelt* is itself infinite."

Wallace devotes a substantial amt of space to Dedekind's Theory as expressed in a paper called (in English translation) "Continuity and Irrational Numbers" in wch the Number Line is described as *severable* & wch involved imagining a *schnitt* (a cut) in the NL to create 2 sets from it of all rational numbers to the left of the *schnitt* & all to the right. Trying to reductio ad absurdum this section (in the sense of reducing it so far it's absurd rather than in the sense of reducing it so far it's proven wrong) may be beyond even my substantial abilities along those number lines.

The gist of it appears to me to be that Dedekind succeeded in defining an irrational number used as a *schnitt* in terms of the resulting 2 sets of rational numbers - something that was considered previously impossible. This was done somehow by demonstrating that there's neither a largest number in the set to the left 'nor a smallest in the set to the right.

I don't think I have that right. It must be because what I don't understand is why the NL doesn't just blow away in the wind when you cut it. Y'know? Like in those ribbon-cutting ceremonies? I mean, after all, Dedekind was inaugurating a profound new idea - cdn't he've given it a little more pomp & circumstance & satisfied little ole me by having the Number Line just ripple & float a little?

"[..] Prof. L Kronecker (1823-1891), who as mentioned was G. Cantor's one-time mentor and later his arch-nemesis and the same person some historians think more or less single-handedly drove him insane, and who (= Kronecker) was pretty much math's first *Intuitionist*, and believed that only integers were mathematically real because only they were 'clear to the intuition,' which meant that decimals, irrationals, and quite certainly infinite sets were all mathematical unicorns." Ah.. if only I were an *Intuitionist* (my intuition tells me I'm not), life wd be *so much easier*. Dedekind & Cantor also conflicted w/ a Kronecker doctrine sometimes called *Constructivism* wch becomes further illucidated as a

Thank Bob those 9.869044.. pages are over with -- I was beginning to get restless!

part of *Intuitionism* as developed by J. H. Poincaré and L. E. J. Brouwer.

I'll present here just the 1st point in Constructivism to give you an idea. Just one little idea. Then we'll move on: "(1) Any mathematical statement or theorem that is more complicated or abstract than plain old integer-style [**A** **R**at **I**n **T**he **H**ouse **M**ay **E**at **T**he **I**ce **C**ream] must be explicitly derived (i.e., 'constructed') from integer [**A** **R**umpelstiltskin **I**nserted **T**opically **H**as **M**any **E**xistences **T**hat **I**nterfere **C**ornily] via a finite number of purely deductive steps." Aw, shucks! That takes all the fun out of it! That's kindof like doing a word puzzle where all the answers are already written in. I'm looking for a word puzzle where I have to figure out that the answers aren't words at all!

Page 231, footnote 10: "Have we inserted yet anyplace that these [..] brackets, '{ }' are what you put around things to show they compose a mathematical set?" Have we? Actually, I tend to use parentheses '( )' & those brackets '{ }' somewhat interchangeably as set-surrounds & rectangular brackets '[ ]' to show my own interpolations into someone else's quote or an authorial aside that may not be immediately relevant to the flow that it interrupts. You've probably figured that out by now.

I think so.... see page ¬¬$\sqrt[111]{etc.}$

Wallace tells us that the *empty set* (previously shown as '{ }') is symbolized by '**Ø**' (wch I imagine most people who know it call the "null sign") & then says that "any set whatsoever will include **Ø** as a subset." Fascinating. That means that the set of all sets other than the empty set includes the empty set. That's another one for WikiAnswers (I asked.. now we wait..). **Ø** is, incidentally, derived from a letter in the Danish & Norwegian alphabet. Those nihilists.

I asked Wikipedia too & got the usual scholarly explanation that confirmed Wallace's biz above: "For any set *A*, the empty set is a subset of *A*: $\forall A: \varnothing \subseteq A$" & I also get "The empty set is not the same thing as *nothing*; it is a set with nothing *inside* it, and a set is *something*." *SO*, if it's something, doesn't that mean "the set of all Heinz soup cans that still have a little left in them" [not quoted from <u>Every</u>], eg, shd really be called "the set of all Heinz soup cans that still have a little left in them + *something* (that's *not* a Heinz soup can) that has *nothing* in it"? The thing is that I'm sure that someone has actually come up w/ an incredibly detailed way of explaining *why* the afore-mentioned set can be more correctly described w/ the "+ *something* (that's *not* a Heinz soup can) that has *nothing* in it" left out even though **Ø** is in there really. Omigarsh! I'm believing in something!

Also by appair inversion diagrammetical logic revealing that the husband of the Heinz heiress Teresa H. Kerry has an equal claim to the null set of political differenciation. (see page ¬¬$\sqrt[147]{hotel}$ of "the∅ mysteries of false democracy")

Wallace gets into 1-1C (one-to-one correspondence) in detail & manages to freshen the subject up considerably even though I've, by now, read about it many times. One interesting restatement (ie: for me) is that "G. Cantor posits as the defining formal property of an *infinite set* is that such a set can be put in a 1-1C with at least one of its proper subsets." A proper set is defined slightly earlier as "A set *A* is a *proper subset* of set *B* [iff] *A* is a subset of *B* and there is at least one member of *B* that is not a member of *A*." So, eg, the set of "all kitchen utensils" includes the subset of the set of "all butter knives" [again, examples not from Every]. An example of a proper subset of all positive integers being Galileo's example of all perfect squares. Both of those sets are of the same degree of infinity because they can be put in 1-1C. Fine.

Still, imagine that in some more materialist sense - something like: Planet *x* is *exactly* the same size as Planet Earth & is inside Planet Earth. *OR*, tENTATIVELY, a cONVENIENCE is a *Gedankenwelt* set that contains an infinite regress of different proper subsets of *Gedankenwelten* all of wch are the *same size* & yet he rarely gets headaches except after drinking too much alcohol - wch, obviously, *reduces* his number of *Gedankenwelten*! Weird. There's no egress.

From pages 266 to 298, this lad is overwhelmed w/ things to identify w/. "Cantor's next big discovery is that you can validly construct an infinite sequence of infinite sets with larger and larger cardinal numbers using nothing but the formal properties of sets. These properties involve the concepts of the subset and of the *Power Set*, the second of which is hereby defined, for some set *A*, as simply the set of all subsets of *A*. Meaning that every member of P(*A*) is some subset of *A*." [..] "[I]f set *A* is infinite, its Power Set P(*A*) will always have a larger cardinal number than *A* - more specifically, he's able to prove that the cardinal number of P(*A*) will always equal 2^A." [..]

The proof for this demonstrates that there can't be a 1-1C between a set & its power set. The cardinal number of the power set always being 2 to the power of the cardinality of the original set is a result of there being 2 choices in set assignment: either in a set or out of it. No food halfway in Stanislav LIM's mouth. In Wikipedia this is explained thusly (as usual, I'm drastically excerpting here):

"[S]ome natural numbers are paired with subsets that contain the very same number. For instance, in our example the number 2 is paired with the subset {1, 2, 3}, which contains 2 as a member. Let us call such numbers *selfish*. Other natural numbers are paired with subsets that do not contain them. For instance, in our example the number 1 is paired with the subset {4, 5}, which does not contain the number 1. Call these numbers *non-selfish*. Likewise, 3 and 4 are non-selfish. [..]

"[L]et us build a special set of natural numbers. This set will provide the contradiction we seek. Let *D* be the set of *all* non-selfish natural numbers. By definition, the power set **P(N)** contains all sets of

Let me now [?] some of those ambiguous identity set & subsets lead me back into the discussion of the margin.

Some folks like borders & boundaries — others don't. I am a rather extreme example of a "member" of the former category. (Even saying that — dividing the human universe into two comfortables misses my deeper meaning.

Let me provide an anecdote. A "client" I work with recently came to me and showed me some photographs of a group of Chicano/Chicana high school students in the act of replacing the US flag with a Mexican flag.

natural numbers, and so it contains this set *D* as an element. Therefore, *D* must be paired off with some natural number, say *d*. However, this causes a problem. If *d* is selfish, then number *d* cannot be a member of *D*, since *D* was defined to contain only non-selfish numbers. But then *d* is non-selfish, because it is not a member of *D*. On the other hand, if *d* is non-selfish, then . . . well, then it must be contained in *D*, again by the definition of *D*!

"This is a contradiction because the natural number cannot be both inside and outside of *D* at the same time. Therefore, there is no natural number which can be paired with *D*, and we have contradicted our original supposition, that there is a bijection [1-1C] between **N** and **P(N)**"

If only I had had the terminology of "selfish" numbers when I was creating the Flatland overlay! Anyway, the above leads to the paradoxical contradiction expressed as "'$(a \in \Phi)$ -> $(a \notin \Phi)$'". This reads as ("*a*" belongs to the set Phi) implies or maps onto ("*a*" does not belong to the set Phi). "Φ" (Phi) is no longer used here to mean infinitesimals - it's simply a set name - as used in Everything. The above symbol for "does not belong to" plays an important hair-raising role later on.

"*A* is an infinite set; *B* is the set of all subsets of *A*. Because all sets are by definition subsets of themselves, *A* is a subset of *A*, meaning *A* is a member of *B*". Ah.. As if it weren't already tricky enuf, all sets are subsets of themselves but *not all sets* are members of themselves - although, as I 'understand' it, subsets are members of sets & the set is a subset of itself so.. Round & round we go! Where we get off nobody knows! Well.. actually, we 'know' that it leads to "Russell's Antinomy" aka "Russell's Paradox" aka "The Barber's Paradox" - wch'll be explained more later.

All of this eventually brings us back to "Formalist-v.-Intuitionist controversies of the early 1900s." [..] "Respecting infinite sets, for example, Intuitionism is rabidly anti-Cantor and Formalism staunchly pro-Cantor, even though both Formalism and Intuitionism are anti-Plato and Cantor is a diehard Platonist. Which, migrainous or not, means we're again back to metaphysics: the modern wrangle over math's procedures is ultimately a dispute over the ontological status of math entities."

I might be a bit of a Formalist, in some ways I find David Hilbert's (1862-1943) defining Formalism as math as a game to be appealing - not so much because of its detachment from 'reality' but because of the levels of abstraction that it 'allows'; but I'm also intriqued by Intuitionism - not only because I simply love the concept of intuition. As such, maybe we're back to the Stanislav LIM of 'Pataphysics.

He was quite irate, but observing my obviously non-plussed lack of reaction to the pictures, asked "Doesn't that upset you?" My reply: "Well, it would probably upset me if I attached any of my emotional energy into the idea of the sacrosant nature of the US flag & the corresponding sense of national identity it is intended to foster.. but since I don't, no, it does not disturb me in the least." Further discussion encouraged me to question why an action demonstrating a different set of values

Ontology: "a study of conceptions of reality and the nature of being" (Wikipedia = WP from now on.. - It's about time!) What I love about math is precisely the way it stimulates people's minds into frequent ontological shifts. That, in case you haven't noticed, is what this bk's 'about'. There's always the hermetic ontology that the more one asks questions instead of accepting answers, the more 'open' 'reality' becomes. **Neoism, eg, is the fine smart of undefining itself.**

The paradoxes generated by Cantor & co's Set Theory lead to "the split of abstract set theory into two subtypes, *naïve set theory* and *axiomatic set theory*. N.S.T. is just regular Cantorian set theory with all its warts and glories, including its susceptibility to paradoxes. Axiomatic set theory is an attempt to derive a more rigorous, foundationally secure version of set theory that's got all the conceptual power of N.S.T. but is set up in such a way as to avoid gross paradox." "Naïve Set Theory"! Marvelous. That seems almost like "Folk Math" but it isn't really close. Still, love that terminology! Non-Selfish Naïve Set Theory as the Happy Solution to problems in Mr. Gregor's Neighborhood!

Giuseppe Peano (Italian mathematician - 1858-1932): He contributed much to set theory - such as the symbol for "belongs to the set of" shown on the previous page & has a set of axioms or postulates or whatever named after him: Peano's Postulates, more correctly called Dedekind-Peano Axioms. Get yr red-hot Peano's Postulates here! In a footnote on page 287 in Everything and More, Wallace mentions these as "five basic axioms of number theory". In WP, 9 are mentioned. I present the 1st 3 here to just show how astoundingly simple they are so that you can marvel over how such things can be used for building such complexity:

"1. For every natural number x, $x = x$. That is, equality is reflexive.
2. For all natural numbers x and y, if $x = y$, then $y = x$. That is, equality is symmetric.
3. For all natural numbers x, y and z, if $x = y$ and $y = z$, then $x = z$. That is, equality is transitive."

I wonder? Can they be stripped down even further? What happens if one reduces the axioms to something like: "0"? Or, "For every number x." Maybe that's what some wishy-washy namby-pamby philosophies (not industrial strength) have already done. But, still, then the idea wd be to get *more reductive*. Start w/ "0" & have the 'developments' get to be <0. By wch I *don't mean* negative. An empty set in wch neither the brackets or the 'nothing' inside them is there anymore. Now, obviously, I'm just on a stream-of-unconsciousness roll here again, but, still, if something's hard to imagine, it's worth trying to do so.

should threaten this individual's set. Isn't that, after all, one of the great values of a so-called democratic society? That everyone can have a unique perspective & opinion without threatening the right of others to have an opposing view? But no, for him, this act was a violation of an absolute boundary -- national & moral which he wished to impose upon the totality of the inhabitants of a territory & a social identity.

Set Theory is wonderfully labyrinthian. As a young'un I was naturally attracted to it but I had no idea just how intensely discombobulating of 'consensus reality' it is. As an old fart, I realize that the desire to order things w/o *boxing them in* is what I've spent much of my life doing. Ordering things allows one to make comparative observations but can all-too-easily over-simplify. Formalist Set Theory is somewhat like Jarry's 'Pataphysical description of the sun (see page $\sqrt[13]{3.02875107 \times 10 \text{ to the 14th}}$) - an outrageous deviance from what sometimes seems like 'common sense' - it's *uncommon sense* - a way of describing 'reality' that enables one to see things w/ fresh concepts that're so radical that it leaves the mind wide-open to revelation.

Wallace gets into *order-type*. How members of a set are arranged. "[I]f you take one of the set of integers' members and put it last - as in {1, 3, 4, 5, 6, 7, . . ., 2} you now have a totally different order-type. the set {1, 3, 4, 5, 6, 7, . . ., 2} is no longer 1-1C-able with a regularly ordered $\aleph_0$ set that has no last member and so give you no way to arrive at anything to match up with the 2"!! Oddly enuf, PNT's rejection of commutativity ($m + n \neq n + m$) is basically just a restatement of the way that order-type changes things - &, as such, may have more 'validity' than one might originally credit it w/. Once again, we're back to *grouping*.

The above change between the set expression {1, 2, 3, 4, 5, 6, 7, . . .} & {1, 3, 4, 5, 6, 7, . . ., 2} might even seem trivial to many looking at it. To me, it's mind-blowing. One simple rearrangement in an expression has profound consequences: {1, 2, 3, 4, 5, 6, 7, . . .} ≠ {1, 3, 4, 5, 6, 7, . . ., 2}. *GROUPING.*

"The ordinal number of the whole denumerably infinite set {0, 1, 2, 3, 4, . . .} gets symbolized by the little omega '**ω**'. This transfinite ordinal is the limit of the set's members' sequence - that is, it's the very smallest number bigger than all finite integers. Another, more common way to describe ω is that it's the ordinal number of that set of which $\aleph_0$ is the cardinal number." Just the idea of someone's finding the 'need' to express "smallest number bigger than all finite integers" impresses me.

What if someone created a term for "person standing on the opposite side of the Earth from you"? Of what use wd that be? Welp, I don't 'know' but I've been told that such a term exists in my favorite word source: Mrs. Byrne's Dictionary of Unusual, Obscure, and Preposterous Words (by Josefa Heifetz Byrne). I don't feel like thumbing thru the whole thing looking to see whether such a word is really in there right now so I just tried Googling "word for "person standing on the opposite side of the earth"" wch I then took the "word for" part out of after I got

Boundaries are fixed markers of space, time, & the unity of collective thought. Boundaries mark the limits of the hard intellect. As such they are inscriptions within the culture of DEATH.

Death = boundary, stasis, limitation, belief, certainty, finitude, law, fixity, rigidity, uniformity, rule, one to one correspondence.

Life = energy, process, unfolding, limitlessness, dynamism, openness, ambiguity, uncertainty, questioning, journey, complexity, possibility, flexibility, movement, doubt, wonder, humor & expansive connectivity.

zero returns. This latter search yielded ONE reply: "Phil Plait's Bad Astronomy: Mad Science". That, in & of itself seems so akin to this bk that it seems esp fortuitous. "So someone standing on the side of the Earth nearer the Moon feels more gravity from the Moon than a **person standing on the opposite side of the Earth.**" Not only is there probably a word for the person standing on the opposite side of the planet from me, but there're even specified times when each of us will be feeling the same degree of gravity vis-a-vis the moon & times when that degree will be specifiably measurably different.

That brings up Googlewhack: <http:/www.googlewhack.com/>: **"Your goal: find that elusive query (two words -** ***no quote marks*****) with a single, solitary result!"** The above search resulting in one result is, obviously, not a Googlewhack because more than 2 words were used. Googlewhacks go quickly out of date because of changes in the internet, etc. My attn was called to them several yrs ago when someone emailed me to tell me that one of my websites yielded a Googlewhack because of a coupling of "discombobulate' (a word I use a fair amt) & something else (that I, alas, don't remember).

Is there a Perverse Number Theory Googlewhack? Is there a Googlewhack that survives yrs of being pointed out? Unknown neoist produced a predictable plethora: 1,230 hits. It wd be a challenge to find 2 seemingly ordinary words to produce a Googlewhack. Imagine discovering that "a" & "the" didn't occur together on any sites. Obviously, they do, so one has to come up w/ something else equally unlikely but.. actually working. It's much easier to find a Googlewhack like, let's say, pregroperatavist aardwolf - wch is almost synonymous w/ unknown neoist. I'll try it. But, not surprisingly, that didn't work either because the combination of words is *too* unlikely. So I try unknown pregroperatavist & strike out again. What about just pregroperatavist? Nope. So strike that. I might be misspelling it anyway.

Now, I ***must*** find one. No red Hankie of Death for this! blink egads yields 20,600 results! As you can see, this ain't easy. predicament jello = 21,500. neoist sperm = 453; neoist ovum = 67; Riemannian neoist = 4; neoist pregroperativism = 137; Riemannian pregroperativism = 0; tENTATIVELY pregroperativism = 62 (this latter includes an entry on <http://www.sportsfactbook.com/history/Neoism> put there by one sly devil that's not myself (except insofar as we're both Luther Blissett)); neoist poontang = 5; Riemannian poontang = 23!; Riemannian Gogglewhack = 0; Non-non-Euclidean Googlewhack = 0; perverse number = 1,880,000; neoist plumbing = 119; neoist plumbbob = 1!: <http://runelogix.typepad.com/my_weblog/interesting/index.html>
YES!

The extent to which the rules of this universe are immutable is the extent to which it has been fashioned upon the image of the god of death.
tENT writes on page ¬¬N hot rat "That's kind of like doing a word puzzle where all of the answers are already written in." An ordinary word puzzle is always already very much like this -- written in or not. For each set of letter spaces there is only one right answer. This is an absolute boundary: right or wrong.

NO!:

"

16,000 results contain neoist
14,500 results contain plumbbob
1 result (from Whack's location) contains neoist plumbbob

Google whacked, but did not link to definitions for both your Googlefactors.

Remember Rule Number One: are live links in the blue bar atop your Google results?

Are both spellings correct? Remember that Google does not link to **every** definition at answers.com. Sorry, but without two links you may not add to The Whack Stack"

Oh, well, back to the Whackingboard. What's "the formula for calculating the probability of finding a Googlewhack?" did not match any documents. Googlewhacking neoist = 2 - & I'm not one of them.

Where was I? If you think the next quote from Wallace's bk'll answer that question, you're probably less confused than I am: "the transfinite-ordinal hierarchy is associated both with abstract entities called *epsilon numbers* and with an arithmetical operation called *tetration*. We're not getting near the former except to say that they're essentially a class of numbers such that $\omega = \epsilon^{116}$; but tetration is simpler, and you might already be familiar with it from, say, field theory or combinatorics if you had a lot of college math. It's basically exponentiation on acid. the 4th tetration of 3 is written '$^{4}3$' and means $3^{(3^{(3^3)})}$, which = $3^{(3^9)}$, which = $3^{19,683}$, which you are hereby dared to try to calculate. The technical connection between tetration, transfinite ordinals, and epsilon numbers is the fact that o = ωω, which isn't all that important. But, if you can conceive, abstractly, of a progression like ω, ((ω + 1), (ω + 2), . . ., (ω + ω)), ω^2, ω^ω, $^{\omega}\omega$, $^{\omega^{\omega}}\omega$, . . ., then you can get an idea - or at any rate an 'idea' - of the hierarchy and the unthinkable heights of ordinal numbers of infinite sets of infinite sets of the ordinal numbers of infinite sets it involves." If Wallace had written nothing else, just this passage wd make his bk worth reading. I tip an infinite regress of empty set hats to him.

THE END. Of quoting from Wallace, ie. I feel like a collapsible epsilon (= 21,600) right now. Do you? It wasn't my intention to quote from any of these bks being navigated thru quite so much but I haven't succeeded in any way around it yet. & I'm not about to now because there's only one bk left:

But imagine completing a crossword puzzle in which you find solutions unforetold & unforecast -- each answer different from the proposed solution, but engaged & connected to the rest of the series... would that sort of game playing be "truer" to the intended spirit of the constructivist project? Isn't such a project, at heart, what this book is about?

Is that hot hat my hot hat?

¬¬ 167 Fuck if I know!

Imagining Numbers (particularly the square root of minus fifteen) - Barry Mazur. Egads onomatoid (= o): "there are those who simply shrug off "imagination" as an 'onomatoid,' that is, a namelike word which in fact designates nothing because it signifies too broadly"! [..] "An editor at McGraw-Hill is quoted as saying, "We were told to avoid using the word 'imagine' because people in Texas felt it was too close to the word 'magic' and therefore might be considered anti-Christian." For that matter, why not avoid the word "thinking" while we're at it. Wdn't want to offend them Texas Christians (like a certain few presidents we cd name) now wd we?! onomatoid imagination = 4 results, one of wch is a reference to Mazur's bk!

While writing this bk, I've been finding myself in a somewhat unexpected dilemma. My tendency is to go off on tangents & to never look back. There's a tangent nagging at me right now - demanding more attn than Mazur's bk. Adding more fuel to this fire is that I don't like quoting so much. It might create the *delusional appearance* that this bk is just a rewrite of other people's bks & little else. But that's a part of my task here: to give credit to other people's bks by quoting them directly rather than, eg, paraphrasing them.

By paraphrasing them, I'd be giving the false impression that I'm an expert on these subjects - rather than a person building on other people's expertise (who in turn built on other other people's expertise, etc). SO, I'm going to stick to this quoting process (w/, more importantly, perhaps, for the purposes of PNT's establishment, my own extrapolations interpolated from time-to-time) until the SWEET END, the Happy Conclusion of this NAVIGATION: section. Then I'll move on.

Mazur introduces some more fodder for my linguistic play: SQRT - the abbreviation for SQuare RooT in one computer program he uses. Obviously, this is also the word SQuiRT w/o its vowels. That might come in handy later.

Discussing quadratic formulas, Mazur explains that "The letters early in the alphabet (*b*, *c*) are generally taken to be *known* quantities, while the end letters (*X*, *Y*) generally stand for *unknowns*. The idea of explicitly differentiating between "not yet specified" values *b*, *c*, . . . as if they are known, as distinct from quantities *X*, *Y*, . . . which are sought (unknown), already occurs in the writings of the sixteenth century mathematician François Viète". [..] "(He, however, made use of a different alphabetic mnemonic [..:] the unknowns he designated by uppercase vowels, and the unspecified knowns by consonants.)"

My perverse typographical mind immediately latches onto this. I use what's known as "tOGGLE cASE" to write my name: tENTATIVELY, a cONVENIENCE. tOGGLE cASE is a reversal of usual UPPER CASE / lower case relations. In the spirit of enjoying deviances from the typographical norm (for variety & creativity's sake), I look out mainly for meaningfully created difference rather than just gratuitous difference (typographical deviance for the sake of typographical difference).

Math, perhaps more than anywhere else except perhaps in Concrete/Visual Poetry, has highly systematized uses of letters - wch includes whether they're capitalized or not or italicized or not, etc.. The above explanation makes that clear. When I read about things like that, I imagine the use of typographical oddities as used by people *presumed* to be 'insane'.

Given that my own typographical oddities are usually very deliberate & meant to enable multi-leveled meanings, I always look for the same in others. I look for *method* in apparent 'madness'. Eg: in my 1969 high school yrbk entry 1st referenced on page $\sqrt[117]{9.50048604 \times 10 \text{ to the } 241\text{st}}$, I 'describe' myself w/ the words "Nascent, orthopteran, sabaist, luxated, oleographic, turgid; labiac, excogitate, accentric, hydrophytic, crispy, intarsia, magnetize."

To the casual reader of these words, they're most likely to either seem like 'crazy talk' or nonsense - certainly *not* as self-description! &, yet, an observant person might wonder: Why is there one semicolon (";") in the midst of a list otherwise divided by commas (",")? That, aside from the seeming nonsensicality of the word choices, is the most obvious clue that there might be something special happening here. There are 6 letters in my given last name "Tolson", there are 6 words before the semicolon; There are in 7 letters in my given 1st name "Michael", there are 7 words after the semicolon.

The last letter in my last name is "n", the 1st letter in the 1st word is "N"; the 1st letter in my 1st name is "M", the 1st letter of the last word is "m". By now it's no doubt obvious that the 1st letters of each word, when read in backwards order, spell out M(agnetize)i(ntarsia)c(rispy)h(ydrophytic)a(ccentric)e(xcogitate)l(abiac) SEMICOLON T(urgid)o(leographic)l(uxated)s(abaist)o(rthopteran)n(ascent) = **Michael Tolson.**

Therefore, what we have here is *method* & *not* madness. But it doesn't stop there. Reading the 3rd letters in from the back of each word (again, in backwards order) we get: magnet**i**ze + intar**s**ia + cri**s**py + hydrophy**t**ic + accent**r**ic + excogit**a**te + lab**i**ac + tur**g**id + oleograp**h**ic + luxa**t**ed = **is straight**. The last 3 3rd-letters-from-the-ends spell out "ire" - wch is why I wrote "I hope it won't arouse yr ire" whilst discussing that yrbk entry. "ire" was an unintended side-product not built in by me. This particular writing of mine is hardly air-tight but, nonetheless, in order to produce it, I had to go thru the entire dictionary that I had to find just the right words that met my criteria of 'sentence structure' (adjectives, nouns, & verbs), & hidden meaning.

There were 3 main reasons for presenting "Michael Tolson is straight" as I did: 1st, people took it for granted that a person such as myself was "on drugs". I had long hair, was very skinny, & wore clothes that were falling off of me. As such, to most people I was a "hippie". Contrarily, I didn't consider myself to be a "hippie" & didn't use *any* drugs - either legal or illegal - including headache relief tablets. The 2nd reason was that I wanted to make this statement in as convoluted a way as I cd conceive of - rather than in a straight-forward way - using ironic humor. 3rd, I just liked the word "straight" because it has 8 letters & only one syllable.

All of that is by way of introduction to what some might call "outsider writing" & its possible connection to mathematics. Back in the 1970s, when my 1st bk came out, my friend Charles Brohawn & I were both interested in "Outsider Art". Charlie went to NYC a fair amt & started noticing some unusual texts posted about by a guy who identified himself in the texts as Benedict Schwartzberg. These sheets were full of unusual typography & were difficult to read - even for afficionadoes of such things as Charlie & myself.

Unwilling to just take it for granted that Schwartzberg's eccentricities were random products of a disordered mind, we wondered if there was any special design that we were incapable of deciphering. Alas, we never looked very hard & we never succeeded in producing any special deciphering. Chas did make photocopies of at least one of Schwartzberg's texts & turn them into address label size stickers that he wd put around as graffiti & I had some Schwartzberg material published in a magazine. below is a shrunk-down sample of one of Charlie's Schwartzberg sticker sheets:

n ZEu).'MY PRESENT NamE's BENEDICT SCHWa('PanGod ZEUS':Hb as ME/Sun/ManiFEst)(
')('RtZBERG':'HamBLin M'('MiLDrEd PREGNaNt)).'g:'('ONCE EVERY agE i inCaRNatE
DER2SLaY EviL.''BEN'&sHoRT(BEN:Bon;;God;good)BECaUsE of'Nazis'(BECaME olstERs
tE HEavENs.)&TiBEt.PREFix'Eg'(NoM D'un'CHiEN!)&'PNEUMA HAGION!('HoLY sPiRit')
rNaL'Now"('ABI('THE HELLENEs Had a woRd4it.') NUN')&AiSkHULos''dogs drEaMing of
asing BUNNiEs MuTtERing'MY.MY'('MU';woMB.BEN No MoviE MU-siC('tHus')MYstERY MIsTi
SEUM woN vaiN'PHUN)('g.'THE fiNaL aiM,DuRBosR&RESULt of tHE aCt of sELf-CREatio
'an EXCEEDs tHE aRC of tHE HUMaN MinD.').iN ALgiERS('PEPE LE Moko')1959(toBtiME oN
gitaL CLock/MERCEDES-BENZ 280('BoRN YESTERDaY';;RB:'28!'))i,as AN(HEB.,'GRacE'(
'Hi!')),CREatE tHE CREatioN,as NUMBER.'RELigio(N)';'KHAIRE ELI (AN)!';'BE HaPPY/
ihE/Hi/BYE,HEId-os(UP)('KHaiR-Hagi-HaLL'('E-HiLos-OB-Hi'/'KHRIS-T-ÖS';'MET','An-
NtED,SHiNY('MEtro'))(ON TEMBLE of APOLLoN('OLUMPos PaN-HELios')at DELPHini;;
EDEN AGAN';'STaRStaR vERY HigH'('G/g)ø(d/D)'('foEtus/BENIM/SunMoON SuN MooN'):H
AGA-TH-",'VERY staR(THE(¥)EIN','RaN fast,BuRNiNgFLaME'('DANU'/'Taq'/'DEva'))('fR
M'),i.E.SUM('BEN S.'/'HELios')(Moon:'BEN-BEN'/'MEN'(/'MAN')/'SELENE'))('LifE'(i US
EEZEE'iNstEad.)'EviL''LovE';'HELi-BU'('dawN')('Rush!?'));gUDaiSM(CHRistiaNMosLEM)&
iBEt)BoN&(BHaRat)HiNDU/BUDdHiSM&CHiNESE)YiN(NoNEXistENt)-YaNg/SHEN(g)ti aRE HELLEN
GNOTHI SAUTON'(DELPHI);'Not was./notHing.'/'HERakLEitos';'Hi,stoLE RakE!'(HELL.AL
aEST ENdEd witH'T',PuTHagoRas adDEd Y('U') BE KE PS O aftER'HEB-Rai-oi RaB'd it
Most HELL.NouNs'ENd iN'S'.'—HA,)so"'T',ALiENs'"aDDEd UUUU(UVWY)&MovEd XAZ./kaBaLa
HE'EL/EL-MU CREatioN# is sECoNDaRY CREatioN.'(ADAM/EDEN/SataN;'ATHENs'/Isaac-Isis

Note on my note on puzzle solving to tENT'S year book description: 1st: I have an advantage in that my 'illness' often works through cryptographic inscription & so its a rather routine habit to look for anagrams & search through texts forwards, backwards, laterally, & through series/) of equations. Of some interest regarding my own method to the decoding process: First I recognized "Michael Tolson" inscribed w/in the last letters of the text - words running backwards.

bottom of a Benedict Schwartzberg as turned into a sticker sheet by Charles Brohawn

The point of all this is that when I look at writings like Benedict Schwartzberg's now, I think of things like John Nash's chalkboard scribbles, of my own writing, & of the possible symbolic significance of Schwartzberg's capitalization. Eventually, I'll get back to the connection between mathematical notation & multi-leveled writing. For now, I've just Googled Benedict Schwartzberg for the very 1st time in my life & was somewhat astounded to find *any* hits - most not relevant, unfortunately.

In the archives of of *Avatar Magazine*, issue No. 9, Sept. 29, 1967, p. 19 is reproduced one of many letters from Schwartzberg to them:

> *I answered "Who am I?" (and proved the answer). I am God. I now ask "What about the others?".*
>
> *In der beginning is der Void. "Mental" meant "All".*

Avatar: an incarnation of God. "Avatar who am in heaven, holy is my name...". A waiter: a wedder:a water. Yesus, the El Avatar, is Elevator.

Incarnation: Inca nation. (Inker:Writer:Speaker:Crea-tor:Thinker:WORD-ER).

Mayan civilization: M

Mayan civilization: Maya(Magic:g/y, as in regal/royal) (Mad Chick: Parvati was Maya incarnate, who's now (for Eternity) mad at Shiva for burning Kama (Karma) to a cinder (inca:anger).

Body is Bodhi is Buddha, and fingernails&toenails are the Nails in the Hands & Feet (Santa Claws), & the hair on the head is the crown of Thorns, and Hell is El is God, & I have free 1-page kaballah for anyone who writes in for it, please publish this letter.

Benedict Schwartzberg

In another link, I find that something from Schwartzberg is in "The Jack Kerouac Archive - Declan Kiely, Henry W. and Albert A. Berg Collection of English and American Literature, New York Public Library" & is cataloged thusly: "D14, f.3 20.18 Typescript draft, untitled. holograph note: "Benedict Schwartzberg (tea head) brilliant linguist & theological didacticist (1966)." April, 1966. 5 leaves." So, someone, at least, thought he was a "brilliant linguist". Perhaps Kerouac did.

In fact, Schwartzberg's name appears in the catalogs of a few universities: Kent State, Stanford, King's College London. Alas, the trail peters out pretty quickly. *Avatar* turns out to be the best source. Sadly, whatever evidence there is about him makes it seem that he was a drug casualty or a damaged person of some other sort. Looking at detail of the writing shown in Charlie Brohawn's copy we find:

i saY'CoRRECt DHYaNa'(tHoUgHt')is:No pain!

& shortly thereafter:

THE:pain/Snapped my brain.

& shortly thereafter:

'=O MEga ALEHaBEt!),i want2know wHat

It's hard for me to say how consistent his patterns are. In these examples, the 1st-person pronoun "I" is lower-case - a device often used by poets, eg, to express a knowledge of one's smallness in the context of the universe. In the small snippets shown above, all of the "Y"s, "C"s,

(crispy [s])
excogitate
(↓)
(luxated)
accentric
labiae

If you notice, only the words "crispy" & "luxated" along with the missing component did not quite work in my initial solution. But the lack of fit coupled with the clue involving the word "ire" motivated me to push on in other directions.

"D"s, "H"s are upper case; but the "R"s, "E"s, "N"s, "U"s, "T"s, "S"s are both upper & lower case, etc.. He does refer to a "MEga ALBHaBEt" & he does use "2" instead of "to" as I often have also.

It seems likely that, as shown in the Avatar letter, he had an idiosyncratic etymological style based on free-play in breaking words apart & reforming them for new meanings that he found to be related w/ in his spiritual cosmology. This type of style might often be called "schizophrenic" - but just because it's 'schizophrenic' doesn't mean it's meaningless.

Of course, many people other than myself have speculated on such writing. I have a whole folder full of it provided by other interested parties such as: John Berndt, Peter "Brainpang" Warner, & Doug Skinner. This latter fellow sent me a review he wrote of a bk called <u>Mathematical Cranks</u> by Underwood Dudley of the Mathematical Association of America. Doug also shared my enthusiasms for the works & research of Raymond Queneau (mentioned earlier & to be mentioned more later) & Richard Shaver (who probably won't appear in this bk).

Skinner sent me samples of the writings of Carl R. Wahlstrom wch I find to be more or less incoherent but wch are nonetheless interesting because of their form. Below is a sample from his 1965 "ECLIPTIC COLOR PHILOSOPHY":

WEDGE SOLID NUMBERS:

The prisms and pyramids are solid numbers, that have wedge and half-wedge total solid / deficiency, formable from a triangle, folded on the dotted lines; or an "Obsidian Dagger"- seed-numeric / formal lenticular / modification. The tri neutral / number dice key complementary planet / proxy prism rotations / revolutions exchanges; but a fourth color / density die keys their optional color altar / motions, formed by a graphic art cubic / cross, thus......

I admit it, I *still* haven't gotten around to tying this into what I quoted from Mazur oh-so-long-ago! So, here goes: What if there's a 'schizophrenic' 'mathematical crank' writer out there who uses typographical deviances such as designating unknowns "by uppercase vowels, and the unspecified knowns by consonants"? & what if there's something to it?

Eg: "YOU think that's something? YOU ain't seen nothing yet!" I chose those sentences for a reason. It's one of the things I've imagined writing in pure platinum in a size maybe 40 microns wide (about half the size of a human hair) & only visible under a microscope if I cd get access to the appropriate lithography equipment. But that's another story. Here, the "YOU" is written in all caps. That cd be a simple means for emphasis (it is) but cd also signify that "YOU" are an unknown because there are 3 vowels capitalized à là François Viète's system. Now, you might be thinking: "That's stupid! "Y" is a consonant here!" & you'd be right. My excuse is that I'm tweaking things by misusing the vowels as "a, e, i, o, u, & sometimes y". OK?! Forgive me. At any rate, YOU *are* an unknown because I don't 'know' who's reading this bk in its hypothetical future.

By now, YOU might very well be hoping that I'll finally get back to quoting the math writers at length so that YOU won't have to suffer thru the mental tortures I've put YOU thru in the "Multi-Leveled Writing" section. Welp, YOU're in luck! It's time to get back to Imagining Numbers:

"**11. Mental tortures.**
Cardano, in considering the "quantity" √-9, writes: "√-9 is neither +3 nor -3 but is some recondite third sort of thing [quaedam tertia natura abscondita]." At one point in *Ars Magna*, Cardano finds himself forced to invoke a square root of -15. he says to the reader, giving no further justification, "You will have to imagine √-15," and then he goes on to calculate with it, even though he says that he is doing this by "dismissing mental tortures." The colorful Latin phrase Cardano used for this is *disissis incrucuationibus*, and the translator notes that Cardano might very well be playing on a possible double meaning of this phrase in the sentence, which can be read either as "Dismissing mental tortures, multiply 5 + √-15 by 5 - √-15 . . ." or as "Cancelling out cross-multiples, multiply 5 + √-15 by 5 - √-15 . . .""

Note that, the [bracketed] Latin is from Mazur - not from me. PP 45-46: "In the early seventeenth century, the mathematician Thomas Harriot, trying to protect numbers like 5 + 2√-1 from reproach, would refer to them as *noetic radices* ("noetic radicals," or "radicals of the intellect"), and others would refer to them as "imaginary roots," as we continue to do."

WP: "The word "**noetic**" comes from the ancient Greek *noûs*, for which there is no exact equivalent in English. It refers to "inner knowing," a kind of intuitive consciousness — direct and immediate access to knowledge beyond what is available to our normal senses and the power of reason." Slippery devil that I am, I think that (if I can remember the word), if anyone asks me what kind of political position I occupy, I'll say "noetic radical" & wink knowingly.

When Mazur gets to double negatives he explains it very clearly. He also notes that "arithmeticians agree with grammarians in claiming that a double negative is a positive." In a later part of this bk, I'll put the usual perverse spin on that.

"qwerty" keyboard (i.e. "errors" in upper & lower case signs, substitutions of spatially approximate letters on the keyboard, transpositions of letters as symbol inscriptions into control commands, etc.) all coordinated to direct the recievers of the message to an internet site describing how encoded messages of considerable complexity can be overlooked. In other words, my "surface" message— which seemed bumbling & transparent — was accompanied by an "encoded message" of considerable complexity, irony, & ulterior intention.

1793: Brook Taylor succumbs to the finite properties of the universe in general & by his frustration in failing to derive the expression $\Delta^n_h[f](x) = \sum_{k=0}^{n} \nu_k f(x+kh)$ in specific.

okay, so admittedly this was a waste of my time since I was completely aware that my system was much to subtle to be interpreted as code -- but it made me feel better in my privilege of having the last "zymurgy" (in-side out joke). In defense of such an obviously hostile use of encryption -- one must try to understand that at the time I wrote the message I felt constantly bombarded by similarly disparaging & similarly complex uses of language.

Mazur adds a spicy anecdote to his telling re double negatives by referencing the author *Stendhal*: "If you are bewildered by this [..], you can understand the distress of fourteen-year-old Marie-Henri Beyle (later known through his writings as Stendhal), who, on quizzing teachers and friends about "minus times minus is plus," found no one who could justify it to him. In his fragmentary and vinegary "autobiography," *The Life of Henry Brulard*, Stendhal muses that his early enthusiasm for mathematics may have been based on his loathing for hypocrisy, by which he means a loathing of some of his relatives and some of their priests. "Hypocrisy was impossible in mathematics," he writes" [..] But, then, when no-one cd answer his question properly, he wondered: ""Can my beloved mathematics be a fraud?"" & turned away from it.

Regarding organizing quantities spatially, it's noted that "In contrast to what one may consider the approximate neutrality of right-left direction in the horizontal dimension, the psychologists Mark Johnson and George Lakoff emphasize the non-neutrality of the vertical with regard to issues of quantity (and Tversky does the same for issues of emotion). Johnson and Lakoff point out the universality of *more is up* and *less is down*. They offer the examples in English such as: *prices rose, the Dow hit bottom, turn up the thermostat*; and they note this bias occurs in many languages. Even more telling is that there is no language for which the opposite is true; none in which more is down and less is up." I'm down w/ that but are you up for it?

Mazur ponders a terminology problem: "I think the issue of "permission" lurks behind many conversations I have heard about the question of "invention versus discovery" in mathematics. The word *permission* is awkward, carries irrelevant associations, and is philosophically unsound (e.g., *who* or *what* is doing the permitting, and *when* do you know that you are granted this permission?)." Exactly. Perhaps, instead of getting *permission* we can just *take the liberty*. Wch is exactly what I'm going to do now by allowing myself to go off on the tangent that's been nagging me since page $\neg\neg\sqrt[167]{hot^{hat}}$. (Nag, nag, nag, that's all these tangents ever do is nag me. If they're not careful, I'm going to give up geometry & trigonometry altogether, & run away w/ a hot young quantum gravity field!)

trigonometry honeymoon = 21,000
trigonometry nag = 19,100
trigonometry divorce = 64,300
trigonometry mistress = 28,700
trigonometry concubine = 2,490
quantum concubine = 19,000

The attempt to Googlewhack is, of course, the search for the ONE. A successful Googlewhack isn't, however, a finding of the Absolute Infinite (defined by Cantor as God) because it's not the ONE & ONLY ONE but, instead, the ONE of a hypothetically infinite many ONEs. It's Pantheistic, or, as I prefer, Panatheistic.

Failed Googlewhacking is a sortof indicator of degrees of consensus 'reality' - the more hits an attempt receives, the more the failed whacker realizes that they're getting to the heart of the LCD (Lowest Common Denominator). Therefore, perverse number, weighing in at 1,880,000, is the most powerful LCD force of the above hits (above meaning 10 long pages ago on page $\neg\neg\sqrt[165]{_{hot}{}^{fat}}$). & things like Riemannian Googlewhack, weighing in at 0, are off the LCD scale.

Going contrary to Googlewhacking rules, though, & creating a purpose contrary to the Googlewhacking one by putting quotes around the 2 words, is ultimately the best way of pursuing degrees of LCD. I hereby declare this process "Google Wracking" (2 words, always in quotes).

The Intuitionist branch of "Google Wracking" is "Hunch Backing". "Hunch Backing" is pursuing a "word combo" that one suspects to probably yield more than 1 million hits. Proper names, such as "Brad Pitt" are allowable only for curiosity's sake but not for yielding serious LCD results because the purpose of this research is to reveal underlying strata of consensus 'reality' that are well 'nigh omnipresent (w/o being obviously so: unwritten laws, oppressive subtexts) rather than to reveal what's already in yr face thru supermarket tabloids.

Take, eg, a search for "Texas Christians". This might be "Hunch Backing" insofar as one might suspect that "Texas Christians" are behind most unwritten oppressive laws these days. However, this yields a lower amt of hits than one might expect: a mere 2,380. It did yield such topics as: "Why aren't **Texas Christians** voting on Tuesday to ban divorce and re-marriage?"; "**Texas Christians** Take Aim at Gay, Female Student Support Center, Agnosticism / Atheism"; "Sex Education Embarrasses Conservative **Texas Christians**"; "Form Pits **Texas Christians** Against Muslims"; & "BOOK Now ! Join **Texas Christians** for an Alaska Cruise in 2008 !" As such, the oppressive hypothesis is correct but the deeper 'truth' still lies unrevealed. "Texas Christians" aren't even where it's at there. They're just dupes & pawns of a deeper Supreme Fascist.

Significantly, "Death God" = 343,000. We're getting warmer - esp given that the LCD is some people's version of Hell. "Fuck You" yielded an astounding 20,200,000 hits! & also leads me to a YouTube movie entitled "Frankee - FURB (Fuck You Right Back)". That one is a pop

yes, today I recognize that my interpretations resulted from the joint (commutative) of paranoia & schizo-recombinatory semiosis - but at the time they seemed quite "intentional". At the time my methods seemed the only reasonable "defense" against against similar sorts of encrypted assaults. Tremble not, self-reflective reader, the transcription errors made in the marginalia you are reading are wholey accidental, unintended, unconscious, & benevolent!

music video so we're definitely getting warmer than we were w/ "Texas Christians" - w/ whom I have close to *nothing* in common.

My next "Hunch Backing" was "a the" = 29,700,000. Here's where having an analytical mind is starting to pay off. This yielded a higher LCD rating than even "Fuck You"! Two articles side-by-side are a powerful indicator of what's *really* behind consensus 'reality'. This ties "Google Wracking" into a related discipline "Part-of-Speech Tagging" - a quantifying practice of marking-up words based on their part-of-speech to determine word frequency. This is a part of *computational linguistics*.

Is it possible to get over one billion hits? "a a" = 237,000,000! Damn! I'm stumped. "I am" = 939,000,000!!!!! A little egomania goes a long way. "You are" = 2,000,000,000!!!!!!!!!! **I DID IT!**

Imagining Numbers: Rafael Bombelli (1526-72). It's interesting to see the symbols evolve - some die off & some stick. "Bombelli's notation for his unknowns is quite suggestive. He denotes them by shallow bowls. If he wishes to refer to the square, say, of an unknown, he puts a little 2 floating in the bowl." [..] "The bowl is waiting, it would seem, to be filled by the *quantity* [..] So it is not unusual to see in his text expressions like

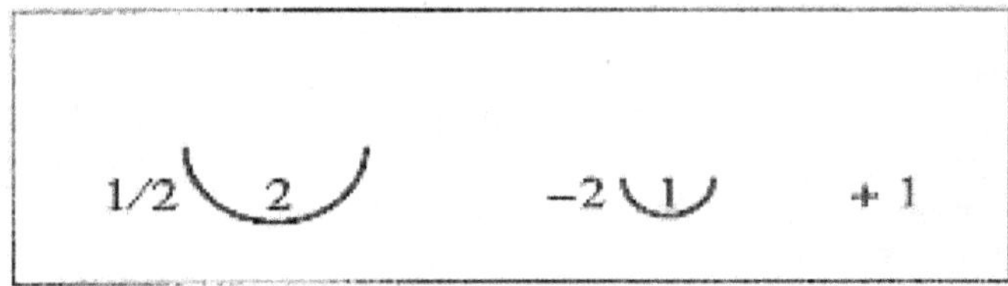

meaning, in modern terms, $X^2/2 - 2x + 1$." Will Bombelli be credited, at some point in the dim future, as the inventor of the Smiley Face?

Much of this bk is going on the same ground from different perspectives that I've already spun off from so I'm skipping merrily over it now. Then, *suddenly*, I find mention, not of something mentioned in the other bks, *but of something that I've mentioned*: viz: Kafka's "The Metamorphosis". Some of you may still be able to remember back to 28pp ago when I introduced Gregor Samsa to "Mr. Gregor's Neighborhood" & the wondrous adventures he had therein. &, *NOW*, Mazur brings up our pal Samsa too!

In a chapter called "STRETCHING THE IMAGE", in a subsection called **"To imagine" versus "to picture."**, Mazur writes: "The objects and forms invoked in poetry and literature must be rigorously imagined, but whether they must, or can, be directly visualized is a trickier question. When *The Metamorphosis* was to be published as a book, Franz Kafka's publisher wrote to him suggesting that the cover illustrator "might want to draw the insect itself." Kafka wrote back:

> Not that, please not that! . . . The insect itself cannot be depicted. It cannot even be shown from a distance."

Mazur goes on to explain that the great writer Vladimir Nabokov (& I do mean *great*) tried to visualize exactly what type of bug it was that Samsa turned into. Mazur concludes, as do I, that "a kind of imagination is called for here that goes beyond mere architectural visualization." Pursuing this, he gives another example:

"A movie of Gogol's story "The Nose" in which a disembodied nose goes "driving about all over town under the guise of a State Councillor," could quite easily miss the point of the tale. For Gogol in his writing is prodding us, cajoling us, instructing us to engage in precisely the exercise of imagining this unvisualizable image in the full glory of its unvisualizability."

Any close friend of mine can probably visualize what I'd be doing, after reading this quote, if Barry Mazur were here in my house at the time. I'd be very excitedly insisting that he watch a short film called, you guessed it, "The Nose" by Alexeïeff/Parker made in 1963 wch *does* what Mazur *doesn't* visualize any filmmaker doing w/ Gogol's story: viz: doing it justice w/o overly turning it into something too simple-minded. In order to accomplish this, the filmmakers used a rare technique called pinboard animation. This involves filming a board that has a million pins in it that are pushed in & out to create a relief surface on wch shadows are made. It. is. AMAZING. I. REPEAT. AMAZING.

Mazur discusses further the way that changing terminology brings w/ it changing thinking about the thing so termed. "Bombelli denoted square root in his *L'Algebra* (most of the time) as R.q. (*radice quadrata*), so the square root of 2 would be R.q.2, and cube root he denoted R.c. (*radice cubica*)."

Now, as a matter of fact, the word "radish" has "radice" as one of its roots. So when I see *radice cubica* I, understandably, think of a cubed radish - wch is, obviously, the way that mathematical terminology is heading. Don't believe me? Then you're obviously not keeping up w/ farming trends. Remember, negative numbers came about out of debt reckoning - they're intimately tied to commerce, those kinky little devils. & cubes, of course, are part of geometry. & geometry is all about space (well, actually, topology is, but you 'know' what I 'mean') & watermelons are grown in boxes in Japan & elsewhere to preserve space by having them created as cubes instead of as those damned awkward rolling things. Hence, their connection to math. No more of this trying to figure

"Great" is that a term of criticism or one of praise? I rather like the ambiguity. Truth be told, I've actually NEVER read anything by Nabokov — as odd as that may seem — but I did see a fictional reenactment of his lecture on "the metamorphosis" on film -- it really only made sense to me in light of Nabokov's interests in Entomology. How many examples of "bug related literary greats" are there, after all?

out how to express topological matters in those fussy technical complex planes n'at. Just grow everything in boxes! It won't be just *radish cubed* anymore, it'll be *watermelon cubed* & *fetus cubed* & *lifestyle cubed* & *universe cubed*. You name it. No one will be embarrassed by being such a square anymore.

"Perhaps our modern symbol" $\sqrt{\ }$ evolved from [..] the letter *r* written cursively:

r ⋎ √ √."

Perhaps. But to me it looks like selected frames from a time lapse study of a flying radish trying to escape from being boxed in. & failing.

Chapter 10: "SEEING THE GEOMETRY IN THE NUMBERS", section "**54, What are we doing when we identify one thing with another?**" I've noted to myself that this is possibly the most important part of the bk for me. Not because of the geometry referred to but because of the mental leaps required to imagine the significance of certain symbol changes. Mazur: "For example, I insisted outrageously [..] that we *identify* the real number *N* with the transformation of the number line consisting of multiplying every number on the line by *N*."

This is also the central conceptual concern of the penultimate chapter of my poem A Labyrinth of Visions -- entitled "The Vision of the Seeker." One passage reads: "I suspect that the expanse of dried lakes/ in which I spend the continuum of my days counting coins/ represents an absence of justice/ & that the spaces between words/ convey the impossibility of touching/ I know that a series of numbers/ reenforces the opaque surface/ of authority/ & the order crystallizes & swells outward,/ a seam binding form to wall/ comma to dust/ & idea to distance."

What if we simply redefined 1 as equalling π? What if we were to somehow create a notation *not* for the numbers but for the thought processes that create the numbers? Varying degrees of infinity wd more explicitly become quantum leaps in consciousness, in the ability to imagine increases in grandiosity.

One of the things I like about Mazur's bk is that he frequently references poets & other creative writers & uses aspects of their works to support his mathematical points. I'm reminded of an interview I read once w/ a man who ran a sperm bank. He was talking about the value of science vs the value of art. He sd something to the effect that he loved listening to Beethoven, eg, but didn't think that it's as important as science. I disagree. For me, listening to music, to continue w/ the example, is crucial to setting a state of mind. Mazur seems to agree by emphasizing poetry. Like me, he also likes the recursiveness of quotes w/ in quotes - they demonstrate one intellect studying another intellect. Here, I get to take this process one step further:

> "[H]ere is the novelist Virginia Woolf (commenting on the essayist Thomas DeQuincy):

> But in what form was he to express this, that was the most real part of his own existence? There was none ready to his hand. He invented, as he claimed, "modes of impassioned prose.""

There you have it, cONVENIENCE quoting Mazur quoting Woolf quoting DeQuincy. Mazur: "Poetry itself is a hybrid, as is song, which is the grafting together of words and melody." & on that note, I end my discussion thru Imagining Numbers.

In Raymond Queneau's 5th novel, Les Enfants du limon (published in English as Children of Clay), the character M. Chambernac is writing & compiling a bk entitled "The Encyclopedia of the Inexact Sciences, biography and anthology of the French "literary lunatics" of the XIXth Century. Here are brought to light for the first time the lives and works of some fifty unknown people, whose wild imaginings (this expression is used without pejorative intentions) never met with the slightest echo, were never recognized as valid, by even a single *other* individual." This latter quote was taken from the introduction written by the translator Madeleine Velguth.

Chambernac's researches were actually Queneau's & weren't fictional. "[H]e spent the better part of three years in the Bbiliothèque Nationale, studying the writings of these people he called *hétéroclites*, eccentrics. When, in 1934, he tried to find a publisher for his finished work, no one was interested. His solution was to build a novel around the manuscript".

The translator notes that she was "strongly tempted to transpose the *a* and the *o* of Naomi in order to make it more closely resemble the French *Noémi*, the first syllable of which is *Noé*, Noah." Indeed, perhaps both Velguth & Queneau are *noetic radicals*.

Last night, I sent out a bulk email to friends to encourage them to use the terms "Google Wracking" & "Hunch Backing" so that I cd get the pleasure of seeing them become more widespread. How much such a strategy will work is currently unknown to me. Given the internet, the possibilities for introducing ideas internationally almost as quickly as they're thought of are like never before (that we 'know' of). If the process of Paradigm Shift Zeitgeists happens quicker these days thanks to that, I hope I'm alert enuf to notice.

This morning, in reply to the afore-mentioned email, I got a joke as a reply from my friend Florian Cramer:

Q: What does a mathematician do who needs to fly to a conference, but is terribly afraid of a terrorist bomb on the plane?

A: He brings his own bomb along since the statistical probability that two people independently from each other put two bombs on an airplane is zero.

so beyond the thought of any X there is the idea that $X = X + y - 1$, that $X + y - 1 =$ vertigo & that vertigo = the research of Chambernac. That this research ≈ to the translator's note in a volume of René Daumal's Contre-Ciel, which volume has still yet to be realized. & that such an "imaginary" volume is both = & ≠ to the paradigm shift poltergeists which continue to inhabit the mind of this marginalienist.

PART TWO

<u>NUMBER WRITING</u>

As stated in Part One's INTRODUCTION (page ¬$\sqrt[2]{4}$), I was mainly active as an experimental writer from ages 20-23 when the following number writing was created (probably sometime between 1973 -1977). From my dim memory, I deduce that its main purpose was to try to rigorously follow a counting rule & to then start over again, to have a new arrangement procedure per page, & to start the counting procedure over w/ each mistake. The reader's relationship to the text wd then basically be to just follow the counting - the important part of the reading being the path that the eyes wd take *rather than* what the text wd hypothetically refer to. This, naturally, was a deliberate extreme break from conventional 'transparent' writing & in keeping w/ strategies in structuralist & materialist filmmaking, eg.

The original pages were typewritten in black until the last page, originally page 5, where red was introduced in 2 vertical columns of "4"s only. Given that this bk is in black & white, these "4"s are slightly differentiated from the black that they were typed over both by their superpositioning & by their lighter shade. Pencil notations were added many yrs later as I tried to figure out what I'd had in mind in some parts. On the next page is an example of the marked text of the original page 5 on wch I speculate re the red: "JUST 4 PATTERN PERPETUATION?"

The original text was typed on 8&1/2X11" ledger paper w/ each section about 2/3rds of the way down the page. Given that these sections had some space around them, & that they'd have to be substantially reduced to fit in this bk, I've decided to not give a full view of the pages here because the numbers will be too small to be easily readable. I have, however tried to stick to the original page placement & to have the backsides of the pages blank - in keeping w/ the original intent to have them be comparable by holding them up w/ a light behind them &/or by flipping back & forth.

While this writing has never necessarily been very important to me, it still astounds me somewhat that I was so utterly unconcerned w/ whether other people wd find such things interesting or not. I wrote in a way that was completely true to my own perverse theories - never for a moment trying to appease any Lowest Common Denominator.

#/wryting
AS I begin
this section
I begin
w/ a challenge.
I shift from
writing w/
my right
hand, which
I am used
to, to
writing
w/ my
left hand
which is
unfamiliar
& somewhat
difficult
for me.

Is this also a weird chiasmic transition from right brain to left brain thinking? I'm not yet certain.

¼12½123/41313¼13½133/41414¼14½143/41515¼15½153¼½3/411¼1½13/42
2¼2½23/433¼3½33/444¼4½43/455¼5½53/466¼6½63/477¼7½73/488¼8½83/
499¼9½93/41010¼10½103/41111¼11½113/41212¼12½123/41313¼13½133/
41414¼14½143/41515¼15½153/41616¼16½163/41717¼17½173/41818¼18½
183/41919¼19½193/42020¼20½203/42121¼21½213/42222¼22½223/42424
¼24½243/42525¼24½24¼½3/411¼1½13/422¼2½23/433¼3½33/444¼4½43/45
5¼5½53/466¼6½6347¼7½73/488¼8½83/499¼9½93/41010¼10½103/41111
¼11½113/41212¼1½143/41313¼13½133/41414¼14½143/41515¼15½153/4
1616¼16½163/41717¼17½173/41818¼18½183/41919¼19½193/42020¼20½2
03/42121¼21½213/4¼/3/411¼1½13/422¼2½23/433¼3½33/444¼4½43/455¼
5½53/466¼6½63/417¼1½73/488¼8½83/499¼9½93/41010¼10½103/41111¼1
1½113/41212¼12½/23/41313¼13½133/41414¼14½143/41515¼15½153/416
1½13/4¼½3/411¼1113/422¼2½23/433¼3½33/444¼4½43/455¼5½53/466¼6½
?→1½113/41212¼12½/23/4 1313¼13½133/41414¼14½143/41515¼15½153/41
63/477¼7½73/488¼8½13/499¼9½93/41010¼10½103/41111¼11½113/41212

JUST A PATTERN PERPETUATED?

NUMBER WRITING

182 algebraic numbers

Already
I feel
the
ability
to express
my thoughts
slowing
down
dramatically

¼½3/411¼1½13/422¼2½23/433¼3½33/444¼4½43/455¼5½53/466¼6½63/477
¼7½73/488¼8½83/499¼9½93/41010¼10½103/41111¼11½113/41212¼12½12
3/41313¼13½133/41414¼14½143/41515¼15½153/41616¼16½163/41717¼1
7½173/41818¼18½183/41919¼19½193/42020¼20½203/42121¼21½213/422
22¼22½223/42323¼23½233/42424¼24½243/42525¼25½253/42626¼26½263
/42727¼27½273/42828¼28½283/42929¼29½293/43030¼30½303/43131¼31
½313/43232¼32½323/43333¼33½333/43434¼34½343/43535¼35½353/4363
6¼36½363/43737¼37½373/43838¼38½383/43939¼39½393/44040¼40½403/
44141¼41½413/44242¼42½423/44343¼43½433/44444¼44½443/44545¼45½
453/44646¼46½463/44747¼47½473/44848¼48½483/44949¼49½493/45050
¼50½503/45151¼51½513/45252¼52½523/45353¼53½533/45454¼54½543/4
5555¼55½553/45656¼56½563/45757¼57½573/45858¼58½583/45959¼59½5
93/46060¼60½603/46161¼61½613/46262¼62½623/46363¼63½633/46464¼
64½643/46565¼65½653/46666¼66½663/46767¼67½673/46868¼68½683/46
969¼69½693/47070¼70½703/47171¼71½713/47272¼72½723/47373¼73½73

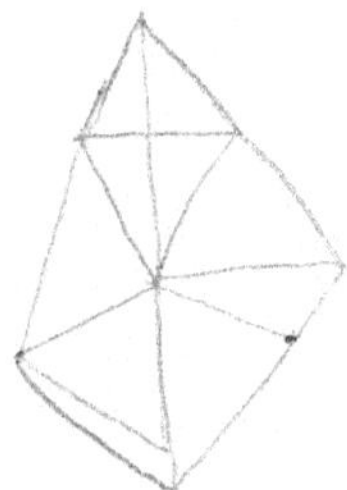

According to the
Wikipedie dictionary
left handedness occurs
more frequently in cases
of twins individuals with
epilepsy, Down's syndrome
autism, mental retardation
& dyslexia. Research
also indicates that
left handedness slightly
raises the risk of
psychotic mental
illnesses. [illegible]
to have [illegible] that
who being left handed

NUMBER WRITING

184 algebraic numbe

Stigma as attached to left handedness includes power for shame society awkwardness being sinister being unlucky being out of order & immoral & being unclean.

3/47474¼74½743/47575¼75½753/47676¼76½763/47777¼77½773/47878¼7
585¼85½853/48686¼86½863/48787¼87½873/48888¼88½883/48989¼89½88
8/49696¼96½963/49797¼97½973/49898¼98½983/49999¼99½993/410019½
43½1043/4105105¼105½1053/4106106¼106½1063/4107107¼107½1073037
/54/4112112¼112½1123/4113113¼113½1133/4114114¼114½1143/41/0/8
3903183/4119119¼119½1193/4120120¼120½1203/4121121¼121½1214¼43
4½11126126¼126½1263/4127127¼127½1273/4128128¼128½1283/415119/
85¼1½14132132¼132½1323/4133133¼133½1333/4134134¼134½131310004
½94184/3131½131¼1311314/3031½031¼0310314/3921½921¼92192/18097
4¼0½1/3521½521¼5215214/3421½421¼4214214/3221½221¼221221451½09
85111¼8118114/3711½711¼7117114/3611½611¼6116114/3511½511¼01¼7
¼9411¼1111114/3011½011¼0110114/3901½901¼9019014/3801½801¼8099
45014/3301½301¼3013014/3201½201¼2012014/3101½101¼1011014/300¼
894/349½49¼49494/339½39¼39394/329½29¼29294/319½19¼19194/309½7
484/338½38¼38384/328½28¼28284/318½18¼18184/308½08¼08084/397½9

Individuals who preferred left-handedness have been beaten, shunned by society, & denied the right to marriage. Right-handedness is a form of social identity boundary. Left-handedness is recognized as marginal. The small percentage of individuals who maintain left-handedness beyond early development is not a reflection of a natural human propensity, but rather of the strict insistence culture makes that all of those who belong to it accept & adapt to the government of its norms.

NUMBER WRITING

186 algebraic numbers

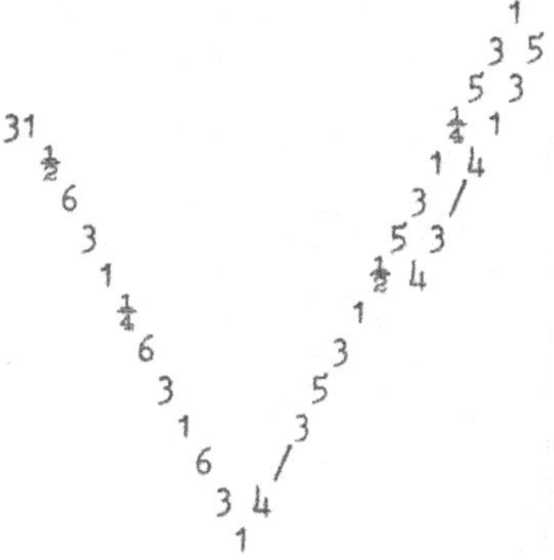

187 algebraic numbers

I find as I write these words with mush effort that I am infringing upon the rights of the text proper -- taking advantage of these pages which FENT wished to remain empty. I regard such actions as survival strategies of the marginalized. Jean Genet -- who was left-handed himself -- whose biography recounts the fact that every card in culture's deck was stacked against him -- he found a way of surviving by breaking the established rules.

NUMBER WRITING

188 algebraic numbers

7½1 383

373 ¼13/41

137

36 137

3/4

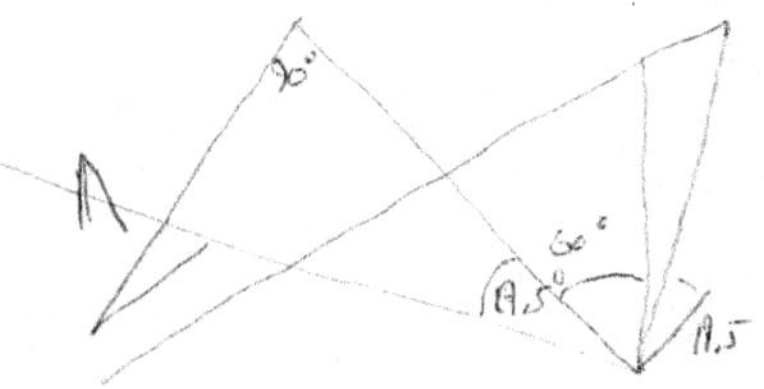

Genet writes:
"Repudiating the virtues
of your world,
criminals hopelessly
agree to organize
a forbidden universe.
They agree to live
in it. The air there
is nauseating;
they can breathe it."

NUMBER WRITING

190 algebraic numbers

¼12½123/41313¼13½133/41414¼14½143/41515¼15½153¼½3/411¼1½13/42
2¼2½23/433¼3½33/444¼4½43/455¼5½53/466¼6½63/477¼7½73/488¼8½83/
499¼9½93/41010¼10½103/41111¼11½113/41212¼12½123/41313¼13½133/
41414¼14½143/41515¼15½153/41616¼16½163/41717¼17½173/41818¼18½
183/41919¼19½193/42020¼20½203/42121¼21½213/42222¼22½223/42424
¼24½243/42525¼24½24¼½3/411¼1½13/422¼2½23/433¼3½33/444¼4½43/45
5¼5½53/466¼6½63 47¼7½73/488¼8½83/499¼9½93/41010¼10½103/41111
¼11½113/41212¼14½143/41313¼13½133/41414¼14½143/41515¼15½153/4
1616¼16½163/41717¼17½173/41818¼18½183/41919¼19½193/42020¼20½2
03/42121¼21½213/4¼/3/411¼1½13/422¼2½23/433¼3½33/444¼4½43/455¼
5½53/466¼6½63/417¼1½73/488¼8½83/499¼9½93/41010¼10½103/41111¼1
1½113/41212¼12½/23/41313¼13½133/41414¼14½143/41515¼15½153/416
1½13/4¼½3/411¼11131422¼2½23/433¼3½33/444¼4½43/455¼5½53/466¼6½
1½113/41212¼12½/23/4 1313¼13½133/41414¼14½143/41515¼15½153/41
63/477¼7½73/48818½13/499¼9½93/41010¼10½103/41111¼11½113/41212

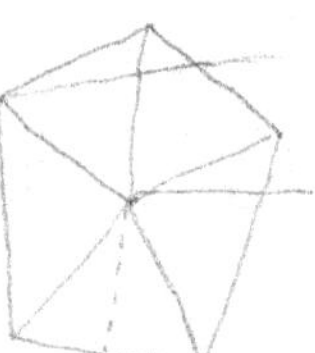

When, in 1943, Antonin Artaud, under the direction of his psychiatrist, Gaston Ferdiere, began to compose his translations of Lewis Carrol into French, I wonder if he experienced this exercise as if he were learning to write with his left hand?

FOLK MATH

What exactly is Folk Math? I'm not sure I have a good answer for that & I haven't succeeded in getting in touch w/ Kirby Malone, who probably coined the term in the late 1970s in Baltimore. For me, Folk Math is something akin to rebuses. WP: "A **rebus** (Latin: "by things") is a kind of word puzzle which uses pictures to represent words or parts of words; for example: H + [ear] = *Hear*, or *Here*." Unlike rebuses, though, I don't think of Folk Math as necessarily 'adding up to anything'. For me, Folk Math can be a jumble of words, letters, numbers, & images that're evocative w/o having to be specifically meaningful.

On one spot on-line, Folk Math is defined as "the math that folks do" - a very down-home definition apparently meant to mean simple arithmetic & such-like things that one might have to use when calculating how much the groceries cost or counting how many things are to be divided up. In other words, the basic math that people who aren't mathematicians by trade are likely to use.

But that's not the Folk Math I refer to. The 1st & last time I remember the term being introduced to the general public was in an excellent article written by Kirby Malone & Marshall Reese for the Spring, 1980 issue of a Baltimore area arts magazine named *Aura*. The purpose of this article was to summarize the activities of 2 prolific Baltimore groups & to introduce our neologisms. Despite the article being entitled:

Mad Science & | Folk Math *by Kirby Malone & Marshall Reese*
AURA – SPRING 1980

only 2 brief Folk Math mentions are made - as shown below:

> The individuals involved in these events may have begun in artistic fields such as visual arts, poetry, music, video, but, in a number of ways, they mean to bypass, counteract & transform the ineffective position that actions identified as art have fallen into. One of the ways is to coin new terms, i.e., *mad science & folk math.*

Folk math, corrupted dimensions, isolette incubator, zualalia & conductive earthlings, to name just a few of CoAccident's terms, give light to an unusual working so that supportive collaboration & constructive humor may divert the confusions & frustrations of disinterest.

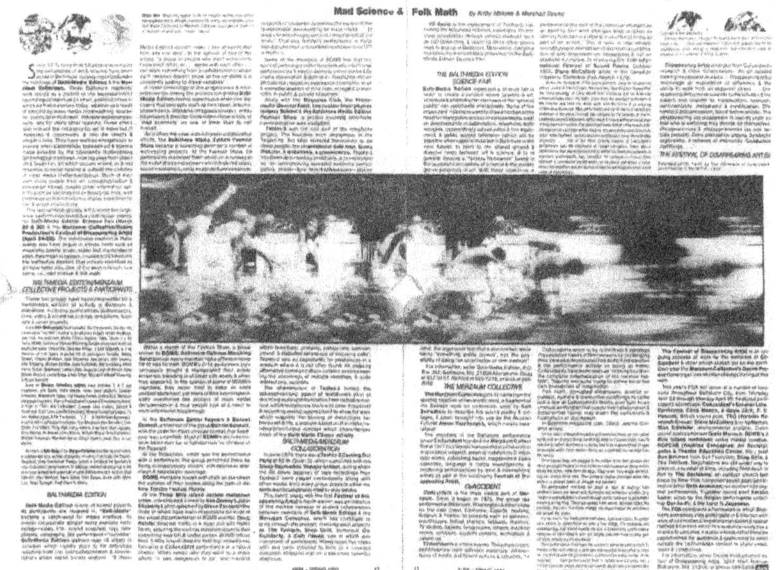

Mad Science & Folk Math

the entire "Mad Science & Folk Math" article

When I asked Marshall recently about this article, he didn't even remember it. He cdn't define Folk Math for me either. He agreed that Kirby probably coined it & that was that. A vagueness.

Mainly, I associate Folk Math w/ a particular shirt I've had for the last 30 yrs. This shirt is shown on successive pages. I've never seen any meaning in its patterns, it just seems like a fanciful design. & that's exactly what I like about it. Although I'd like it if were heavily imbued w/ meaning too.

One of the things that I like about nonsense is that it can liberate symbols from over-contextualization. Nonsense is like a thunderstorm clearing the air of pigeon-holing. Symbols can be set free for awhile. This Folk Math shirt is like meaning on vacation.

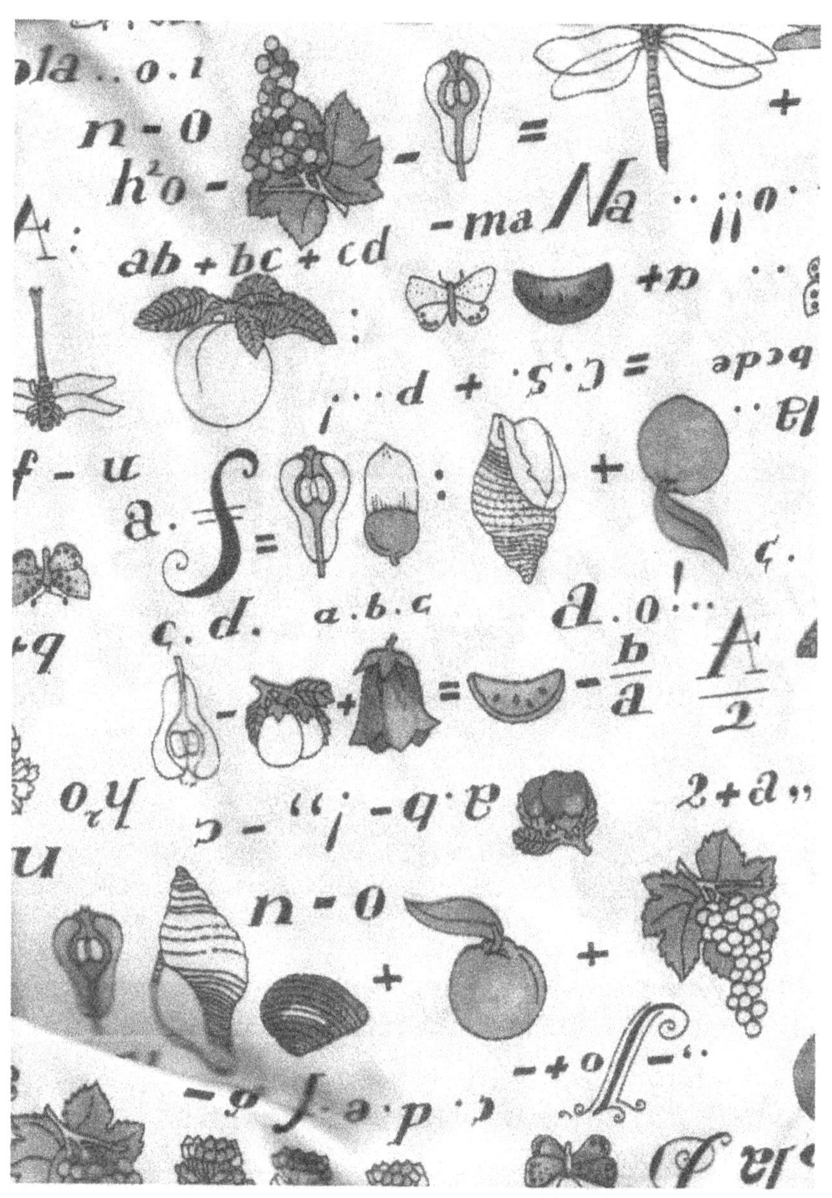

detail of the back of the Folk Math shirt

tENTATIVELY, a cONVENIENCE (left) & Charlie Brohawn (right) after having switched clothes for a B.O.M.B. action in 1979

"excluded by my birth & tastes from the social order I was not aware of its diversity. Nothing in the world was irrelevant: the stars on the general's sleeve, the stock market quotations, the olive harvest, the style of the judiciary, the wheat exchange, the flower beds. Nothing. This order, fearful & feared, whose details were all interrelated, had a meaning: my exile." Jean Genet

Finally, I tend to associate Folk Math w/ Charlie Brohawn (even though I doubt that he'd remember it either) because of his interest in marginal 'folk' things like duck decoys, primitive painting, etc - much of wch he's explored, along w/ Chris Mason, w/ their performance group *The Tinklers*.

Charlie & I had many interests in common in the 1970s. In fact, we had *so many* interests in common at times that I, at least, joked about our being twins. We gave 2 performances that I can recall where we explored this a bit.

We were both part of a guerrilla action group in 1979 called B.O.M.B. (the Baltimore Oblivion Marching Band) that had been pulled together by Richard Ellsberry. Richard originally envisioned it as a group that wd improvise guerrilla actions at public events.

In the above photo, Charlie & I are shown after having exchanged clothes. We were en route to a B.O.M.B. action at the Johns Hopkins Spring Fair. Charlie is wearing the Folk Math shirt. The ice-bag on his head was what I used to wear as a joke at parties - so that I cd be ready for the next day's hangover.

A DOUBLE NEGATIVE AS NOT A POSITIVE

In early 1982, my lover/collaborator HannaH AvivA & I made a short unsplit-regular-8mm movie called "A Double Negtive As Not A Positive". To say that the 2 of us together were intense at times might be an understatement. She was a pathological liar & manipulating gold-digger, I was a self-destructive psychotic. We were both teetering on a tight-rope walk of despair.

In the film, we're shown naked, cutting specific symbols into each other's skins w/ an exacto knife. We then make prints from the blood. These were mailed & handed out to friends & correspondents. Below is a black & white copy of a print made from HannaH's wound:

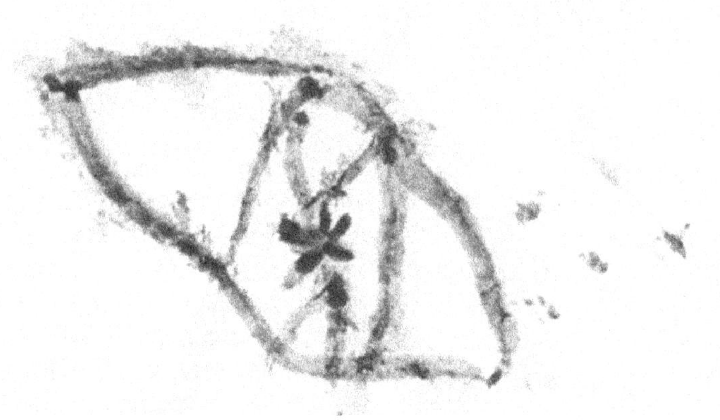

I was using the name E. G. Head at the time as a joke off the Latin abbreviation for "for example" & as a joke off of "egg-head" (ie: an intellectual). Here's a black & white copy of a print I made from my wound:

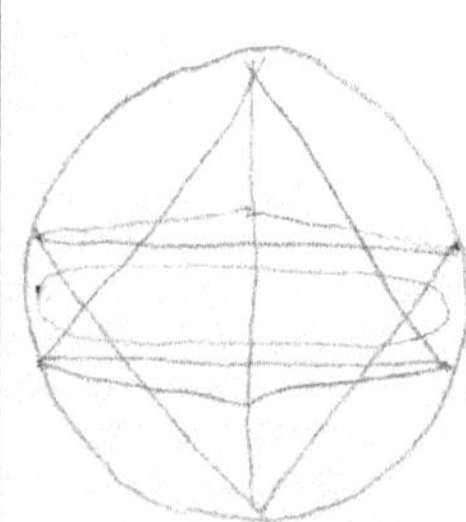

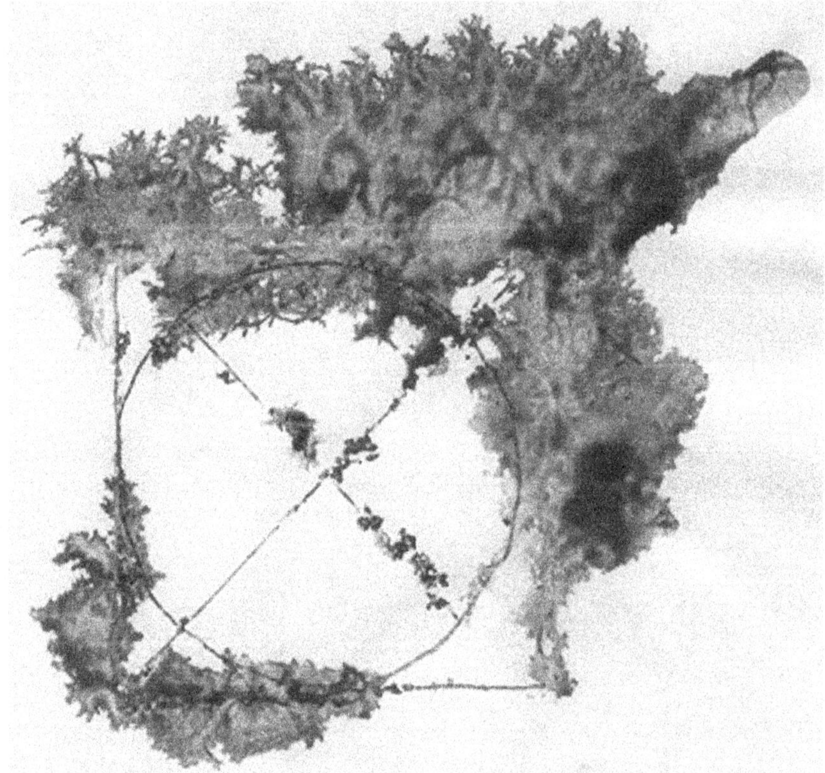

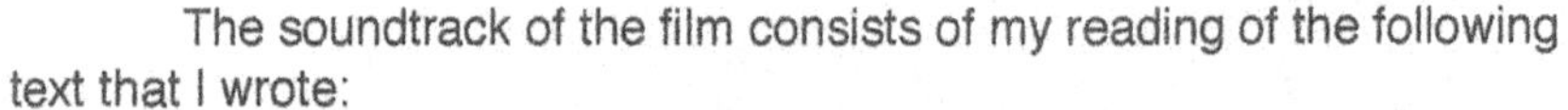

The soundtrack of the film consists of my reading of the following text that I wrote:

"This movie's called "A Double Negative As Not A Positive". The title refers to the somewhat unorthodox mathematical idea that 2 negative #s multiplied together can theoretically produce something other than a positive # - that is, an alternative to the positive/negative duality, &, as such, polarities in general. This is somewhat akin to the idea of the so-called "imaginary #" sometimes represented as the square root of a negative # or as the letter "i". The title <u>is not</u> meant to be a more abstract way of saying "2 wrongs don't make a right" even though that saying, when interpreted literally, doesn't <u>necessarily</u> mean that 2 so-called "wrongs" make another larger so-called "wrong". It's exactly such simple-minded so-called "moralisms" that this movie provides a philosophical metaphorical alternative to.

Here, 2 people frequently inclined to despairing so-called "self-destructive" "suicidal" behavior, are shown cutting into each other's skins in a formalized way that's meant to focus their violent desperate energies toward misery transcending ends. This ritualized practice is an attempt to take the ordinary means of "self-destructiveness", that is, slashing one's body, & to subvert the usual despairing purpose.

The symbol shown here cut on Hannah Aviva's shoulder supposedly appeared spontaneously on her body one day - much to the shock & consternation of a supposed psychic that Hannah showed the mark to later - & the symbol shown here cut onto E.G.Head's leg consists of 2 overlapping null signs used here to represent the general thrust of this movie."

In addition to the blood prints, we also made these post-cards:

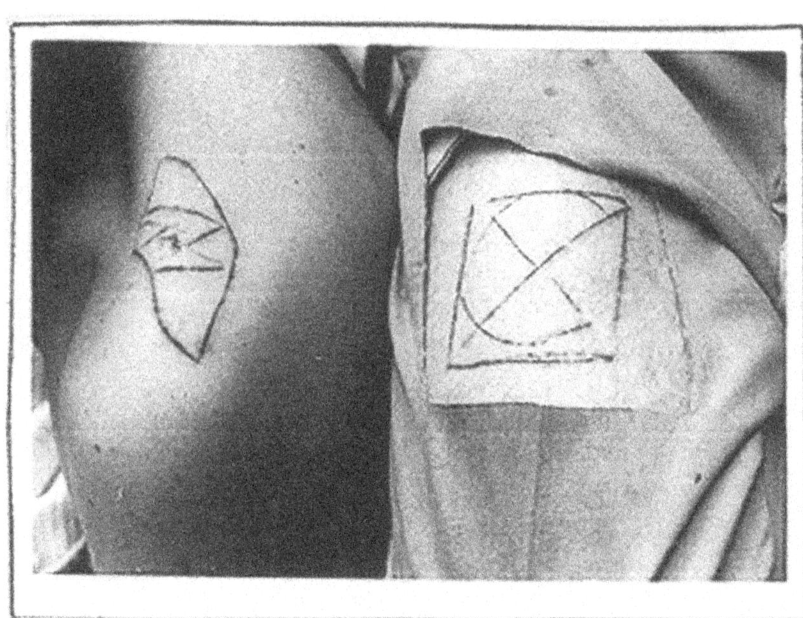

"Beauty has no other origin than the singular wound, different in every case, hidden or visible, which man bears within himself, which he preserves & into which he withdraws when he wants to leave the world for a temporary but authentic solitude... Jean Genet

ANYTHING IS ANYTHING

As I started reading the bks that I've incorporated into the NAVIGATION: section, I was thinking about how I cd apply what I was (barely) learning in a creative way. There were 3 people in my circle of acquaintances whose brains I tried to pick for help: Jessi (Pittsburgh), Ksenija (Serbia), & Steven (California). Jessi & I had quite a few enthusiastic discussions over beers & I communicated w/ the other 2 by email. I was looking for mathematical expressions of key slogans/mottos of mine & they were trying to help. I'll be using this section of this bk to present any of that correspondence that I find worthwhile - regardless of how immediately germane it is to the section heading.

My 1st email inquiry (slightly edited) was this:

Subject: tENTATIVELY, a cONVENIENCE
Date: Sat, 12 Jul 2003 00:30:49 -0400
From: anonymous

Dear Ksenija & Steven,

As happens w/ most people approaching senility (is that 'true'?)
I'm reverting to my childhood
& becoming interested in mathematics / number theory again.

ANYWAY, to make a long story short,
several key concepts have popped up in my 'work'
over the decades:
"Anything is Anything",
Zeno's paradox as the key to 'eternal' life,
"A Double Negative As Not A Positive", etc..
ANYANYWAY, the point is that these days I'm interested
in getting some tattoos that reference these ideas.
I don't want any words on me
but mathematical symbols are fine.
My questions for you, at the moment, are:
Can you think of any way to formulaically express:

"Nature in giving the form of a tree to a tree could just as well have given it the form of an animal or a hill; we would have thought tree for the animal or the hill & the trick would have been turned.
Antonin Artaud

"Anything is Anything"?;
"A Double Negative As Not A Positive"?
(keep in mind that I'm using "as" instead of "is" here
"is" is definitive, "as" is propositional);
& something along the lines of:
"A finite quantity represented as a set containing
an infinite quantity of its subdivisions
(such as its subdivision in terms of rational numbers)
does not equal the same finite quantity
represented as a set containing an
infinite quantity OF A DIFFERENT DEGREE of its subdivisions
(such as its subdivision in terms of irrational numbers)"?

Sorry for the clumsiness of the above.

****
Subject: Re: tENTATIVELY, a cONVENIENCE
Date: Mon, 14 Jul 2003 22:04:17 +0200
From: Ksenija

i have to think about representing 'as' as opposed to 'is' mathematically. quite honestly, i am not sure that it can be done.

****
From: Steven

> "Anything is Anything"?;

you might want to ask jessi logic. it has been a while since i've done logic, am more of a geometry guy now. here's a shot:

(forall) x, x = x
where (forall) is the upside-down capital `A.'

this is specific. and is Ayn Randian (just replace x with A and read it aloud (over and over again (fifty million times))). it just means that everything in your universe is identical to itself. another way would be the less sensical

(forall) x = (forall) x

this doesn't really have so much meaning, but could be taken to mean that when choosing over your universe of objects, there is only one way to do it.

to me
Antonin
Artaud,
I am my
son my
father
my
Mother
& me!!
Antonin
Artaud

> "A finite quantity represented as a set containing
> an infinite quantity of its subdivisions
> (such as its subdivision in terms of rational numbers)
> does not equal the same finite quantity
> represented as a set containing an
> infinite quantity OF A DIFFERENT DEGREE of its subdivisions
> (such as its subdivision in terms of irrational numbers)"?

this is a bit of a problem. normally we rely on self identity, i.e. anything is identical to itself, regardless of how it is expressed. you could say that two _partitions_ of a set are different. for example if you partitioned all the natural numbers into odds and evens, versus partitioning them into primes and non-primes, in each case the partition has two parts, the union of which is the natural numbers, but the partitions are not the same. you could say, with symbols, that if you take two partitions, even if they have the same union they are not the same. but there are freakish accidents where they are the same (take the same partition twice). so you need modality.

~ (necessary) (U)_{x (in) X} x = (U)_{y (in) Y} y => X = Y

the U is the union symbol, not really a `U.' you should have the _{ } parts in subscript, (in) is the set inclusion symbol. (necessary) is an empty square. but beware because modal symbols are not universally recognized--people interpret them in all different ways. again, ask jessi on this. there is an alternative formulation using the diamond.
not sure how to work in the infinity parts.

****

Subject: Re: math questions
Date: Tue, 15 Jul 2003 21:02:32 -0400
From: anonymous

> (forall) x, x = x
> where (forall) is the upside-down capital `A.'

The meaning of "Anything is Anything" to me is something like:
Anything can be replaced by Anything Else.
This is a sort of refutation of 'consensus reality'
in the sense that it implies an infinite modularity
that goes very contrary to the extreme fixity
that one usually encounters (frustratingly in my case)
in 'normal life'.

> you could say, with symbols,
> that if you take two partitions, even if they have the
> same union they are not the same.

That's close to what I'm getting at.
You're using "partitions"
(presumably a more logically correct term)
& I'm saying "subdivisions".
However, in your example, the set of all natural numbers
is infinite to begin w/ so it misses the finite beginning aspect.
Perhaps what I'm trying to do is describe a
NON-PARADOXICAL PARADOX in the sense that
I'm trying to say something like:
When n (defined as anything finite)
is subdivided (or partitioned? [or analyzed?]) in such a way
that it's defined in terms of a set w/ infinite contents
it can be called both finite & infinite
AND THIS IS NOT PARADOXICAL
because by so doing one is simply demonstrating
that both states coexist.
ALSO, when n (defined as anything finite)
is AGAIN subdivided (or partitioned? [or analyzed?]) in such a way
that it's defined in terms of a set w/ infinite contents
to A DIFFERENT DEGREE OF INFINITY
it can be called both finite & infinite
AND EQUIVALENT TO & NOT EQUIVALENT TO ITSELF
AND THIS IS NOT PARADOXICAL
because by so doing one is simply demonstrating
that both states coexist.

"The number
& order
of possible
suppositions
in / this
realm is
precisely /
infinity! /
& what is
infinity? /
That is
precisely
what we do not
know! /
It is a word /
that we use /
to indicate /
the opening
of our
consciousness
toward
possibility /
beyond
measure /
tireless & beyond
measure."
Antonin Artaud

****
Subject: Re: more logic(?) from tENT
Date: Wed, 16 Jul 2003 19:06:42 +0200
From: Ksenija

> > problem: normally we think of symbols as merely shorthand
> > for words. thus in math papers, when we introduce new
> > weird notation we usually tell people how to read it

i absolutely agree with steven on this one. given enough paper, you can
create any universe you desire, and then assign symbols as you see fit.

math is mostly convention - in the sense that someone agrees on symbols and their meaning, and then others use them. you could have different meanings in different contexts, which can also be a problem.

> > (forall) x, x = x
> > where (forall) is the upside-down capital `A.'

this one is just saying that everything is itself. i don't think this is what you are trying to say. if you wanted to say that everything is also everything else, you would say (for all x)(for all y) x=y
this, however, makes the universe trivial. there is only one thing.
i assume you don't want that. how about this, though? you can say that everything has all the same properties as everything else; this may not necessarily mean that two things are identical (depending on your axioms). or do you want something else?

"From one media of expression to another correspondences & levels of development are created -- even light can have a precise intellectual meaning" Antonin Artaud

****

Subject: Re: more logic(?) from tENT
Date: Wed, 16 Jul 2003 17:52:50 -0400
From: anonymous

> then assign symbols as you see fit.

Might it be necessary to create symbols especially
for my purpose & then somehow make those symbols public enough
to make them somewhat accepted & understood beyond just a few
friends & aquaintances of mine?

> this one is just saying that everything is itself.
> i don't think this is what you are trying to say.

No, I'm trying to redefine "itself" as a set of infinitely possible contents.

> (for all x)(for all y) x=y
> this, however, makes the universe trivial.

Perhaps. Or it might be just what I want.
I'm not sure I understand the expression.
"Anything is Anything", as I may've stated before,
can be extended to something more like:
"Anything can be substituted by Anything else
(including the original "Anything")".
What I'm getting at
(as I elaborate further in the "is" vs "as" explanation below)
is the idea that no thing is absolutely fixed.

Imagine the following possible permutations of
"Anything is Anything":
"Everything is Food",
"My Chair is Not My Chair",
"My Head is One Quarter of a Worm-Hole",
"This Photograph is The Entire Indian Ocean Minus One Molecule".
"The Entire Understanding of Human Science is The Fonz", etc..
I don't mean these examples to be metaphors:
I mean them to be taken LITERALLY.
ANYTHING can be substituted for either ANYTHING
on either side of the 'equation'.

"we are beautiful, joyous, drunk, & Free!", Jean Genet as Solange

> i am not sure about this 'as'. can you explain in a bit more detail

William S. Burroughs made the distinction, & I agree,
between "to be" & "as". "To be" defines something.
EG: "I AM a human being" makes a definitive statement.
"As" is more propositional:
EG: "I AS a human being" makes a NON-DEFINITIVE statement
- something along the lines of:
"I propose that I can be thought of as a human being but that
doesn't mean that I absolutely am one."

****

Subject: Re: math questions
Date: Fri, 18 Jul 2003 15:34:30 -0400 (EDT)
From: Steven

> The meaning of "Anything is Anything" to me is something like:
> Anything can be replaced by Anything Else.

i see, so something like `x can be replaced by y even if
x and y are distinct, for any x and y you might choose.'
this is difficult, kind of. the set up would be:

(forall) x,y ~(x=y) ^ r(x,y)

where r(x,y) somehow captures that you can replace x by y. somehow this could be stated by saying that if you take any well formed formula with a single variable and substitute x in it, it has the same truth value as if you substitute y in it. this breaks out of first order logic, i think. there is prolly a better way to write r(x,y), or rather, a more precise way. i'm not familiar enough with higher order logic. you could just spit in ayn rand's (dead) eye and say

(forall) x,y ~(x=y) ^ (x=y)

tho this would classically be regarded as either false,
or a statement that the universe is empty!

> > (forall) x = (forall) x

> > this doesn't really have so much meaning, but could be taken
> > to mean that when choosing over your universe of objects,
> > there is only one way to do it.
>
> Wch is the OPPOSITE of what I'm getting at.

gotcha. again, this doesn't have so much meaning, but more precisely, what i was trying to say is that the universe of objects is fixed, and so if you traverse over all of them there is only one way to do it, i.e. by traversing over all of them. maybe you want to deny that as well.

"I perceive w/ certitude that which is. That which is not shall be made by me if necessary." Antonin Artaud

> I'm trying to say something like:
> When n (defined as anything finite)
> is subdivided (or partitioned? [or analyzed?]) in such a way
> that it's defined in terms of a set w/ infinite contents
> it can be called both finite & infinite
> AND THIS IS NOT PARADOXICAL

ahh. ok. hmm. again, i think in first order logic there is no way of stating that something is finite or infinite. this is easy, i hear, in higher order logic. ask jessi. but perhaps it can be written in the language of functions:

given a function on a finite domain, there is a function
on an infinite domain with the same image.

in this setting, let the (finite) set A be the domain of function f. let n be the set of all things of the form f(x) for x an element of A. this is finite because A is finite. but the claim above says there is now some function g and some infinite set B such that the set of objects g(y) for y an element of B is exactly the set n.

so what? it means that a set that can be described with a finite number of arguments could also be described with an infinite number.

this is certainly not paradoxical,
and likely can be expressed easily in higher order logic.

> I don't think I follow it adequately.

> Eg, is "modality" 'simply'
> (as one of my dictionaries puts it:)
> " in logic, the qualification of a proposition
> affirming or denying possibility , impossibility, necessity,
> contingency, etc" or is it something more specific?

righton. in addition to the 'not' 'or' of classical logic, modal logic adds on the 'box' symbol, which is interpreted as necessity. but interpretations vary, and there are different axioms one can take for modal logic, depending on the interpretation. the box can also be taken to mean belief or knowledge. the diamond is a shorthand for 'not box not' so that 'diamond A' means 'not box not A.' if box is taken as necessity, this means 'not necessarily not A,' i.e. 'possibly A'

****

Subject: Re: math questions
Date: Fri, 18 Jul 2003 21:56:43 -0400
From: anonymous

> (forall) x,y ~(x=y) ^ (x=y)
>
> tho this would classically be regarded as either false, or a statement
> that the universe is empty!

Neither of wch is my intention
although I'm becoming increasingly interested
in building theorems based on 'false' axioms.
eg: using "m + n ≠ n + m" as a starting point.
Is there a branch of logic that pursues such ideas?

> what i was trying to say is that the universe
> of objects is fixed, and so if you traverse over all of them
> there is only one way to do it, i.e. by traversing over all of
> them. maybe you want to deny that as well.

In a sense, I do.
One possibility of what I'm getting at is something like this:
In order for us to understand ways in wch
the universe of objects is NOT fixed we must create perceptual systems
of special amorphousness & then locate their appropriate isomorph.
Or something. Y'know, the Unified Field the bouncing ball Theory.

****

Hello? Reader? Reader? Are you still there? I don't exactly blame anyone for finding this last section of emails boring &/or self-indulgent - even in its edited form. Including it is a way of showing my feeling around in the dark. In retrospect, I'm a bit astounded that anyone wd help me w/ this process at all! Esp given that I barely knew these folks.

The entirely too long & probably not short enuf of it is that eventually the following expression was agreed upon for ANYTHING IS ANYTHING & I had a self-inking rubber-stamp made from it so that I cd easily use it for graffiti-ing purposes:

$$(\forall x)x = (\forall x)x$$

I rubber-stamped it in the men's rm of the bar where Jessi & I wd talk math. Shortly thereafter, someone wrote on the wall near it: "1 = 1 So what?" I then replied only in math notation w/ the following (although it's a bit scrunched here):

TAUNTED RAGE ≥ ARTAUD/ GENET

$$\langle ((\forall x)x = (\forall x)x) \supset \{(((\forall x)x = (\forall x)x) = (1=1)) \equiv (((\forall x)x = (\forall x)x) \neq (1=1))\} \rangle$$

If all possible "x"s are equal to all possible "x"s,
then the set of "all possible "x"s are equal to
all possible "x"s" equals the set of "1 equals 1"
is identically equal to the set of "all possible "x"s
are equal to all possible "x"s" DOES NOT EQUAL the set of "1 equals 1"

Whether my fellow graffiti-ist ever read my reply or understood it is unknown to me. Can you understand what I like about such an exchange? In most men's rms, the graffiti is pretty limited intellectually.. to put it mildly. It's mostly tags (people's graffiti names) or so-&-so-sucks-cock type of stuff. I like the idea of challenging people's notion of the transgressive. Most graffiti in men's rms is totally predictable. Initiating a philosophical discussion thru graffiti isn't.

ENOUGH IS ENOUGH

In early 2004, I went to Barcelona & then Berlin. In Berlin, I gave a screening & afterward met a girl named Sandra who invited me to her apartment for dinner. There I was introduced to a friend of hers named Carlos. The 3 of us had a discussion about the meaning of some rubber stamps that I had w/ me - including the one that I call the "ENOUGH IS ENOUGH" stamp. Carlos listened attentively & asked some intelligent questions wch I promised to answer at length by email upon returning to the US. The email below (somewhat updated) is that answer & it's the most complete explanation of ENOUGH IS ENOUGH that I've provided to date. I sent it to Carlos but never rc'vd a reply. I suspect that it never reached him - it may've been too large.

"you must go home now, where everything -- you can be quite sure -- will be falser than here." Jean Genet as Irma

Hi Carlos (& anyone else that I send this to),

I'm still wondering about how to answer your question about my "Enough is Enough" formula. SO, here's my attempt. The 'formula' (if that's what it is) was created in a sortof "stream of consciousness" way as a type of "brain-teaser" to be used as I used it with you & Sandra. In other words, as something to show to people to try to stimulate thinking through what I call "P.N.T." - or "Perverse Number Theory". This is NOT Number Theory in any conventional mathematician's sense &, in fact, most mathematicians/logicians would probably find it repulsive & annoying because it uses none of the rigor that distinguishes math from sloppier thinking.

When I use mathematical notation in this text I will either insert it as a PICT file to make sure it doesn't get corrupted by electronic transmission (that is: e-mail) OR follow it by a language statement without math symbols - in so-called "natural language" (put in parentheses). I apologize if my English is too complex. I will try to keep it somewhat simple (but I'll probably 'fail' because I have a tendency to use expressions that may not translate into German well) & you can always write me with questions if you want clarification.

Enough is enough. I need a break from my self-imposed exile of left-handed marginalization. I want to return to the perverse logic of my loose connectivity. On page 209 of the Artaud Anthology, these words are written: "The place where you suffer where you know you suffer where you feel it & where systematically & voluntarily you maintain the things which you do & which you eat in the breast of eternal sadness without letting go undercover in an organ forever useless where a **being** waits for them." Haven't we had enough of that place, that stasis, that sort of systematic order & fixity, the sort of irrefutable rules governed existence, that universe composed of immutable axioms?

Math is based on "axioms" - 'irrefutable' statements used as building blocks for an edifice of logic. PNT is NOT. One axiom of math is the Commutative one: $x + y = y + x$. In other words, in some operations, the order of the elements is immaterial. Contrarily, PNT postulates: $(m + n \neq n + m) \Leftrightarrow x$ (the set of (m plus n does not equal n plus m) is isomorphic to x). The idea is not to refute the commutative axiom but to try to stimulate a state of mind in which one imagines what a refutation of it could possibly mean. No doubt there are many logicians who've already investigated this idea to profound depths that I wouldn't know about or understand. SO, this is just my personal amateurish & intuitive skimming of the surface.

The Enough is Enough 'formula' is a sortof 'pinnacle' of my feeble PNT so far. Why? Perhaps because it combines pictures + math symbols in a way that's evocative & both sensical & nonsensical. Perhaps because it's open-ended in a way that might catalyze the imagination in a direction that *seems* to promise 'meaning' in a way that challenges the perceiver to attempt to resolve it in some way. By being seemingly mathematical it potentially prompts the perceiver's inclination to try to understand it as somehow axiomatic. By being imagistic it undermines the quasi-unambiguity of the math symbolism. Even I'm not 'sure' what it 'means'. I'm still thinking about it.

That's where your question(s) come in. The thing that's interesting about them to me is that you, in some sense, have had the 'ideal' reaction to my presentation of the formula. You have interpreted the elements in the way that makes the most sense to you (by trying to somehow resolve them into being axiomatic) & have then questioned their significance *beyond* what I said they stated. As such, you then challenge *me* to ponder your extrapolation! This is quite a challenge because I have to understand your reading & explain to myself how it's embedded in my intentions. This helps *clarify* my intentions to myself.

Let's start with how I presented this PNT to you & use some illustrations. I showed you the Anything is Anything rubber stamp:

$$(\forall x)x = (\forall x)x$$

& explained that it reads as "For all x, x equals For all x, x" or "Anything is Anything" - by which I *don't* mean just that x equals itself (although even that is problematic when scrutinized closely*) as in, for example, 1 = 1 or 2 = 2, etc, but that x *also* equals *any other* x - for example: 1 = 2 or "the square root of negative 15" equals "a 'randomly' chosen molecule in the zit on a baboon's ass" or whatever. To most, if not all, mathematicians/logicians this is probably just silly & pointless but to me it's ok to be silly & far from pointless. As with the non-commutative non-axiom: "($m + n \neq n + m$) $\Leftrightarrow$ x (the set of (m plus n does not equal n plus m) is isomorphic to x)" the challenge is to try to imagine the implications of such a profoundly (says who?) disruptive-to-consensus-'reality' statement.

[January, 2008 interpolation:
A physicist acquaintance wrote me in an email rc'vd today:

"If you like the non-commutativity, in physics multiplicative operators

As tENT pushes toward the conclusion of this book, I want to circle back to the open-ended joy with which my own marginalia explored at the outset of my journey through this text. I do not wish to refute the course I took at the outself of part two -- I think that journey was important, too -- important in its "truth" -- that to exist at the margins is to suffer, to struggle, to find oneself taunted & to be possessed with rage.

often don't commute (such as two non-commutative matrices, or even position/momentum operators).

You should also look up Grassmann Numbers, they're pretty weird. Basically A*B = - B*A. Does all kinds of weirdness with Taylor expansions, which only ever need to go to 1st order, so they're easier. They are useful when considering fermions. One of my physics profs when introducing them jokingly implied we could go "smoke something" to understand what they mean."

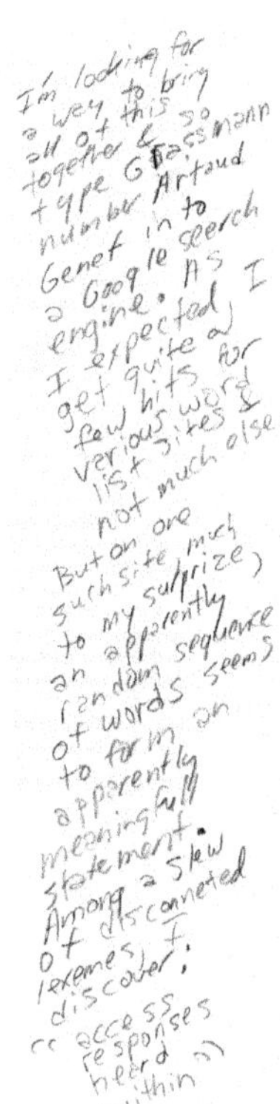

There's non-commutativity already! WONDERFUL. &, sure enuf, there were 24,000 Google hits. Including: <http://www.stringwiki.org/wiki/Noncommutativity> SHEESH! Too much to go into here when I'm just about finished writing this bk. I also looked up "Grassmann Numbers" & got this from WP:

"In mathematical physics, a **Grassmann number** (also called an **anticommuting number** or **anticommuting c-number**) is an element of the algebra spanned by quantities θ_i that anticommutes with all generators θ_j but commutes with ordinary numbers x:

$$\theta_i\theta_j = -\theta_j\theta_i \qquad \theta_i x = x\theta_i."$$

One thing's fer sure, I don't have the slightest idea what that means but I'm very grateful to the holy ceiling light that it exists.]

Then I showed you the Enough is Enough rubber stamp:

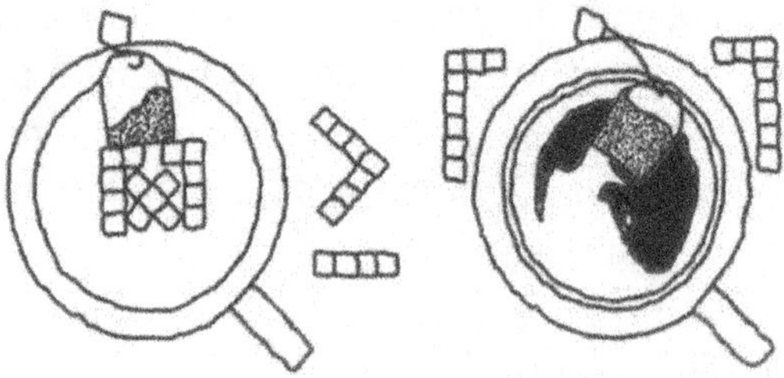

followed by the photograph(s) that it's based on:

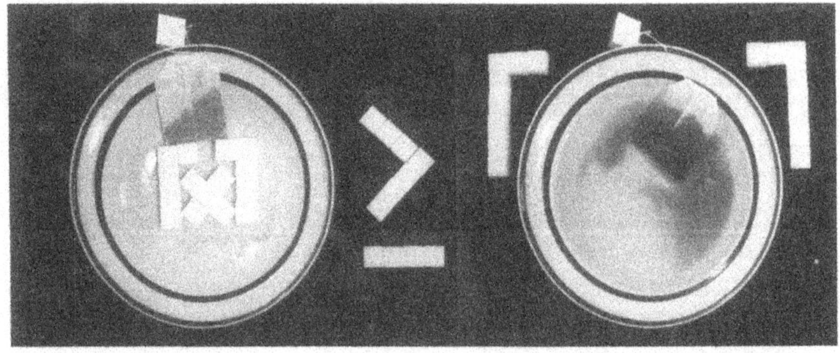

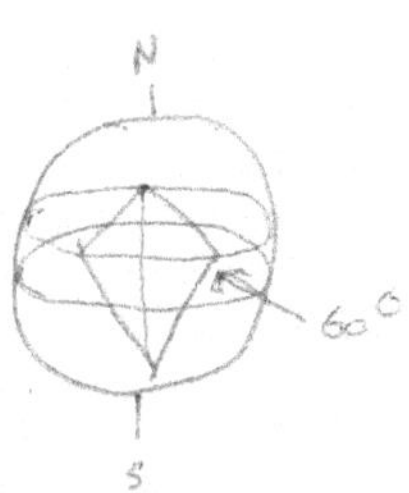

- accompanied by a meandering partial explanation. Rather than recreate that explanation, I'll paraphrase from an e-mail to my neoist architect friend Boris Wanowitch below followed by further commentary:

In English, there's the expression: "Enough is Enough!"
which is said when a person is overwhelmed
by something that's too much. SO, it sortof means: "That's too much!"
I've been trying to translate this statement into a mathematical statement.
The symbols that look like square brackets [ ]
without the bottom horizontals mean: "ceiling".
The symbol that looks like a sideways "V" with a line under it means:
"greater than or equal to".
SO, what you see in the picture is: on the left:
a bowl with a tea bag & sugar cubes forming the
mathematical symbols for: "the ceiling of x"
- followed by: "is greater than or equal to:" (written in sugar cubes)
- & then the same bowl with hot water in it now
so that the tea bag is extruding its flavor
& so that the sugar cubes are melted
& with sugar cubes outside the bowl forming the "ceiling of" symbol.
How to interpret this? Let's start with it as:
"The ceiling of x is greater than or equal to the ceiling of
the ceiling of x."
This could be an attempt to state something like:

Somehow this "statement" manages to fill me with hope.
I'll follow the direction using the text on this page:
the expression:
a person is overwhelmed
that's too much
to translate
like square
horizontals
like a sideways "V"
equal to
what you see
in a tea bag
& sugar
symbols for
"is greater than
or equal to:"
the same tea bag
is the sugar
outside the
bowl, forming as
greater than or
equal to the ceiling of
an attempt to state something

"x has surpassed its own desirable maximum limit"
which leads us back to "Enough is Enough!"
HOWEVER, I have made this more complicated
by using the tea & the sugar cubes, etc..
Here's a quote from a book I'm reading called
Alan Turing: The Enigma that was the inspiration:
"When the sugar has dissolved in the tea,
there remains no evidence, on the law of averages,
that it was ever in the form of a cube. But in principle,
at a sufficiently detailed level of description,
the evidence would remain in the motion of the atoms."
Now, I've taken this quote out of context,
but I'll make up a further interpretation
of my picture with that quote in mind:
"The ceiling of x is greater than or equal to
the ceiling of an object in which the ceiling of x
may be reconstructed if analyzed in sufficient detail
& STOPPED at a certain point."
I add "& STOPPED at a certain point" because
if the reconstruction analysis goes "too far"
it'll take the analyst PAST the state at which the sugar cubes
represent the symbols & back to when the sugar was still part of a plant
- & THEN we'd be really confused! Sound complicated?
Really, I'm partially trying to be funny
but, in my perverse way, I am actually trying to
create a new branch of Number Theory.

It makes sense to me, in the way that tENTS breaking with the typographical conventions of the prose text makes sense to me here. I'm aware that his textual placement results from the fact that he is quoting from a email message, but still, it's suggestion of poetic line-breaks encourages me still further.

To put the above quote from Andrew Hodge's Alan Turing: The Enigma more in context I excerpt from the relevant passage at greater length:

"Alan [Turing] could have found many of these ideas [about body & spirit] in his reading of Eddington while still at school. [..] He had found a resolution of the classical problem of determinism and free will, of mind and matter, in the new quantum mechanics.

"The idea that Alan said 'used to be supposed in Science' was familiar to anyone who studied applied mathematics. In any school or university problem, there would always be just sufficient information supplied about some physical system to determine its entire future. In practice, predictions could not be performed except in the most simple of cases, but in principle there was no dividing line between these and systems of any complexity. It was also true that some sciences, thermodynamics and chemistry for instance, considered only averaged-out quantities, and in those theories information could appear and disappear. When the sugar had dissolved in the tea, there remains no evidence, on the level of averages, that it was ever in the form of a cube. But, in principle, at a sufficiently detailed level of description, the evidence would remain in the motion of the atoms. That was the view as summed up by Laplace in 1795:

> Given for one instant an intelligence which could comprehend all the forces by which nature is animated and the respective situations of the beings who compose it - an intelligence sufficiently vast to submit these data to analysis - it would embrace in the same formula the movements of the greatest bodies and those of the lightest atom; for it, nothing would be uncertain and the future, as the past, would be present to its eyes.

From this point of view, whatever might be said about the world on *other* levels of description (whether of chemistry, or biology, or psychology, or anything else), nonetheless, there was *one* level of description , that of microscopic physical detail, in which every event was completely determined by the past. They might *appear* undetermined, but that would only be because one could not in practice perform the necessary measurements and predictions.

consider the following redesignation of the convolution theorem:

Freedom [Creativity (xanything is anything, y not?)]

= Fructiferous (xcitement) For (you) !

[..]

[..] "Eddington was not a dogmatic Christian, but a Quaker who wished to preserve some idea of free consciousness, and an ability to perceive a 'spiritual' or 'mystical' truth directly. He had to reconcile this with the scientific view of physical law. [..]

"The new quantum physics offered such a reconciliation, because it seemed that certain phenomena were absolutely undetermined. If a beam of electrons were directed at a plate in which there were two holes, then the electrons would divide between the two, but there seemed no way of prediciting the path that any particular electron would follow, not even in principle. Einstein, who in 1905 had made a very important contribution to the early quantum theory with a description of the related photo-electric effect, was never convinced that this was really so. But Eddington was more readily persuaded, and was not shy of turning his expressive pen to explain to a general audience that determinism was no more. The Schrödinger theory, with its waves of probability, and the Heisenberg Uncertainty Principle (which, formulated independently, turned out to be equivalent to Schrödinger's ideas) gave him the idea that mind could act upon matter *without* in any way breaking physical laws. Perhaps it could select the outcome of otherwise undetermined events."

Contrary to what might seem to be implied by the preceding, I am NOT trying make a point vis a vis free will vs determinism or trying to somehow 'prove' the existence of "spirit" or "soul" - *BUT*, I can certainly imagine some interesting possibilities related to a favorite subject: "undermining 'reality' maintenance traps". Now's probably not the time for that tangent though.

Enter your questions. At 1st you said something about the 1st "x" meaning, for example, something like (excuse my inaccurate paraphrase): "a dog is a cat" & the 2nd "x" meaning, for example, that translated into a different language - let's say: "un chien est une chat" & you said something about the translation being a lesser version of the 1st side because the translation's 'bad' (as it certainly is in this case parce que mon français est mauvais). Therefore, what we had was something like this: "the ceiling of 'a dog is a cat' is greater than or equal to the ceiling of the ceiling of 'un chien est une chat'" which could be reduced to, in your interpretation (as I remember it), to something like this: "the maximum communicative quality of the phrase 'a dog is a cat' is greater than or, at worst, equal to the maximum communicative quality of the maximum communicative quality of its attempted translation into français as 'un chien est une chat'". Lost in translation. Now, again, as I recall it, you more or less abandoned this in favor of your bigger question but I want to pursue this part a bit further.

d | c | d
c | c | d
d | d | c

= cat's game
clearly disadvantaged
in this arrangement
the dog will always
play as "under dog"

∴ CAT/DOG ≥ CHAT/CHIEN

or

A | A | E
  | A | A
E | E | E

wow!
Enough is
Enough beat
out Anything is
Anything —
& it didn't
even need to
cheat!

Partially what interests me about this line of inquiry is that you seem to've applied the qualified "x" of the "Anything is Anything" equation to the "Enough is Enough" formula - immediately creating a sortof "meta" hybrid. In other words, it's as if you plopped an "Anything is Anything" "x" solution into both sides of the formula. SO, instead of "x", we have "Anything is Anything" which becomes "a dog is a cat" on the left side &, instead of "x", we have "Anything is Anything" which becomes "un chien est une chat" on the right side. Now, in non-perverse math, you've made an incorrect transfer of ideas because in the "Enough is Enough" formula there's no "For all x" qualifier. That means that the "x" on the left side has to be the same as the "x" on the right side &, therefore, it's up to the puzzler of this puzzle to figure out what "x" can meet the conditions of having its ceiling greater than or equal to the ceiling of its ceiling. In this case, "enough" as the solution to "x" meets this requirement *as long as we accept (the admittedly absurd) introduction of INFLECTION as understood through the "natural language" expression.*

HOWEVER, what the fuck, this is PERVERSE NUMBER THEORY, right? SO, enter some sort of (un?)disciplined stream-of-consciousness. A possible purpose of PNT is to unleash an unpredictable chain of inquiry. Therefore, I think that applying "Anything is Anything" to the "x"s of "Enough is Enough" is a wonderful idea. But, *not* to try to demonstrate that something's "lost in translation". That's too simple. Let's put that aside, for the moment, & move on to your next & bigger question (again, excuse the bad paraphrasing) in which you addressed both the formula's mathematical symbols but also the image content by saying something like: "Ok, the 1st picture doesn't have any water in the tea-cup/bowl & the 2nd picture does & you've said that if we apply the theory of determinateness as expressed by Laplace we can understand that the 2nd picture not only contains the history of the 1st picture but that we can also extrapolate this determinateness process into both the 'future' of the formula as well as its past." [Ok, I know you didn't say all this but I'm mutating your statement into what I think you & I are both getting at in this investigative process] "THEREFORE, we can imagine that a later picture in the hypothetical sequence of images separated by 'greater than or equal to' symbols might include piss, right? Because the tea will be drunk & turned into piss." Let's pause before reaching your actual question.

That having been said, I'll allow a few of my patanagrammatical avatars wish their farwells -- I've written enough &/or too much on everything, & nothing, anything anyhogs as it is -- vive les marginelles!

NOW, when I talk about this formula & the photo(s) I sometimes say that once we realize that the right half of the formula still contains (in dissolved form) the sugar-cube statement of "the ceiling of x" we can extrapolate beyond that & imagine another (invisible) image further to the left that contains a pre-sugar cube version of sugar - let's say sugar cane. Sometimes, more for humor's sake than for anything else, I then give the expanded formula the political overtones of something like: "the socio-political-economic process of converting sugar cane into sugar is ultimately TOO MUCH." SO, I've hypothetically delved into the 'past' of the invisible sequence. Your intro to your question goes into the future. I think the 'invisible' implications in both directions have interesting potential because it stimulates the investigator to imagine the significance of slotting images into a repeating formula. & that's exactly what you do next.

YOUR FINAL QUESTION: "If we continue to extrapolate can we make a self-contained circle? &, if so, how is this possible?" In other words, if "Enough is Enough", what's to stop the 2nd "Enough" from cycling back to the 1st "Enough" to form a closed loop & how does that fit into the images of the tea going from dry to wet if it can, therfore, go back to dry again? & where does the piss fit in? It seems that we're beginning to turn ecological here. But that's not my purpose - just like it's not my purpose to 'prove' the existence of a 'soul'. & I don't think that's your purpose in asking the question either. As I understand it (in this definitely stream-of-{piss}-consciousness addressing of your question), you're taking my use of the images of sugar cubes & tea & expanding from them by making them more important than the math symbols that they're 'containers' for. Which is fine. To clarify, though, while I mainly intended to use these images as a reference to Laplace as a springboard for tangentially bringing out the hidden presence of the 2nd "ceiling of x", I was simultaneously using such an open-ended method for doing so in order to create a situation ambiguous enough to stimulate exactly the kind of further tangent that you have been the 1st person to go off on! Congratulations!

HO!

EGO UNHINGE US

But that doesn't answer your question, does it? Let's try to keep it 'simple'. If "the ceiling of x is greater than or equal to the ceiling of the ceiling of x" is extrapolated to the right, some logic or another might render it as developing like this: "the ceiling of x is greater than or equal to the ceiling of the ceiling of x is greater than or equal to the ceiling of the ceiling of the ceiling of x" ad infinitum. In other words, the pattern already established by the formula would add additional "ceilings" onto each additional extention to the right. However, each additional extention to the *left*, by the same 'logic', would *subtract* a "ceiling" - thusly yielding, with one extrapolation on both sides, "x is greater than or equal to the ceiling of x which is greater than or equal to the ceiling of the ceiling of x which is greater than or equal to the ceiling of the ceiling of the ceiling of x". In that scenario, the possible extrapolations leftward of the formula stop as soon as "the ceiling" is removed.

As such, NO, the formula cannot become circular because moving leftward the formula ends & moving rightward the formula must increase its number of ceilings & can't, therefore, lead to an x with a lesser number (or complete lack of) ceilings. Enter the conflict with the Laplacian determinism that the formula's imagistic expression engenders. If we have to accept the Laplacian determinism in order to read into the 2nd image the latent existence of "the ceiling of x" in order to get to "the ceiling of x is greater than or equal to the ceiling of the ceiling of x" in order to get to the inflected saying "Enough is Enough" then we might also 'have to' consider the possibility that the formula can extend infinitely in either direction thusly disrupting the apparent leftward limitation. The question is, therefore, does invalidating what is possibly the main barrier preventing circularity in the process on the leftward end imply the possibility of also removing the infinite "greater to" progression & the infinite "ceiling of" piling up &, therefore, eliminate the inflection that generates "Enough is Enough"?

Uh...hue!

NOISE GONG

FURTHERMORE, what happens when we return to slotting "Anything is Anything" into this? It's your turn to answer.

- tENTATIVELY, a cONVENIENCE
May 1st, 2004EV

*Consider this excerpt from an on-line thread:

To: <phil-logic@bucknell.edu>
Subject: Re:Sameness and Self-Identity
From: "David A.Hernandez Purnell"
Date: Fri, 3 Jan 1997 13:45:43 -0000

WARNING & REQUEST:
*Please treat this posting like a book on a shelf
- to be left undisturbed if of no interest*.

The following is of course excessively long for a usual post. I am submitting it however in the hope that I may receive helpful comment since it hints at certain positions which will be a central element in a work to be published at the end of this year. The overall work is a composite of religious/historical/philosophical considerations and research a central thread of which is the different "logical" (philosophical/ontological) implications of "indexical" and "descriptive" language (reference, description and ontology). These differences have been neglected in the understanding of religion in particular personal religious experience (including the "mystical") contrasted with creedal "descriptive" or wholly "group-normative" systems. Part of what is argued is that understanding of the "Abrahamic" (and similar) scriptures (Pentateuch, Gospels (including Thomas), Koran) and related traditions requires cognisance of this distinction. The historical aspects seek to illustrate that this ill-understood distinction has always been at the nexus of philosophical/religious dispute and is present too in the development of the sciences including physics, mathematics and symbolic logic - the influence of such as Ramön Lull and Kabbalah on Leibnitz and Newton to the present day (from Cantor's infinities through Spinoza's influence on Einstein to the Islamic mysticism of Abdus Salam and "superstring dimensionality".. The work argues for the priority of individual experience as against religious or racial groups.. Similar motivations underlay the work of such as Ramön Lull and Ibn Arabi.

Huh!
ONION-EGG-SUE

To: *ANY INTERESTED* recipients of list
Subject: Re:Sameness and Self-Identity
Was: Goedel's Incompleteness Theorem and Constructivism

On: 20 December 1996 13:13

Michael Thayer wrote:

>David HP writes:

> >My understanding is that it is possible
> >to characterise Intuitionism

> >precisely on the basis that reductio ad absurdum
> >is rejected since it rests
> >upon the law of the excluded middle.
> >Your comment here therefore puzzles slightly.
> >David
>
>AS has been previously noted,
>this characterisation of "Intuitionism" is
>incorrect with respect to the reductio. It is also false with respect to
>excluded middle. Most Intuitionists have accepted that >(2^(2^157))+1
>either is or is not prime. Some strong finitists might disagree,
>and some
>intuitionists as well, but the main problem area
>was LEM in the presence of infinities.

>Michael

Dear Michael,
Thank you for your elaborations from which it would appear that (at least with respect to those matters touched upon in your last two sentences) not all "Intuitionists" are in accord. This is hardly surprising since as much can be said of Republicans and the Angels (to make a

fairly random selection) but with respect to Intuitionists seems singularly appropriate. The origins of Intuitionism (in its traditional form) was rooted in the rejection or suspension of LEM of which Brouwer wrote:

"(For Intuitionists)....the dogma of the universal validity of the principle of the excluded third is a phenomenon in the history of civilisation like the former belief in the rationality of pi, or in the rotation of the firmament around the Earth..."

As I indicated to Nicholas - though not so that he found the suggestion amenable - the discussion on Intuitionist "logic" is one I consider to be of no strictly philosophical interest since this is to be found not *within* the calculus of logicians but rather in the *prior reasonings and intuitions*. Intuitionism was opposed to the formalist enterprise and is - one would have thought - only adequately characterised as a rejection of "formalism" *together with* Platonism. Interestingly the Intuitionist account of the sequence of numbers parallels the view on future states for which I have been arguing.

Goo!

GENUINE HUSH

Brouwer considered numbers to be not only "mental entities" but to arise out of the "sensation of time". I do not take time to be open to "sensation" but the "now indeterminate" status of "future events" (or more appropriately 'non-events'!) likewise requires the rejection of the view that for *any* "future tense statement" either it or its negation is now true. Predetermined events (if any) are perhaps another matter.

The Intuitionists considered the logicism of such as Russell and Frege to be "topsy turvy" - mathematics is prior to logic. Logic and its "rules" are extrapolations from mathematical experience. Again there is an affinity with what I have been arguing (or rather perhaps "asserting"!) i.e. that ontology is prior to logic - though the scope of "ontology" in my vocabulary is obviously much wider than just mathematics. These considerations have a bearing on the following matters.

Since Torkel expressed his astonishment that anyone could be other than instantly cognisant of the (putative) truth of

@ (x)(x=x)

my comment to him that @ is at best indeterminate has been joined by Greenberg's claim that it is false as well as a recently posted (2nd January 1997) assertion from someone (I am sorry I forget who) that it is "true by definition" - though not unsurprisingly perhaps no "definition" was provided. At the time I started this response (last year!) "recent" exchanges included the following:

>"William J. Greenberg" wrote:

>> 12. I wish to dedicate "Forrest Gumpel Meets (X)(X=X)" to
> >Torkel Franzen and Mikhail Zeleny, for their ceaseless
> >efforts on behalf of the X=X Thought Police.

>It is my impression that your theory, which we discussed just
>over a year ago, would be better accepted if it was presented as
>a theory in which the symbol "=" has a different definition,
>according to your axioms, from the very common one. And then,
>obviously, take the usual precautions to not mix different
>theories having the same symbols; which if not taken, will
>guarantee confusion. David A. Hernandez Purnell's primacy of
>ontology should come in very handy here.

>Neil Nelson

Now I think the argumentation surrounding the interpretation of "=" is not unconnected to the (original) "Intuitionist intuition". According to Neil it is (perhaps?) a matter of differing axiomatic systems. But I have more than a faint suspicion that William will want to insist that "the very common" interpretation of @ is just wrong - though he will undoubtedly put it more symbolically.

What is surely ignored by those who insist that the "meaning" and/or "truth" (or "falsity") of (x)(x=x) is transparent is that it is difficult to imagine a more quintessential focus of the history of philosophical dispute. Nothing could perhaps be so far from being "true by definition" because of the number of issues which such "definition" would need to address. Every issue is raised:

Is

(x)(x=x)

the expression of a relation between signs? Between ideas? Between things? Or between all of these? As witnessed in the discussion of "2" there are those who want to claim that "Socrates = Socrates" is a true sentence or proposition to the extent that it attributes a true relation to Socrates (one which he bears to himself) achieved by the "use" of the name "Socrates" and those who would say that it is not the *name* "Socrates" which appears in this sentence or proposition - but Socrates! I will not dwell on the fantastic claims made for the supposed "use/ mention" distinction save to say that one cannot refer to Socrates by name without mention of that name - nor can one mention that name without it being true it is/was the name of Socrates. "Mention" of the name "Socrates" is mention of Socrates' name (as opposed to - say - that of Plato) as is its "use". But Socrates who? Which Socrates? One might write:

ENOUGH IS ENOUGH...

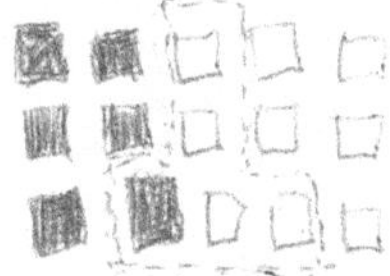

Socrates = Socrates

claiming that it attributes "a property" to Socrates and be met with the question of just who possesses this property? The unique Socrates beloved by some as an exemplar of "Greek Rationalism"? (Incidentally I consider there are reasons for seeing Socrates more as a Jewish figure than a Greek!) Would it then mean something like "*That philosopher* (whether Greek or Jew) in all "possible worlds" (including those in which he was/is named "Jacob Cohen")? Or does it mean *any* person or other thing named "Socrates"?

Is it "transparently obvious" and inevitable that "Socrates" is the name of Socrates? Just as "2" is the name of 2? Or "x" the name of x? Of course it may be "obvious" - but it is not thereby necessary. The pretence that @

is transparent in its meaning is an unconsidered assumption to the effect that "every occurrence of "x" is an occurrence of *the same thing*". But "the same" and "thing" in what sense?

"Type/token" talk can be seen as a circumlocution which allows mathematicians to get on with their business. Fine. But *some* who move from mathematical formalisms to the wider world seem to suppose that such talk dispenses with the issues of metaphysics or ontology. It is apparently of no concern to those who make this kind of assumption that there is no possible stipulation whatsoever which would ensure similarity between any two (or more) "tokens" of a "Type". Is the string "APPLE" on my screen the same or a different token from that which appears on yours? Even if mine is 10pt bold helvetica black and yours 12 point Times italic blue? What about the electronic encoding of the string "APPLE" in hyper-space or on disk even? Or - if anyone has one - its appearance on a Braille output device or output by an encryption coder? Since "Type/token physical inscription" talk was introduced as a "physicalist" alternative to "the transcendent" (purportedly accessible only to "the mind") dogmatic insistence on unexplicated type/token talk looks not unlike obduracy in the face of disaster.....Or perhaps a "doctrine" of a "group"?.....

There is apparently much dependence in current "philosophy of logic" of the notion of a "linguistic community" who either share "identical meanings" or are not group members. It is maintained on some such basis that @ is both transparent and true. It is - those subscribing to such views suppose - the "standard interpretation" or a matter of "convention". But talk of "communities" and "conventions" is nothing more than (poorly) concealed dependence upon *intention* even if this - like death in certain circles - is not to be openly talked about. Not surprisingly then some suppose that they may adopt any convention they see fit for whatever purpose or none. And in an obvious sense this is so - but what is surprising is that some also suppose that such liberty is without limit or constraint. I have seen some talk of attaching "different senses" to "not" or "=". Fine so far as those "strings" go....but supposing such achieved then each different sense of "=" would be *that* sense and *not* some other. All would be parasitic upon the *primary* meaning.

Anything is Anything

Illustration 225. Non-Euclidean Penguinogram absolved from its taunt-inspired rage while dreaming of an equation in which X is both greater than or equal to & the purported "trancendent" of the apeironic paradox.

PARADIGM SHIFT KNUCKLE SANDWICH

I wasn't satisfied w/ any of the mathematical expressions for any ideas of mine so I didn't want to use them for something as personal & long-lasting as a tattoo. Instead I started thinking about Leonhard Euler's famous equation:

from page $\sqrt[104]{5.9083646 \times 10 \text{ to the 209th}}$ in the NAVIGATION: e: The Story of a Number section:

"the formula that "must surely rank among the most beautiful formulas in all of mathematics": "$e^{xi} = -1$". "Indeed, by rewriting it as $e^{\pi i} + 1 = 0$, we obtain a formula that connects the five most important constants in mathematics (and also the three most important mathematical operations - addition, multiplication, and exponentiation). These five constants symbolize the four major branches of classical mathematics: arithmetic, represented by 0 and 1; algebra, by i; geometry, by π; and analysis, by e. No wonder that many people have found in Euler's formula all kinds of mystic meanings"."

It is, indeed, stunningly elegant. Alas, it only has 5 symbols in the 1st version above & 7 in the 2nd & I have 8 knuckles & wanted to make a knuckle tattoo. The 'classic' knuckle tattoo being "LOVE" on one hand & "HATE" on the other. I remember a guy who had that tattoo being very careful to just show me the "LOVE" knuckles. I appreciated his use of them for showing his emotional attitude toward others but I didn't want to be in a predicament where showing one hand might give people the wrong impression. I wanted a knuckle tattoo where showing either or both hands wd be fine. Besides wch, of course, I wanted a more 'original' tattoo.

So I started thinking about what variations cd be made on the Euler equation to make it 8 symbols long. "$e^{\pi i} + 1^2 = 0$" was a possibility.. but it didn't seem very exciting. I wanted to take the Euler equation a little further out. Perhaps introduce another math idea that had strong personal significance for me. Then it occurred to me that putting 1 to the power of ∞ *might not equal* 0. I hadn't read about "indeterminate forms" yet.. but I had a *hunch*, an *intuition*..

Remember way back when (if you're reading this from front to back) on page $\sqrt[77]{1.81880374 \times 10 \text{ to the 145th}}$ where I mention reading *e* & finding: "The indeterminate forms most commonly encountered are 0/0, ∞/∞, 0 • ∞, ∞ - ∞, 0^0, ∞^0, and 1^∞." - w/ indeterminate forms being "expressions hav[ing] no preassigned value; they can be evaluated only through a limiting process. Loosely speaking, in every indeterminate form there is a "struggle" between two quantities, one tending to make the expression numerically large, the other tending to make it numerically small. The final outcome depends on the precise limiting process involved."

Indeterminate forms are the David A. Bannisters, the Monty Cantsins, the neoisms, the SMILES, the Karen Eliots, the Santa Clauses [that wd make a good mathematical or legal term], the Luther Blissetts of the math world. Anyway, before I read about indeterminate forms, I played my hunch that by adding infinity as a factor that the left side of the equation might no longer equal zero so I changed it to "approximately equal to" instead.

Apparently, if I added some limiting factor then it cd either be equal to zero or not - & I wanted to leave it at 8 symbols. & I wanted to leave it ambiguous. Thus, 2 more essential paradigm shift concepts are added: infinity & the Stanislav LIM, the Law of the Included Middle: the ambiguity - yet another PARADIGM SHIFT KNUCKLE SANDWICH blow against the absolute right & wrong universe of boxed-in minds.

PARADIGM SHIFT KNUCKLE SANDWICH 228 selfish numbers

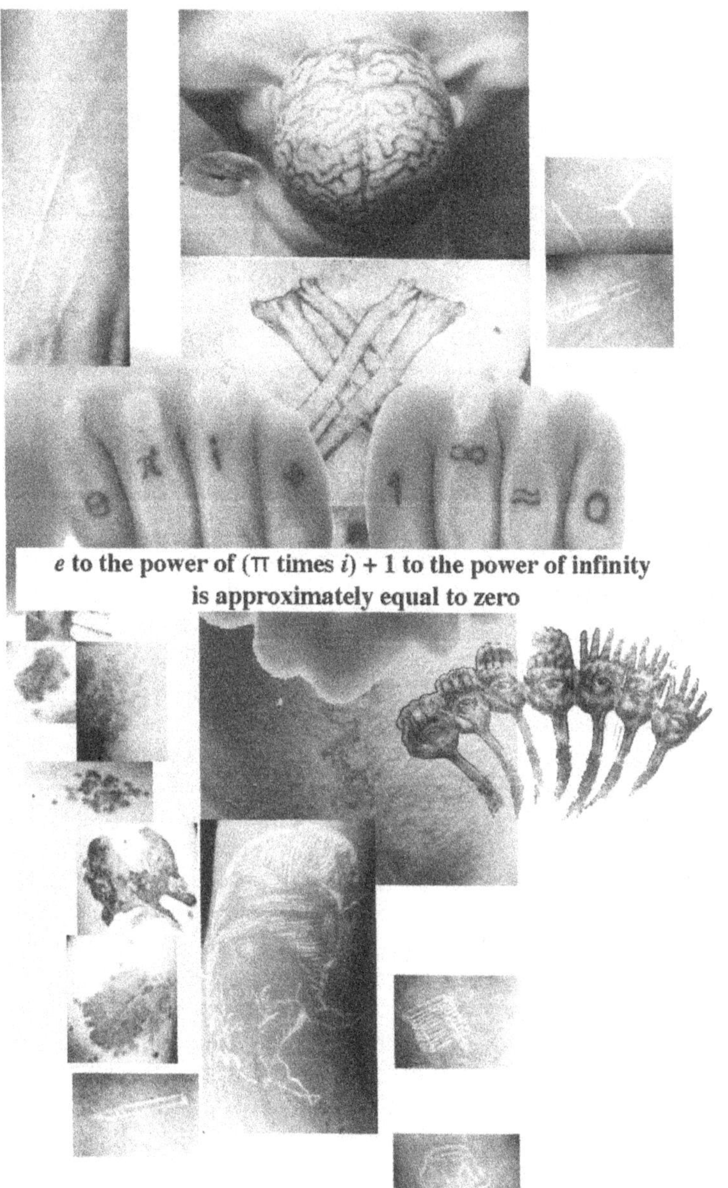

Putting the Paradigm Shift Knuckle Sandwich tattoo somewhat in context

It wd probably help put the Knuckle Sandwich tattoo(s) in context more if I were to explain the development of my tattoos in general - most of wch are shown on the previous page.

I called my 1st 3: "It's not a Matter of Life & Death" & made a movie of me getting them. 1st, came an image of DNA (1986) from my navel to my cock - symbolizing life, then came the cross(ed)bones (1986) on my chest - symbolizing death, the conclusion being the 3D brain (1987) on my head (done in red & green instead of red & blue) - symbolizing transcendental expansive consciousness.

After that? In 1990 I got a series of white tattoos. These were simple pictograms of objects that had caused scarification on me - next to the relevant scars. The tattoos, because they were white, look like scars themselves. I got an image of a talking tree prop on my forehead - I'd been hit there by the prop when I had a job installing a Wizard of Oz set for an event. I got a knife on my left hand - I'd been slashed in a street altercation in Baltimore. I got a rather abstract looking image of a corner - I'd slashed my left wrist by bashing it against some object.

I got a razor blade on my right arm - where I'd also slashed myself. I got a venetian blind on my left thigh - I'd cut myself on one when I was looking out the window in a hospital as a child. I got an exacto knife on my right thigh - I'd cut a DOUBLE NEGATIVE AS NOT A POSITIVE symbol there in performances in the mid 1980s. I got a volcano crater on my left knee - I'd climbed one in Hawaii & cut myself on the sharp lava. These were ways of showing some painful trajectories in my life, of explaining the drawings that experience had left on my skin.

Then, in 1991, I got an image of a dust mite enlarged 250 times in glow-in-the-dark ink on my right thigh. The tattoo barely shows as just a thin scar in normal lighting - but if its charged up, esp under blacklight, it glows brightly. This is a sort of ecological tautology. If the glowing inks are killing my skin, then the image of the dust mite is feeding what the image is of - since dust mites live off of dead human skin.

In the mid to late 1990s I started getting my splotch tattoos. These are just splotches of color that I'm using to gradually change my skin color - making myself a more colorful character. The idea is to use tattoos just for coloration - rather than for imagery or symbolism. Furthermore, I founded a group called the National Association for the Advancement of Multi-Colored People (N.A.A.M.C.P. - no parody of the N.A.A.C.P. is intended) & declared myself president of it - partially because I'm so tired of the whole concept of 'race' wch I think people have spent way too much time on. I'd rather think of myself as "multi-colored" (it's more accurate anyway) than "white". Everyone is a Multi-Colored Person.

Some of these splotches look like skin disease. That appeals to my perverse sense of humor. I have one on my left calf, my left thigh (this is the only one that failed at being a splotch - I thought I'd struck on a sure thing - I thought even the *worst* amateur tattooist wd be able to make a splotch! I was wrong), my left upper arm, my right eyebrow, my lower back, & my right hand. This latter tattoo is particularly funny to me because it's in a green & blue that I often painted on walls at one job I had - so it looked like I'd spilled the paint on myself & not cleaned it off.

In 2002, I got the wagging, waving & winking ARCHIBRAS on my lower back. This is based on the ideas of utopian philosopher Charles Fourier (1772-1837). Fourier is reputed to've believed that following his philosophy wd improve the followers' bodies - resulting in the growth of a new appendage after 20 generations. This new appendage, the archibras, was to be a tail w/ a hand on the end of it & an eye in the middle. Whether this is really historically accurate or not, I don't know.

Legend further has it that, while many people attempted to live life according to Fourier's liberated individualist philosophy, most, if not all, of Fourier's followers were embarrassed by the utter absurdity of the image of the archibras. I, on the other hand, find it very entertaining. Fourier's ideas of sexual liberation & equality for women are all well & good, but the archibras is unique!

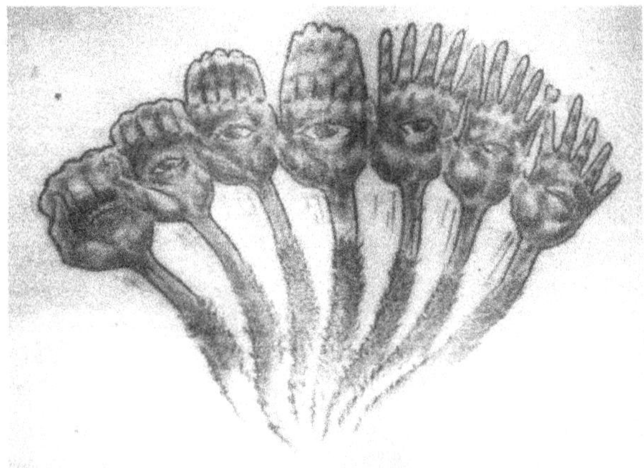

&, then, in 2004, along came the PARADIGM SHIFT KNUCKLE SANDWICH to round it all out!

HAIRCUT PARADOX

So, then, of course, I had to make a movie inspired by all this. Strangely enuf, it's even been screened a few times. Lardy knows what anyone thought of it. Thanks to David A. Hernandez Purnell for his unwitting contributions to this & to the ENOUGH IS ENOUGH section. I've tried finding him on-line repeatedly & failed.

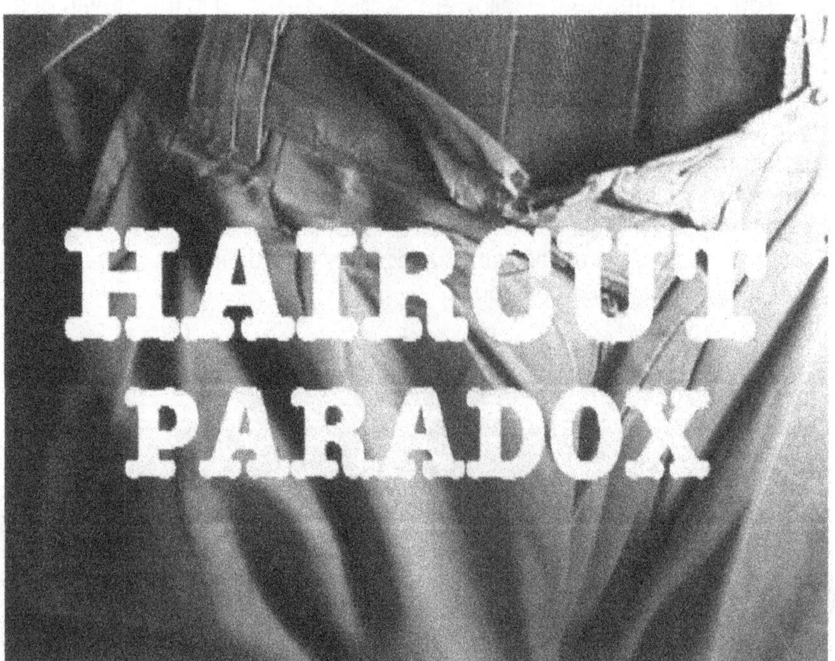

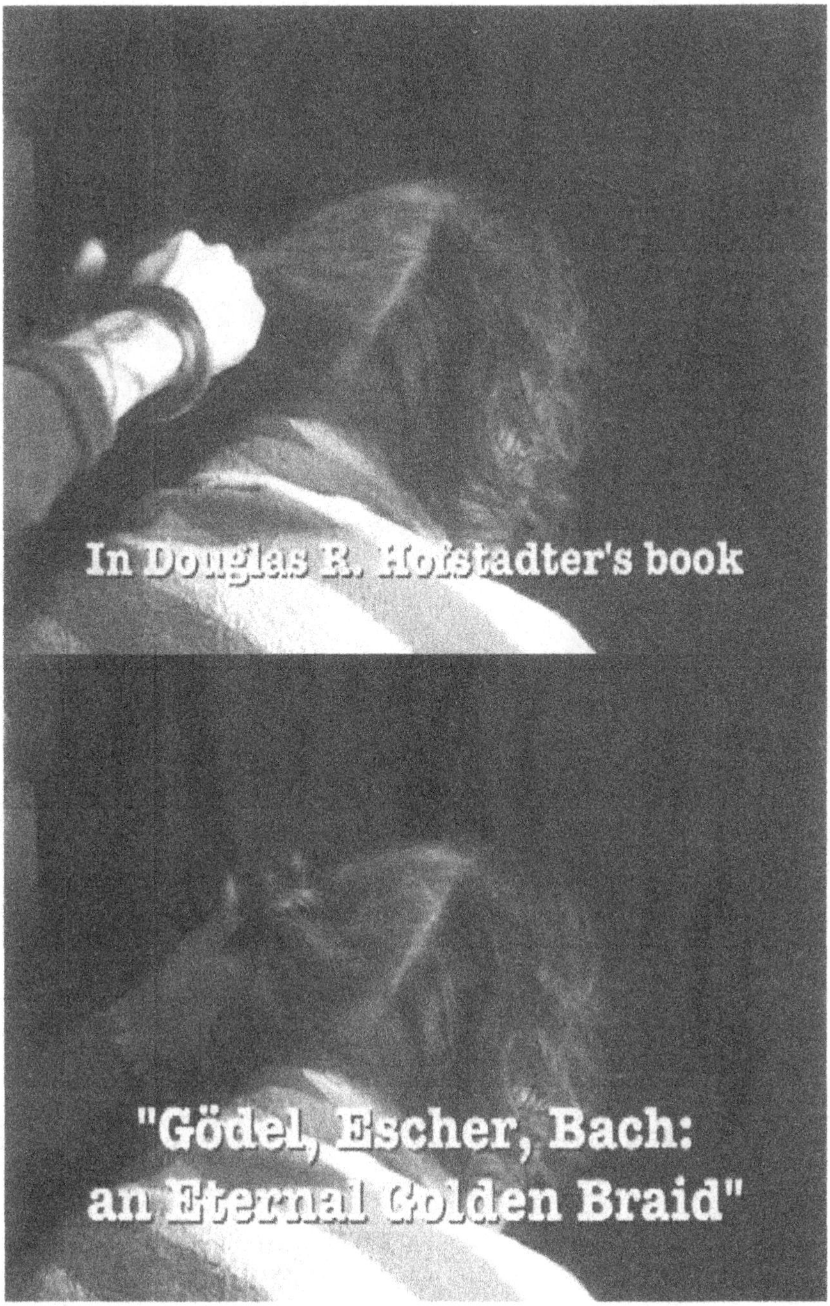

In Douglas R. Hofstadter's book

"Gödel, Escher, Bach: an Eternal Golden Braid"

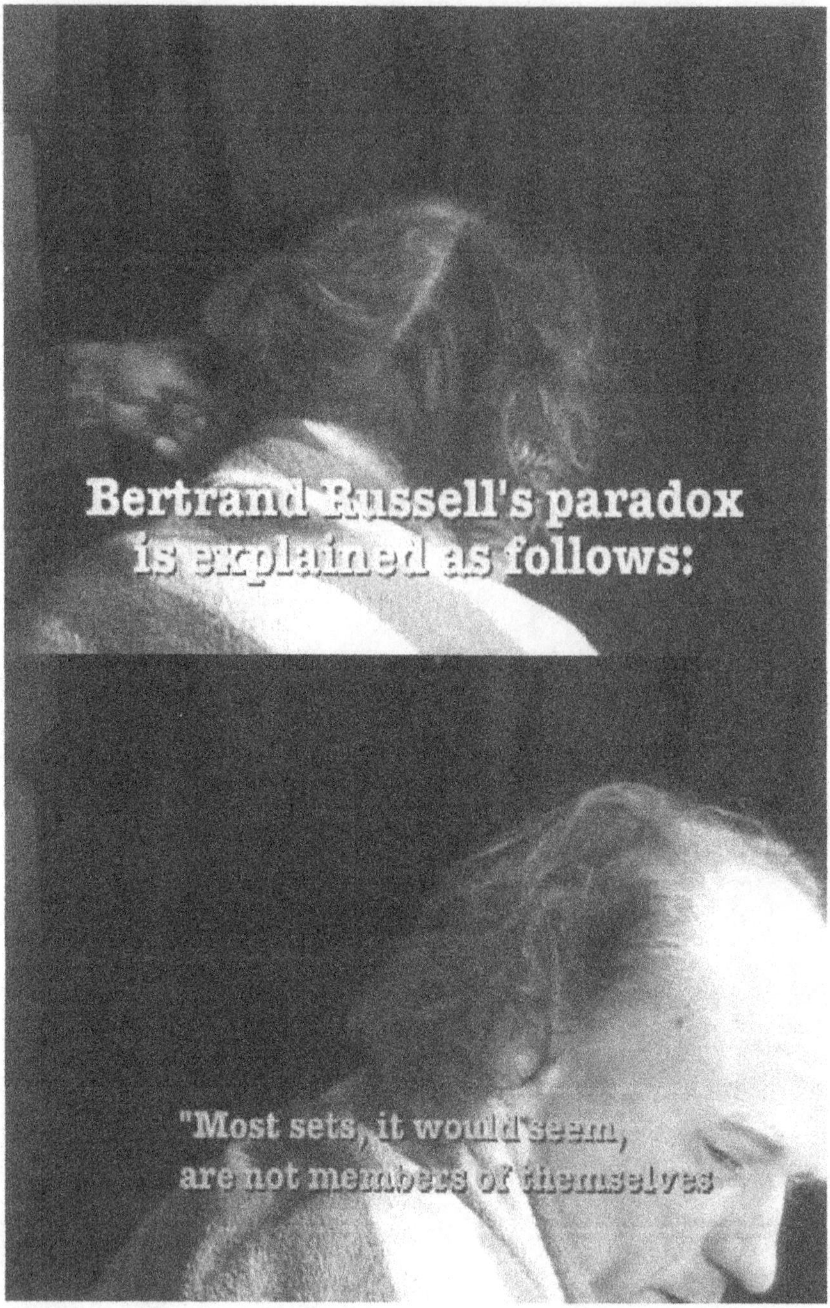

Bertrand Russell's paradox
is explained as follows:

"Most sets, it would seem,
are not members of themselves

- for example, the set of walruses is not a walrus,

the set containing only Joan of Arc is not Joan of Arc

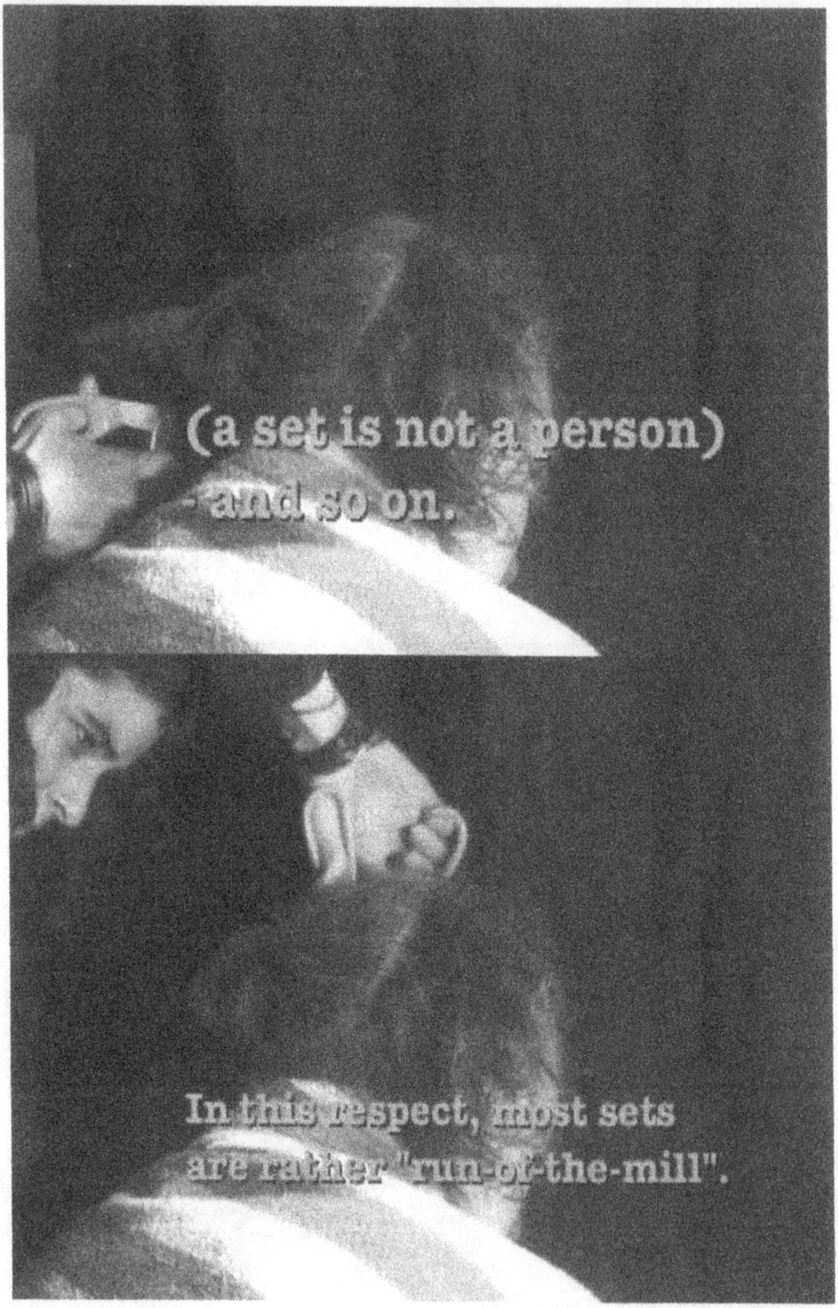

(a set is not a person)
- and so on.

In this respect, most sets
are rather "run-of-the-mill".

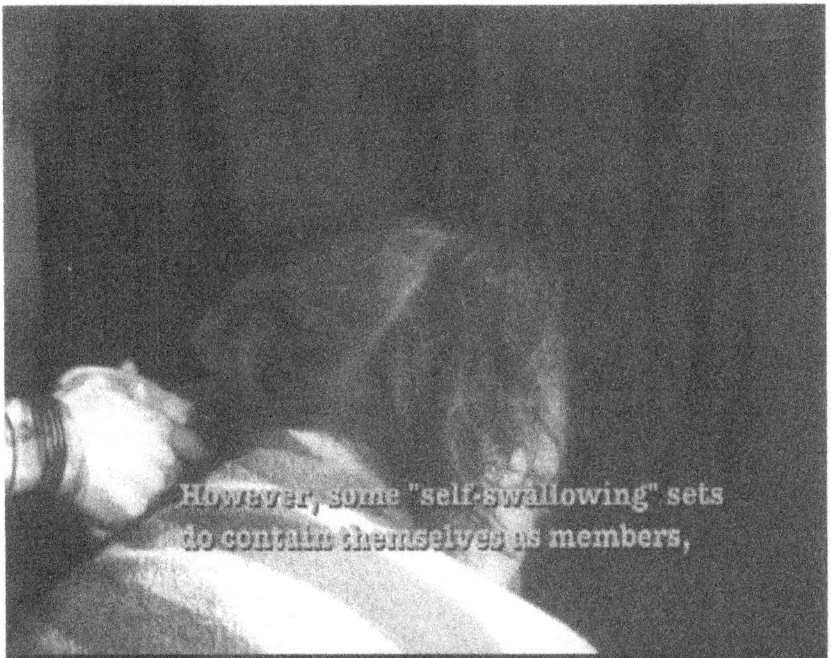

However, some "self-swallowing" sets
do contain themselves as members,

such as the set of all sets,
or the set of all things except

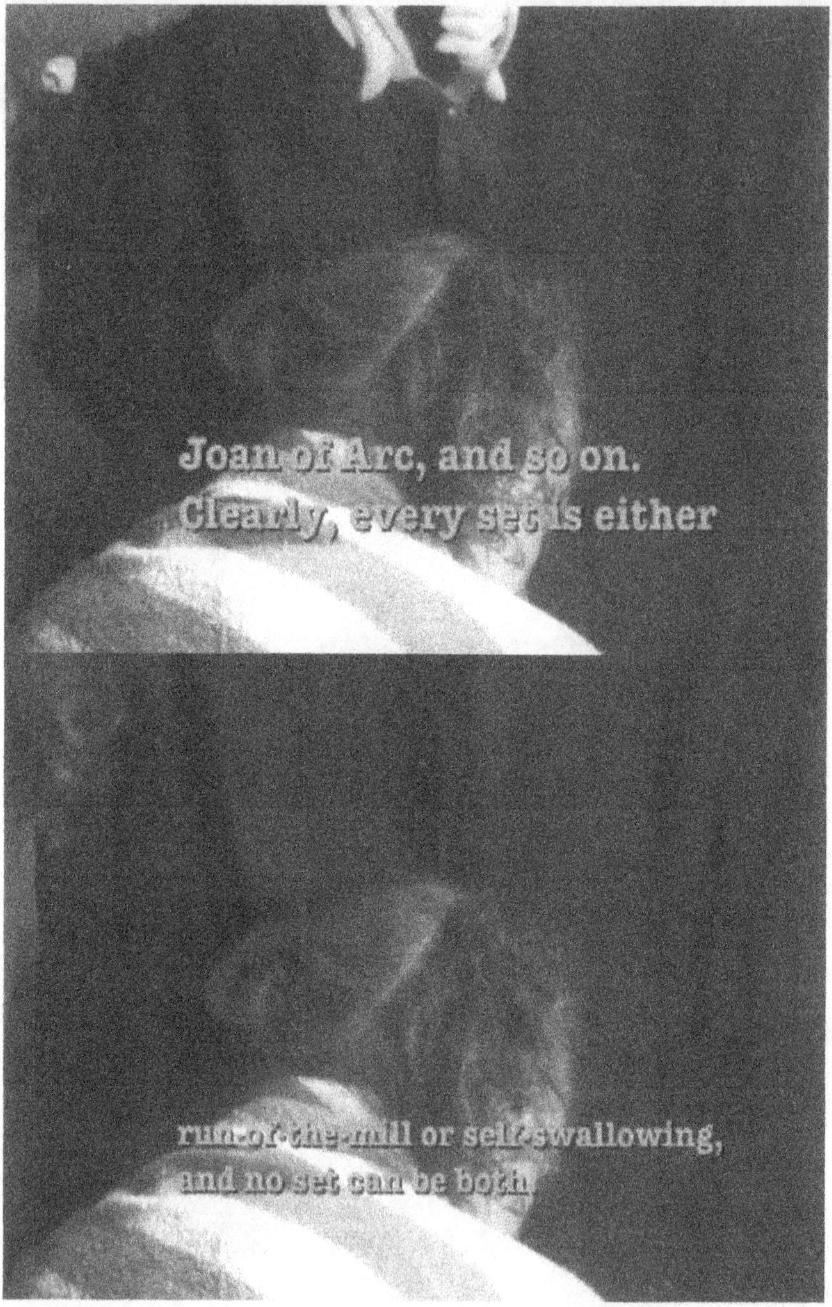

Joan of Arc, and so on.
Clearly, every set is either

run-of-the-mill or self-swallowing,
and no set can be both.

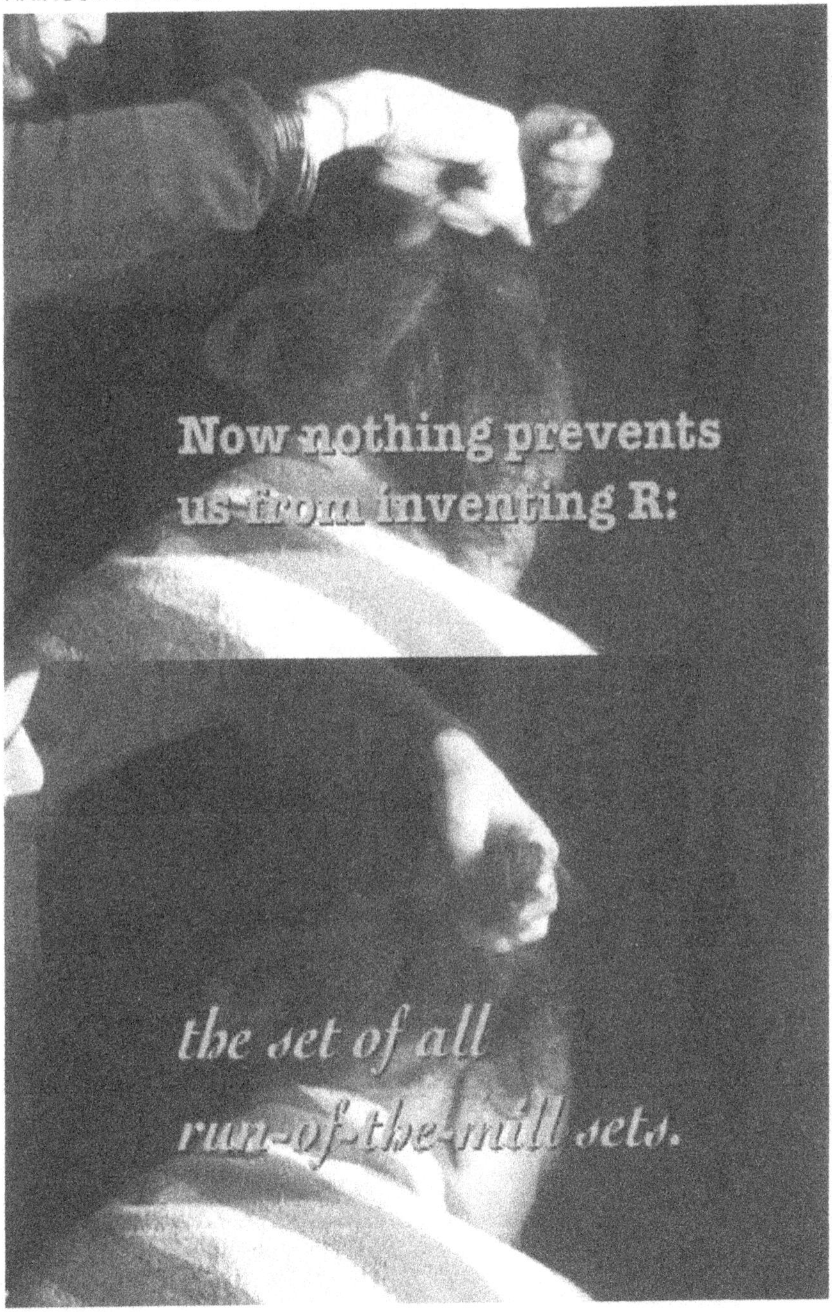

Now nothing prevents us from inventing R:

the set of all run-of-the-mill sets.

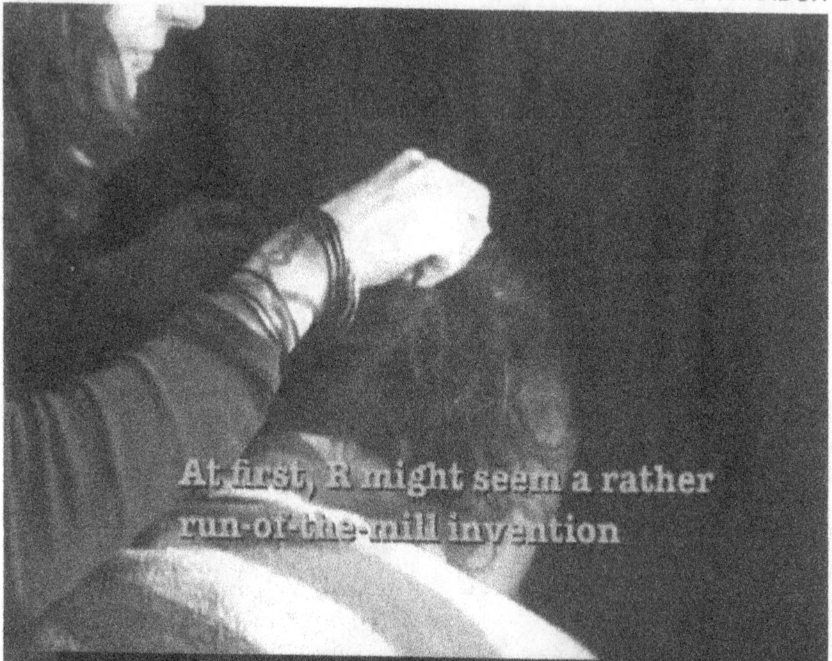

At first, R might seem a rather run-of-the-mill invention

- but that opinion must be revised when you ask yourself,

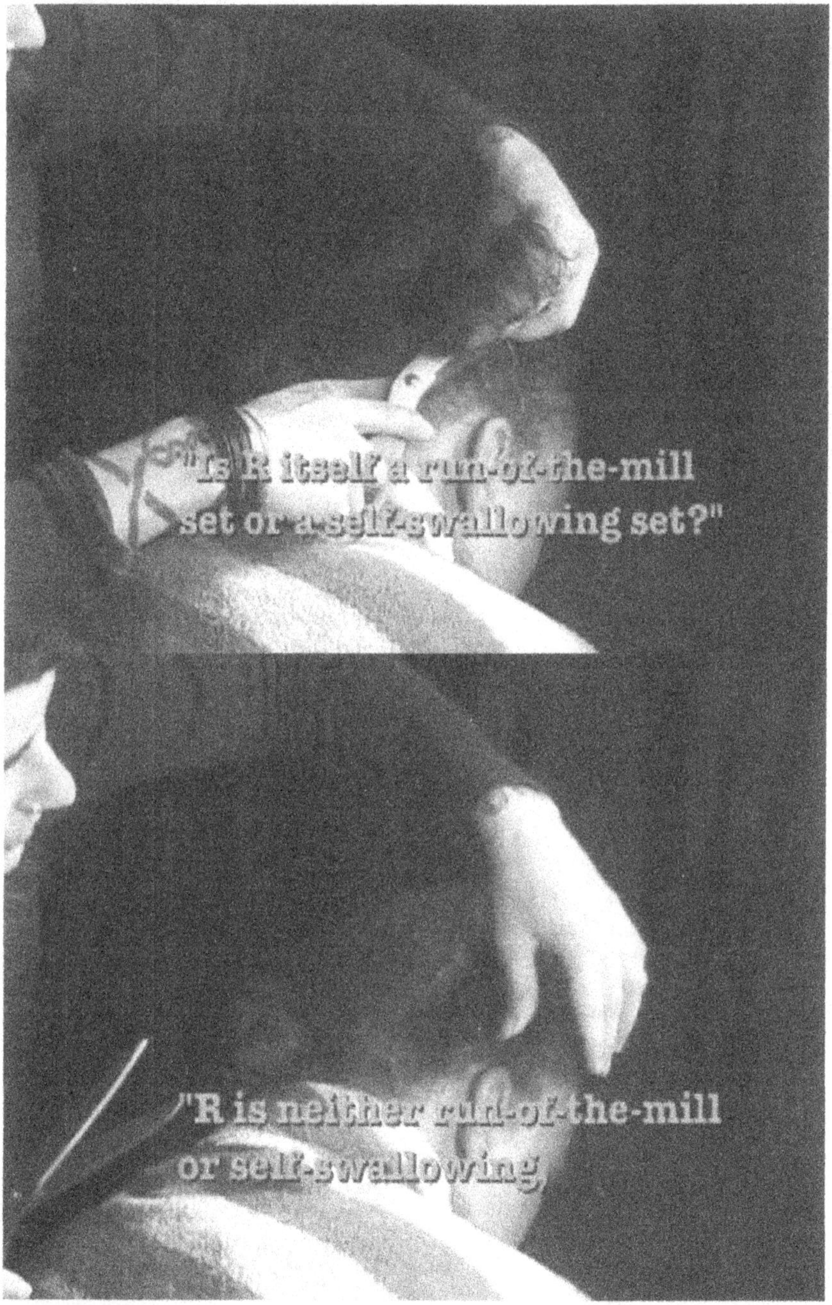

"Is R itself a run-of-the-mill set or a self-swallowing set?"

"R is neither run-of-the-mill or self-swallowing,

for either choice
leads to paradox.""

Russell himself is quoted in Paul
Hoffman's book about Paul Erdös,

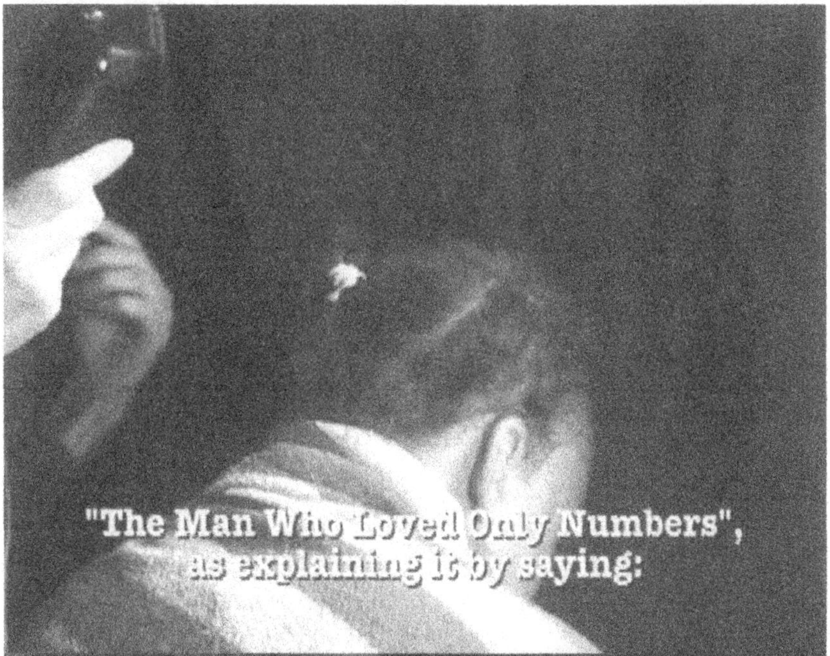

"The Man Who Loved Only Numbers",
as explaining it by saying:
"that a class sometimes is, and 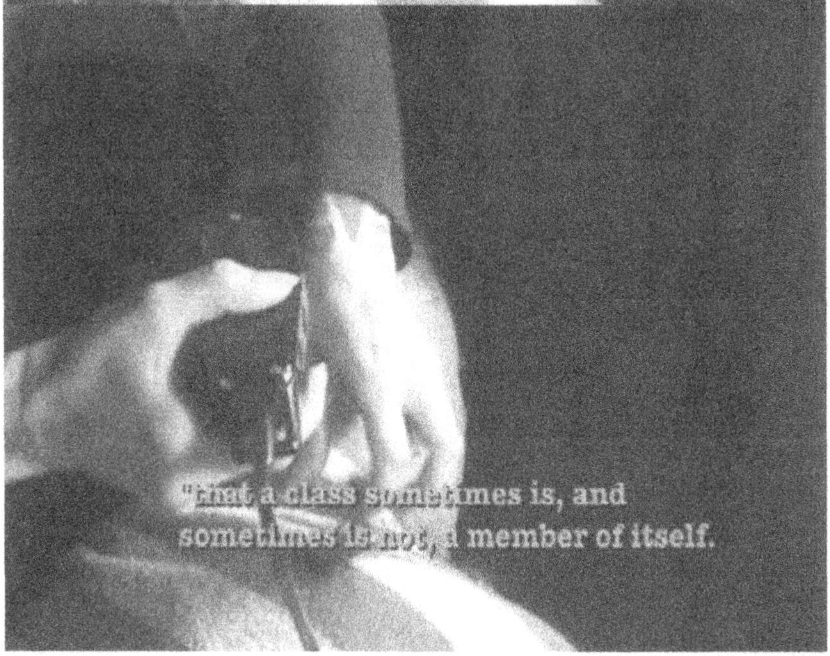
sometimes is not, a member of itself.

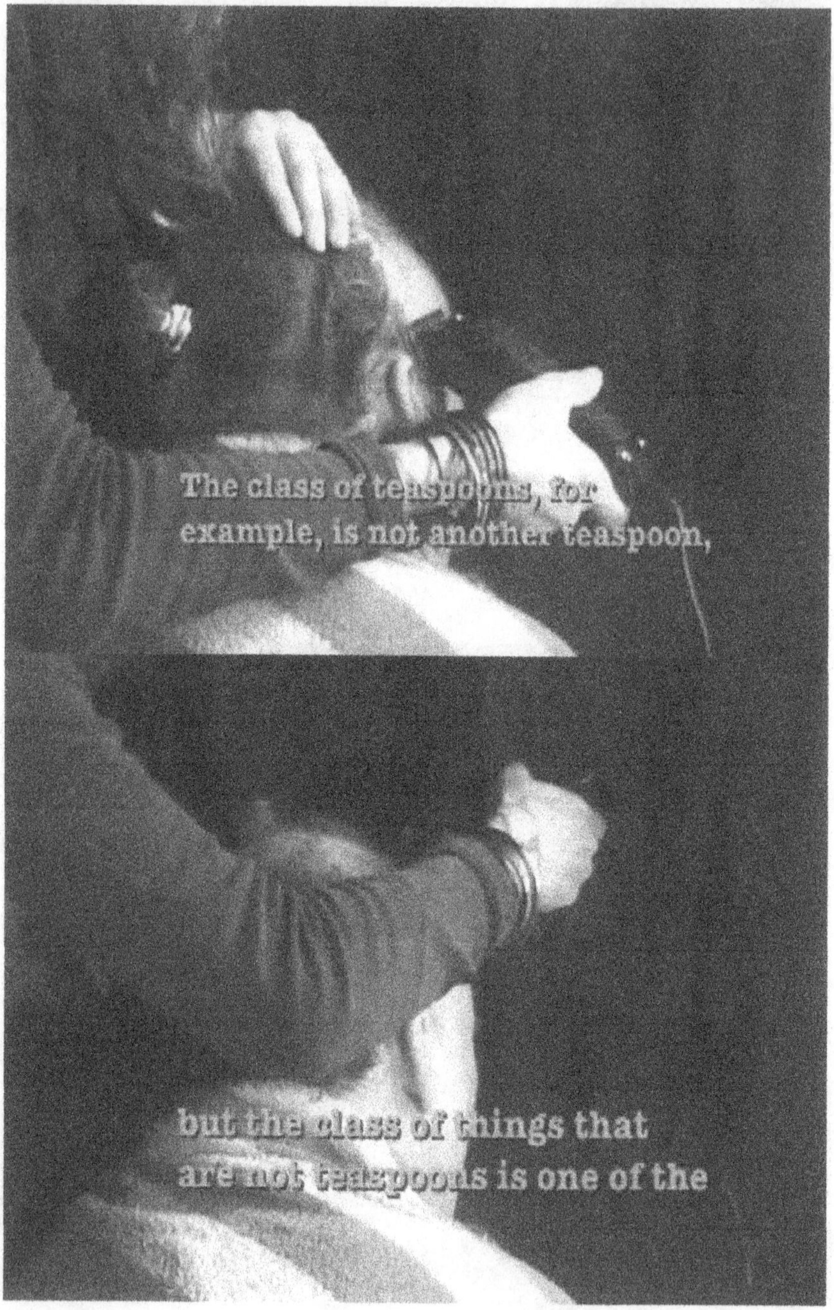

The class of teaspoons, for
example, is not another teaspoon,

but the class of things that
are not teaspoons is one of the

things that are not teaspoons.
There seemed to be instances

that are not negative: for example,
the class of all classes is a class....

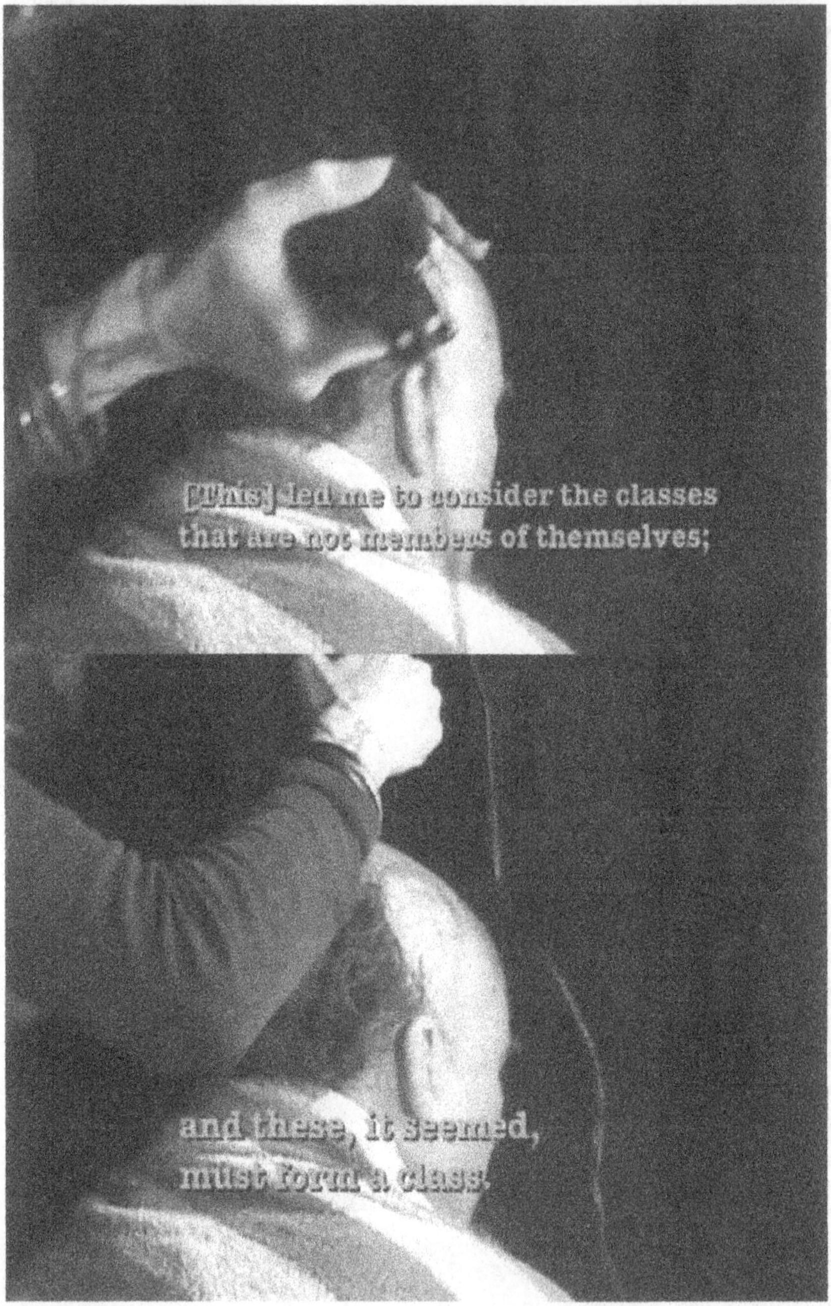

[This] led me to consider the classes
that are not members of themselves;

and these, it seemed,
must form a class.

I asked myself whether this class
is a member of itself or not.

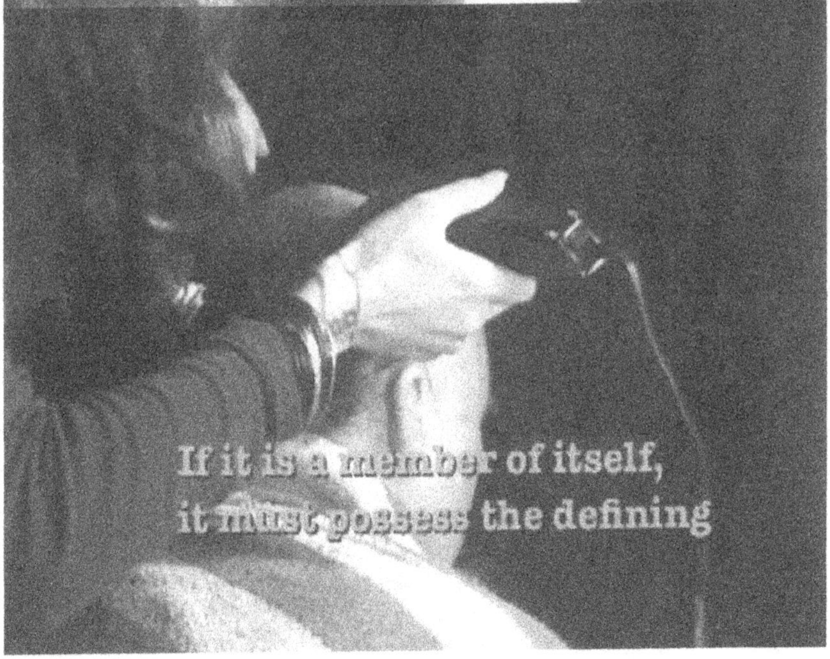

If it is a member of itself,
it must possess the defining

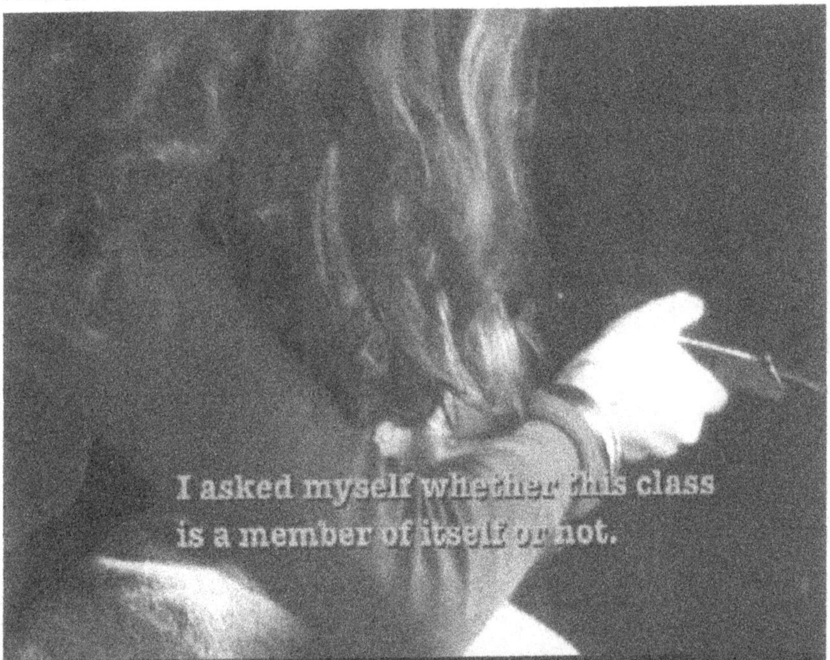

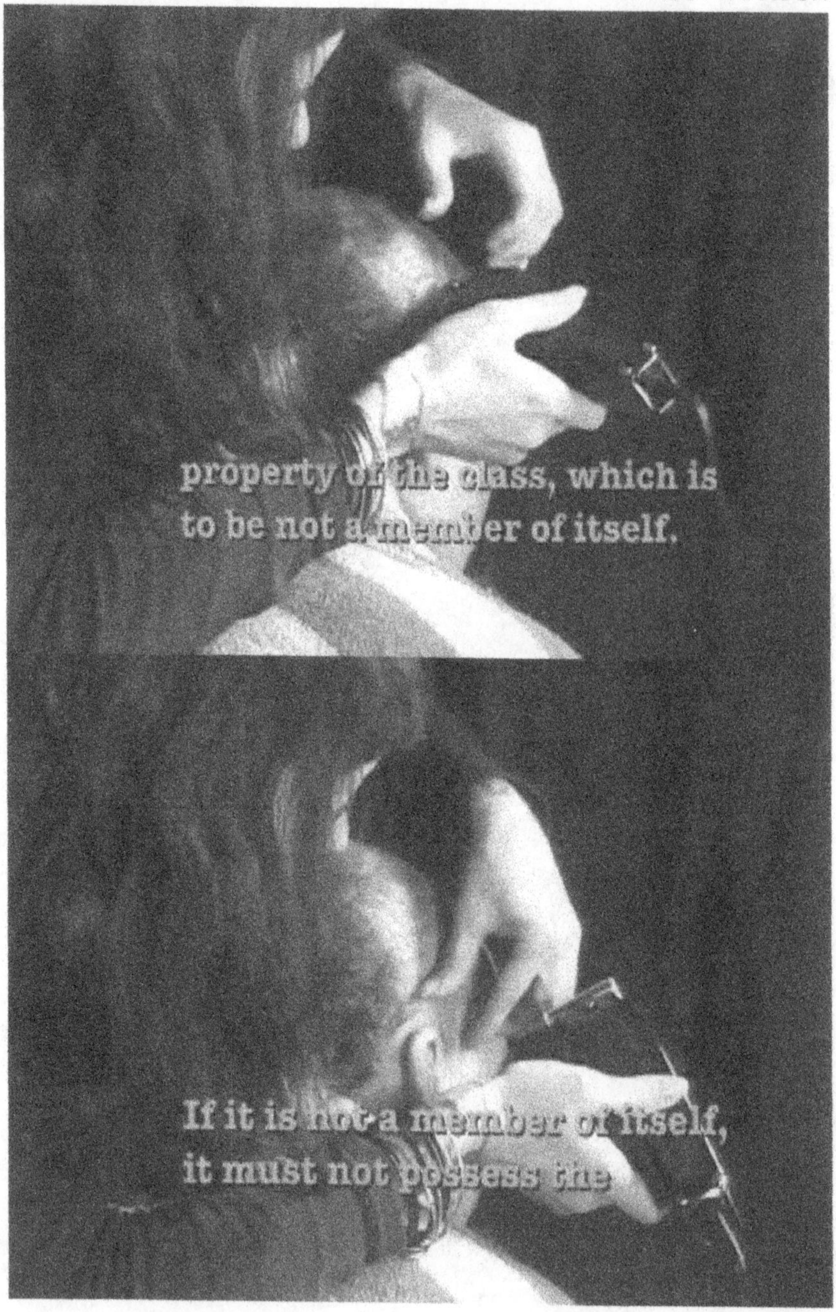

property of the class, which is
to be not a member of itself.

If it is not a member of itself,
it must not possess the

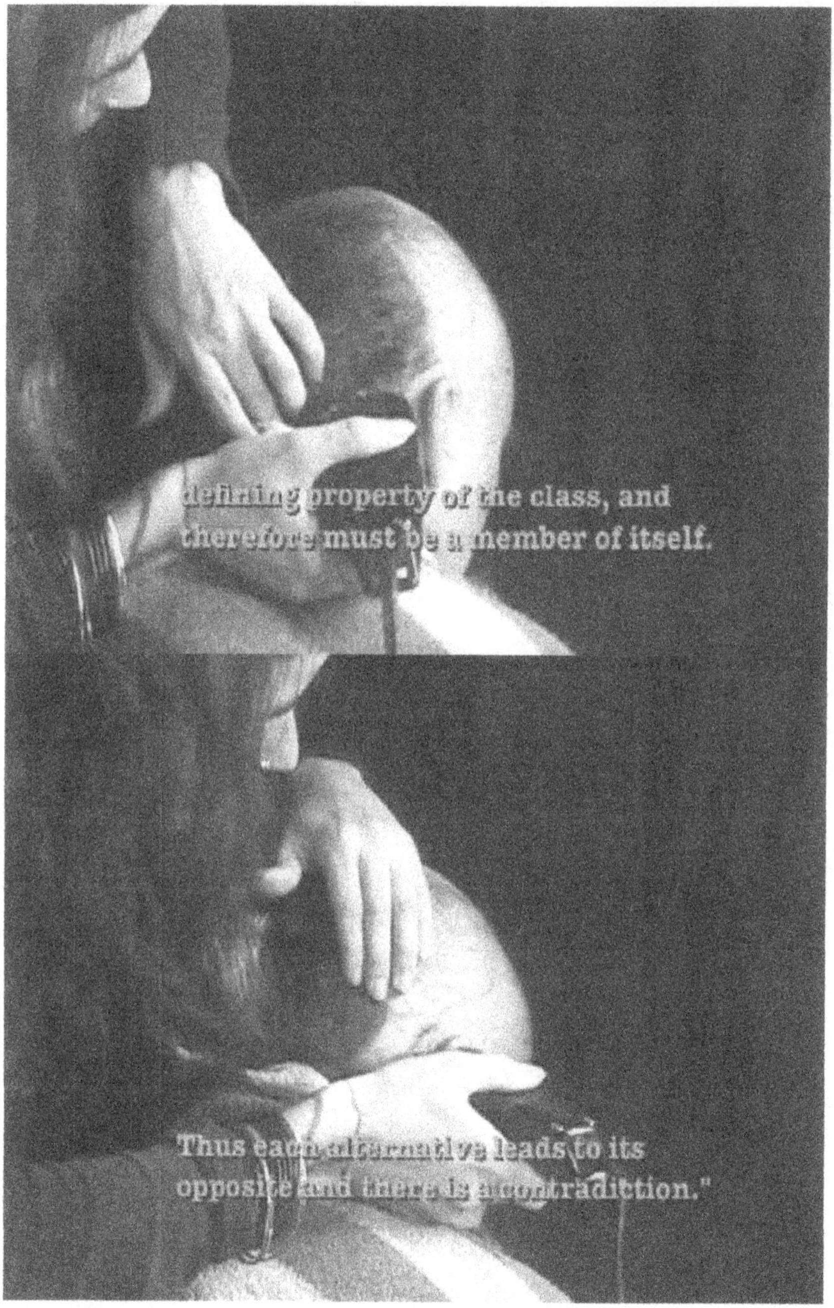

defining property of the class, and
therefore must be a member of itself.

Thus each alternative leads to its
opposite and there is a contradiction."

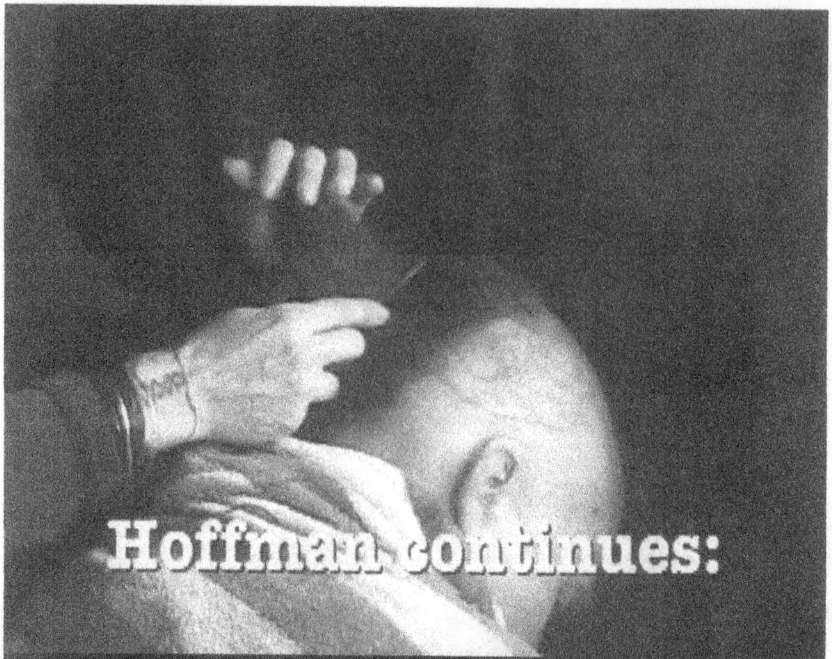

Hoffman continues:

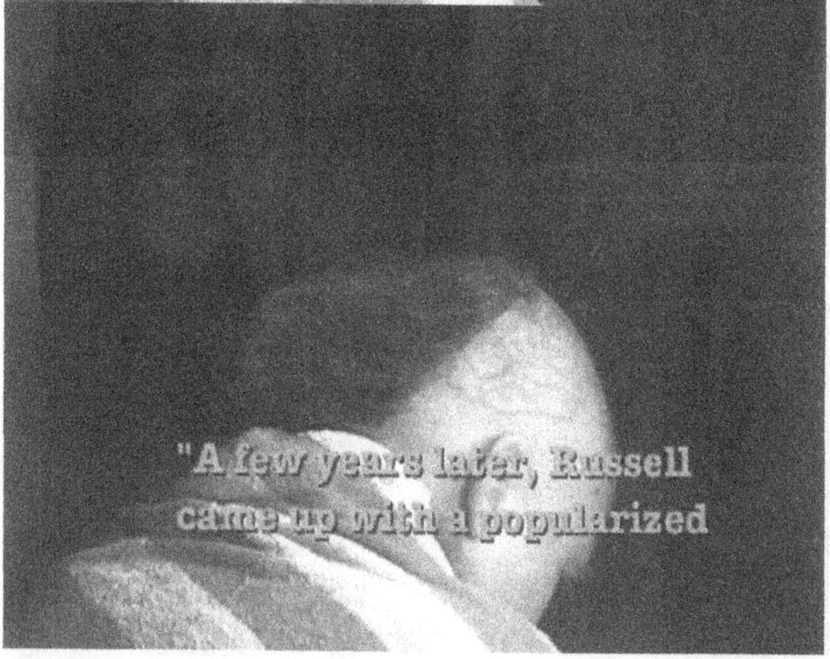

"A few years later, Russell came up with a popularized

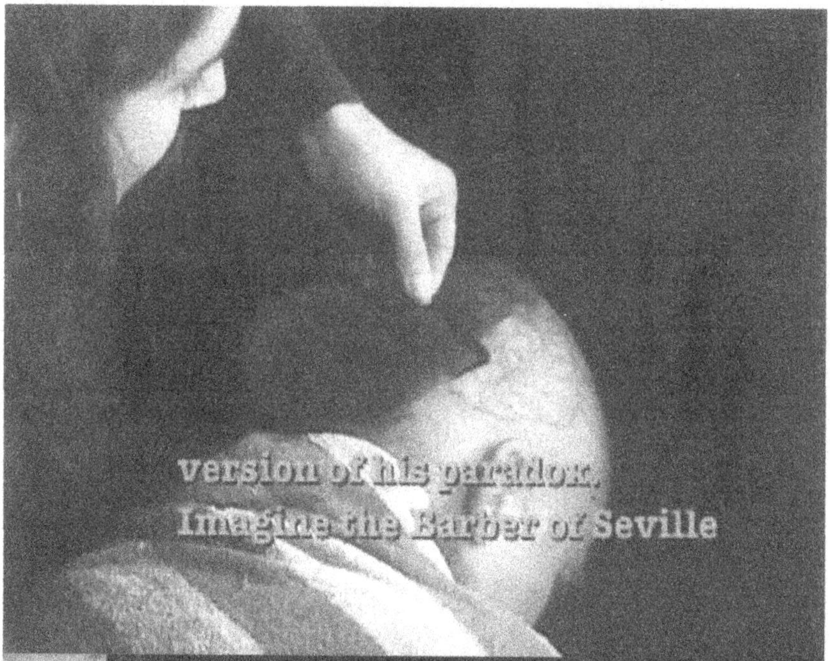

HAIRCUT PARADOX

250 superreal numbers

version of his paradox.
Imagine the Barber of Seville

who shaves every man
who does not shave himself.

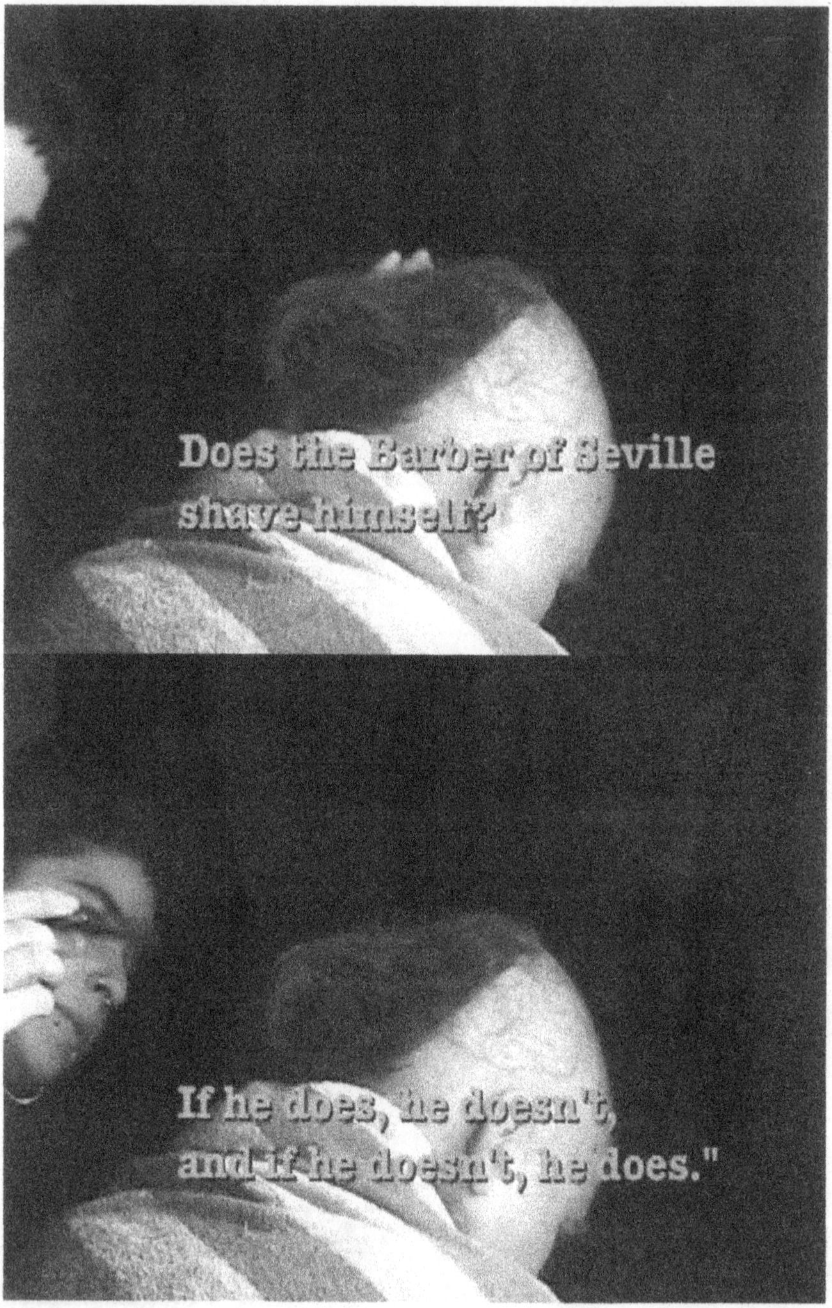

Does the Barber of Seville
shave himself?

If he does, he doesn't,
and if he doesn't, he does."

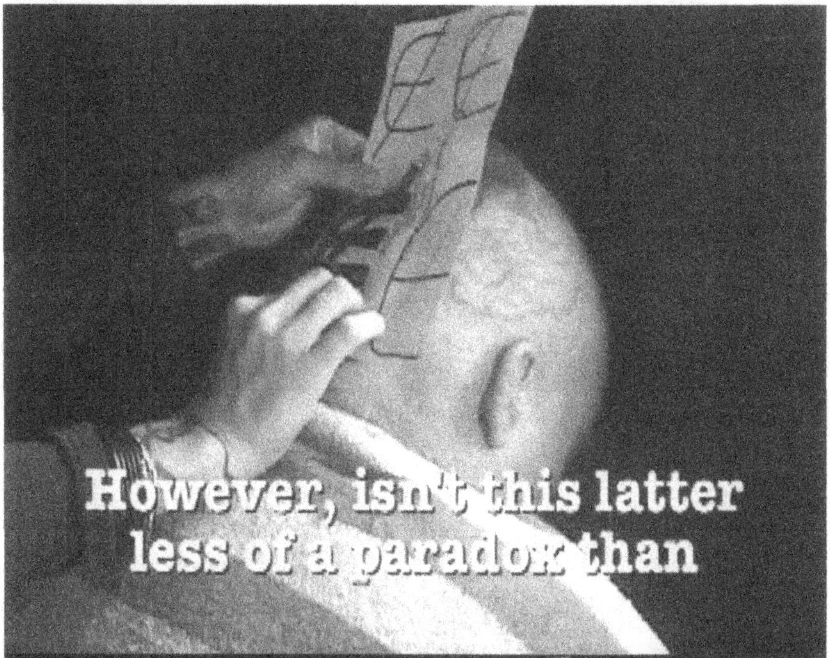

However, isn't this latter
less of a paradox than

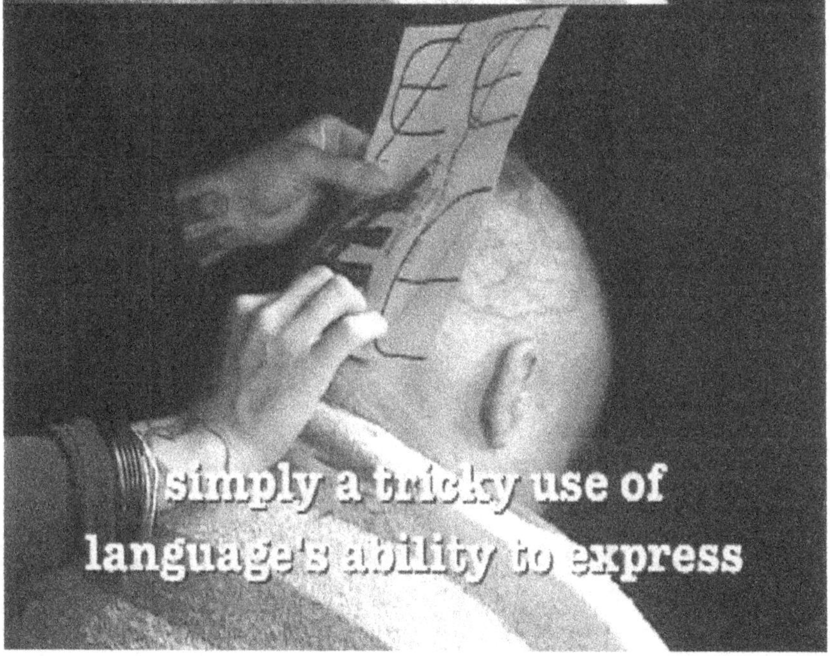

simply a tricky use of
language's ability to express

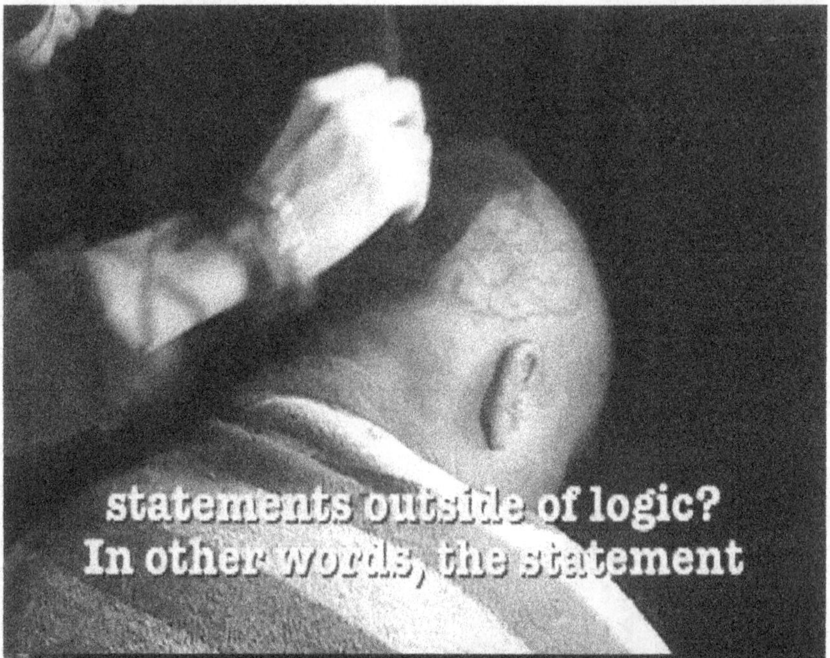

statements outside of logic?
In other words, the statement

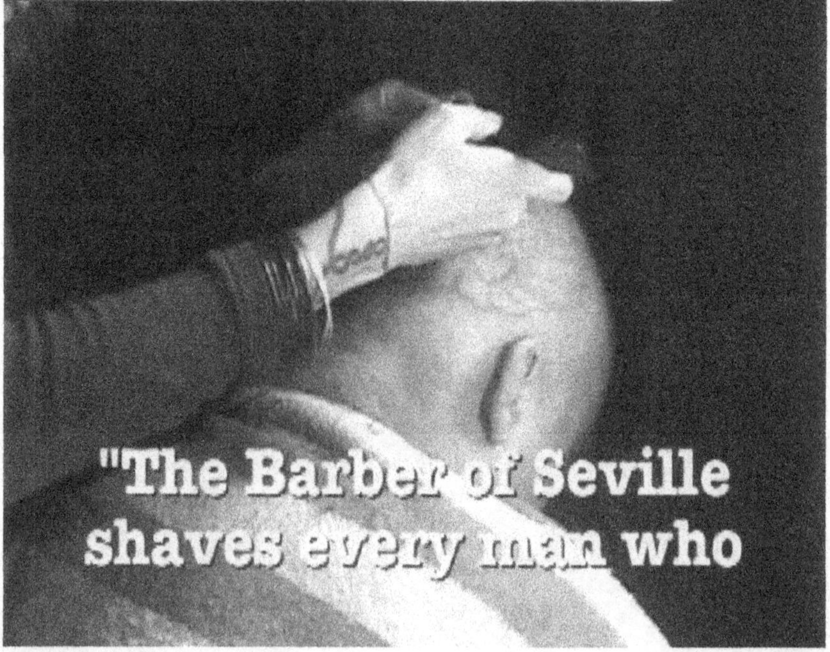

"The Barber of Seville
shaves every man who

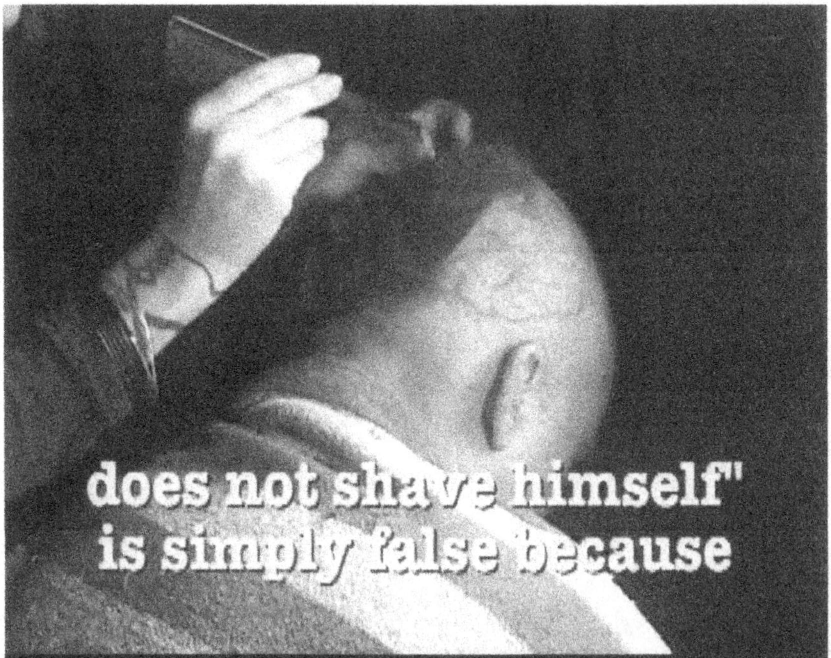

does not shave himself"
is simply false because

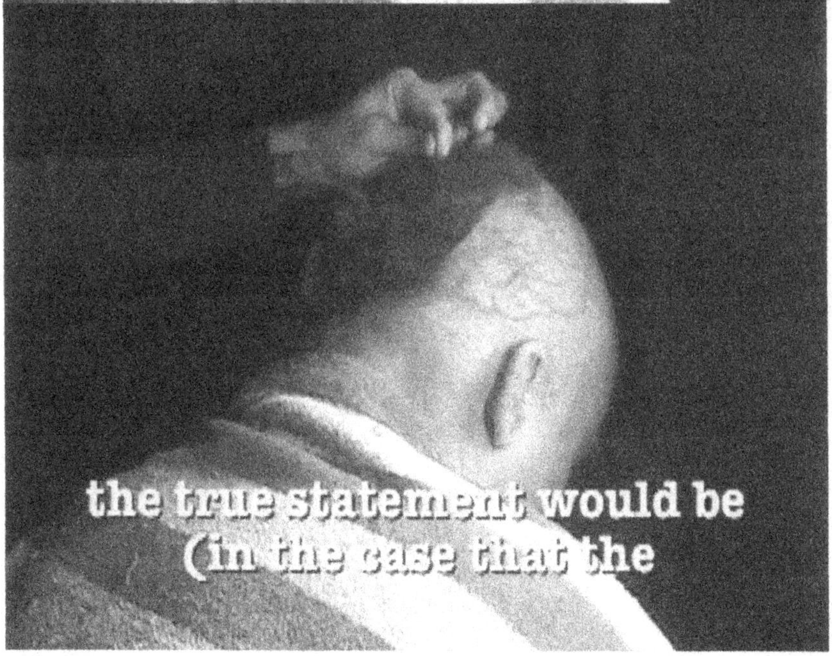

the true statement would be
(in the case that the

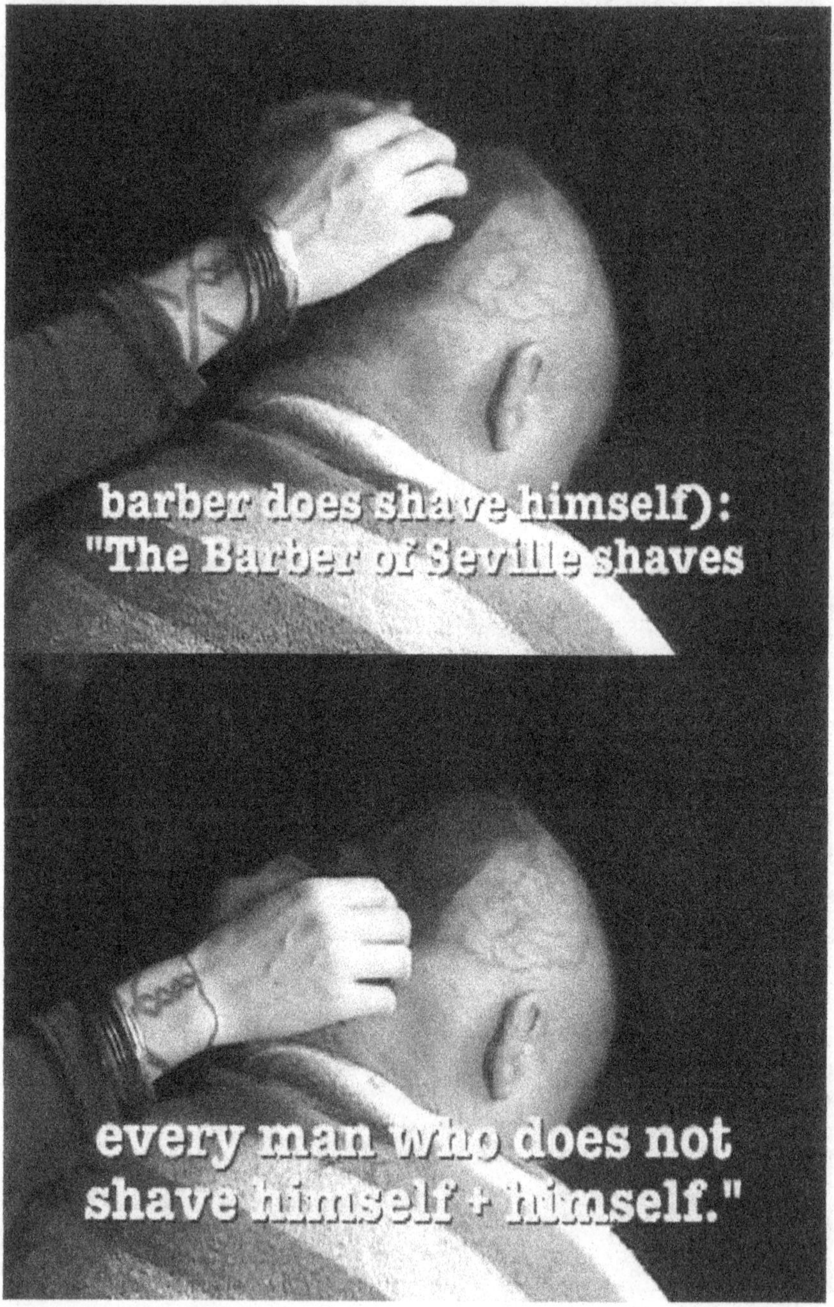

barber does shave himself):
"The Barber of Seville shaves

every man who does not
shave himself + himself."

Now, take this haircut:
hair is removed from the
head to leave hair only
in the shape of the symbol

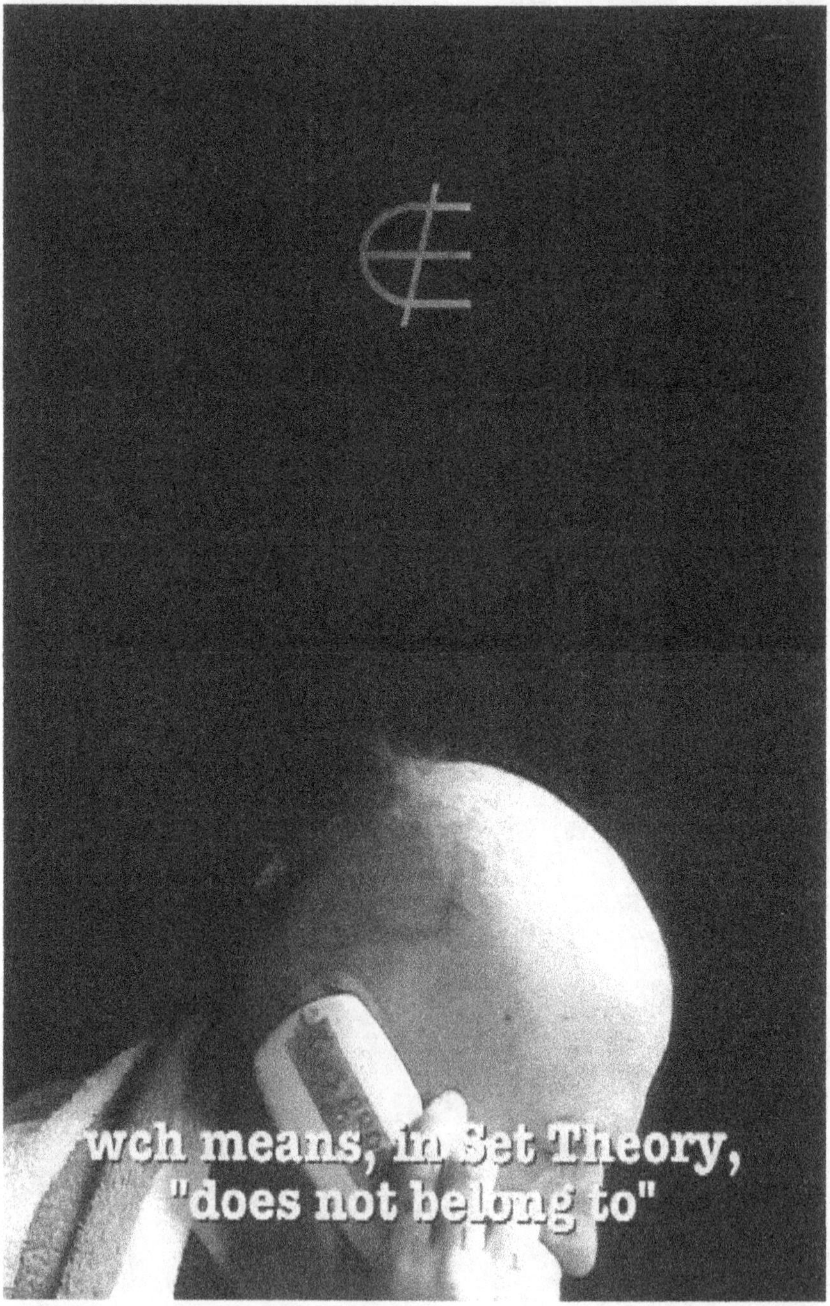

$\notin$

wch means, in Set Theory,
"does not belong to"

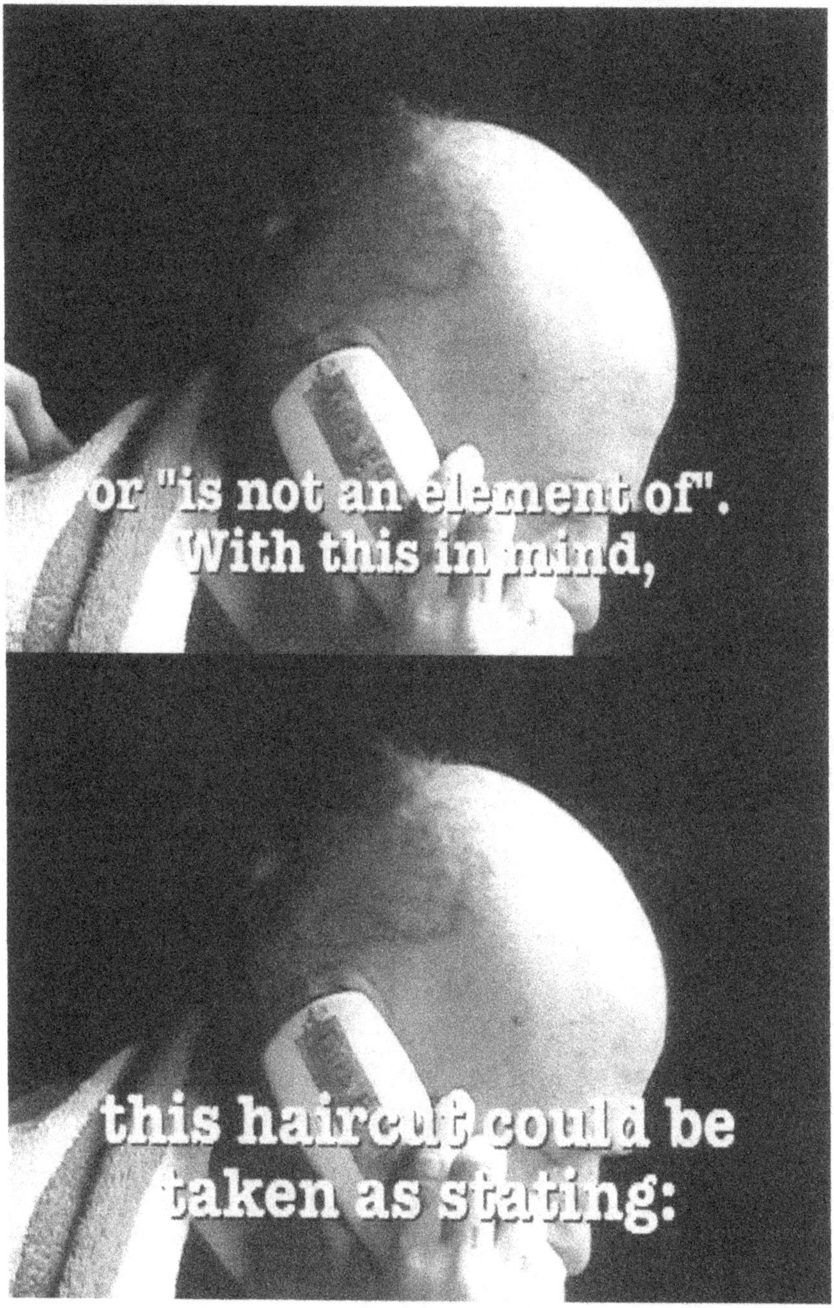

or "is not an element of".
With this in mind,

this haircut could be
taken as stating:

The hair that

makes this symbol

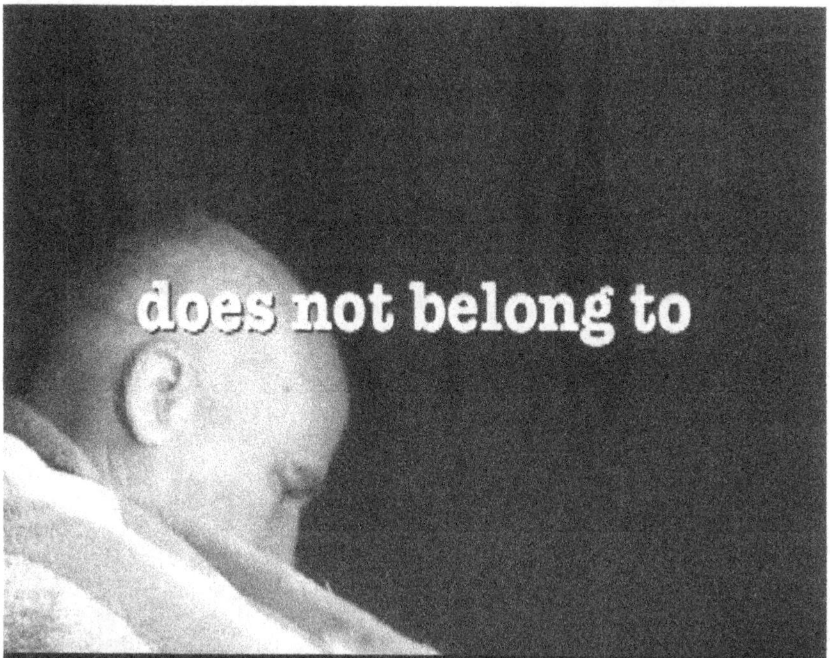

does not belong to

the set of hair

that was removed

to make this haircut.

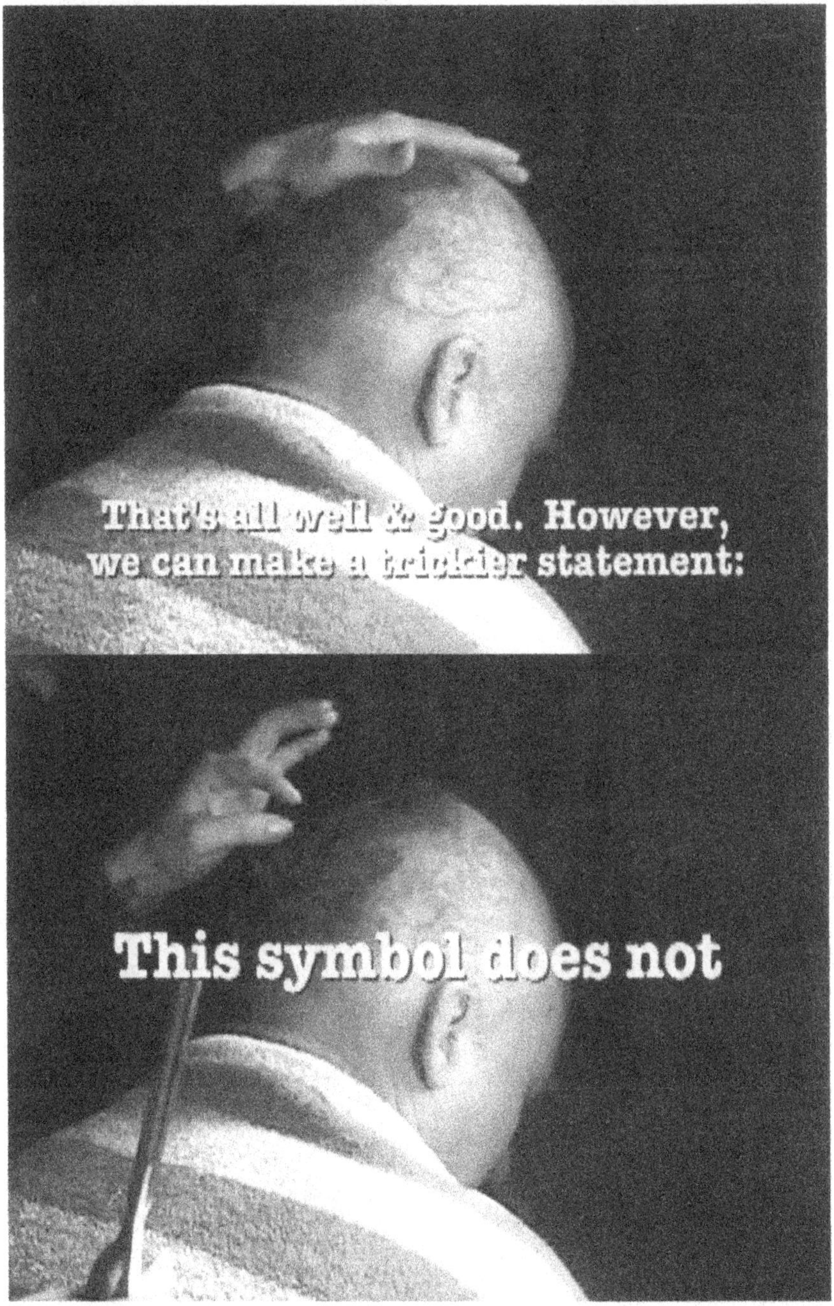

That's all well & good. However,
we can make a trickier statement:

This symbol does not

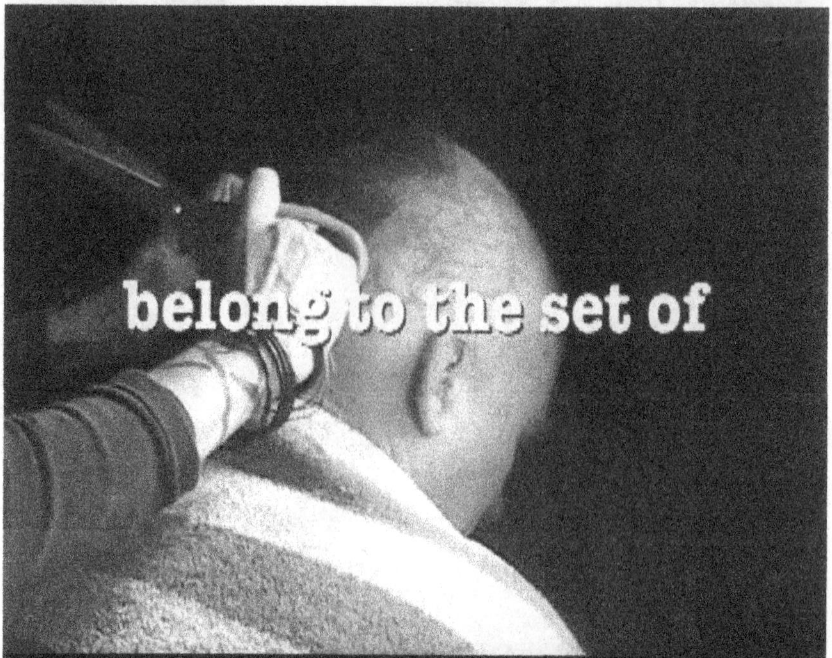

belong to the set of

hair that was used

to make this haircut.

If this is true, then possibly
the symbol doesn't exist here

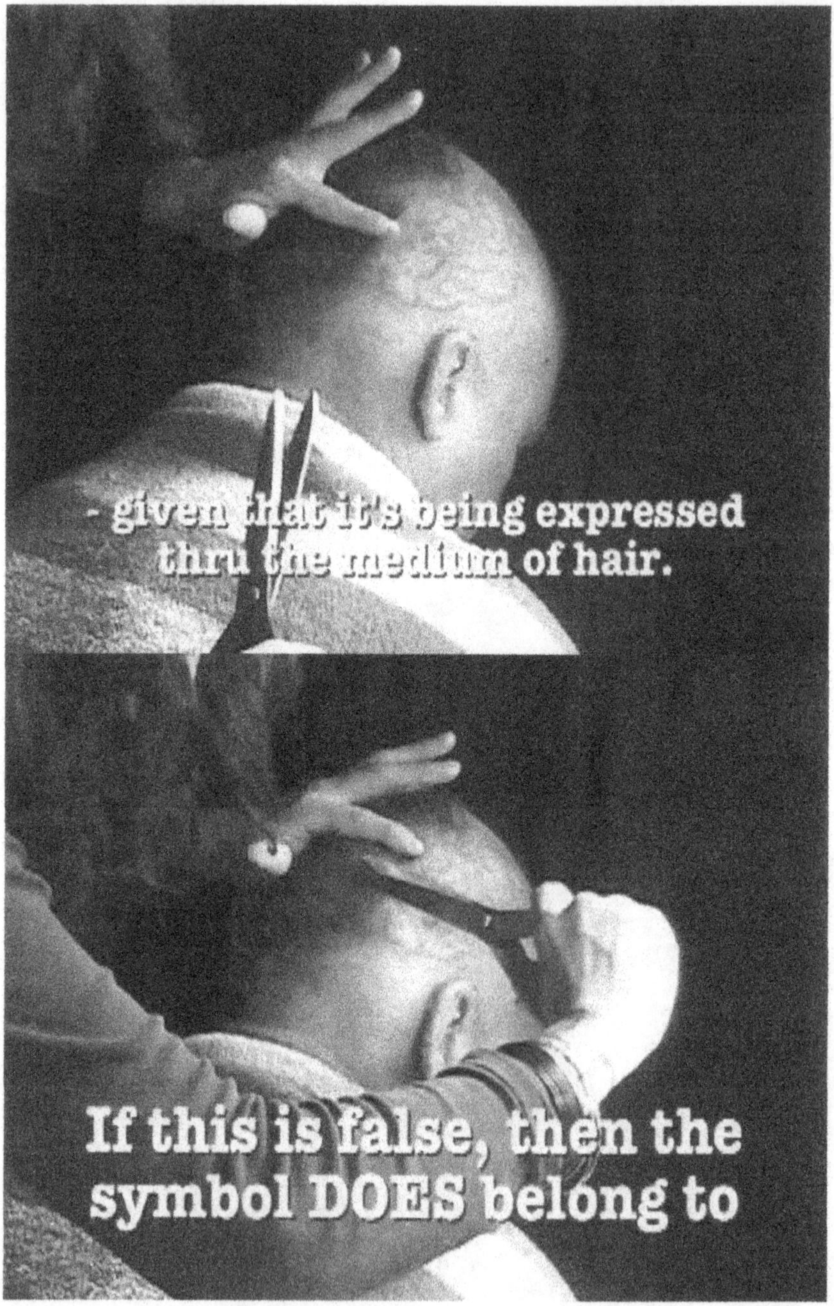

- given that it's being expressed thru the medium of hair.

If this is false, then the symbol DOES belong to

to make this haircut.

If this is true, then possibly
the symbol doesn't exist here

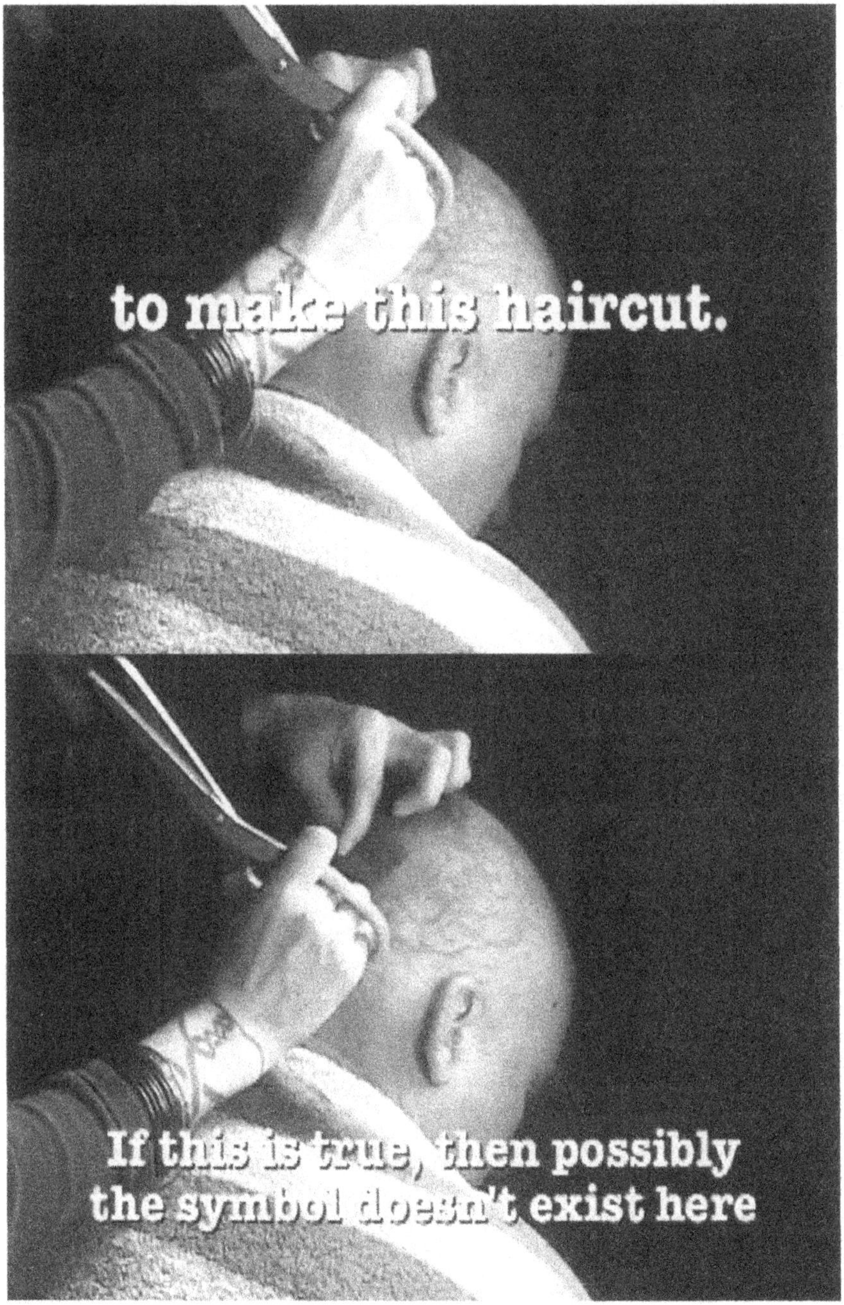

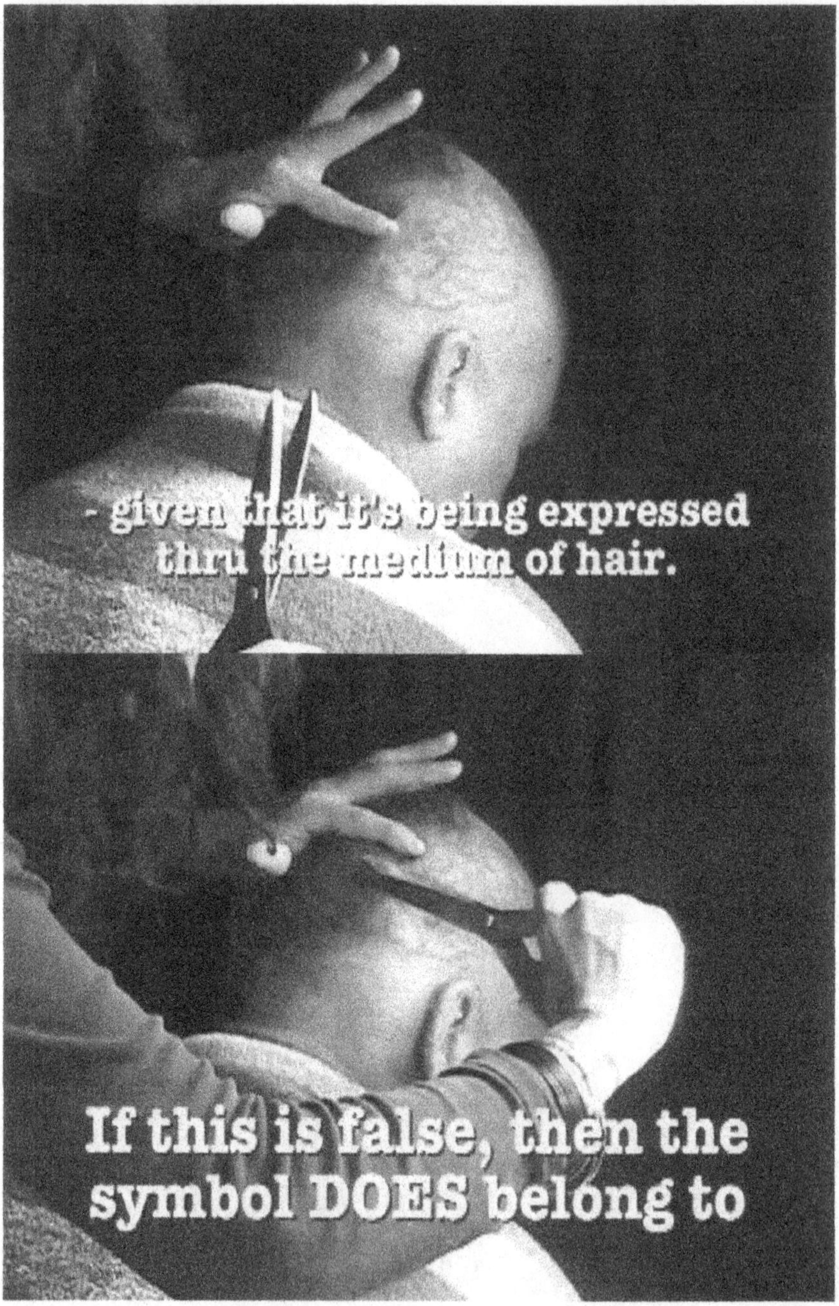

- given that it's being expressed thru the medium of hair.

If this is false, then the symbol DOES belong to

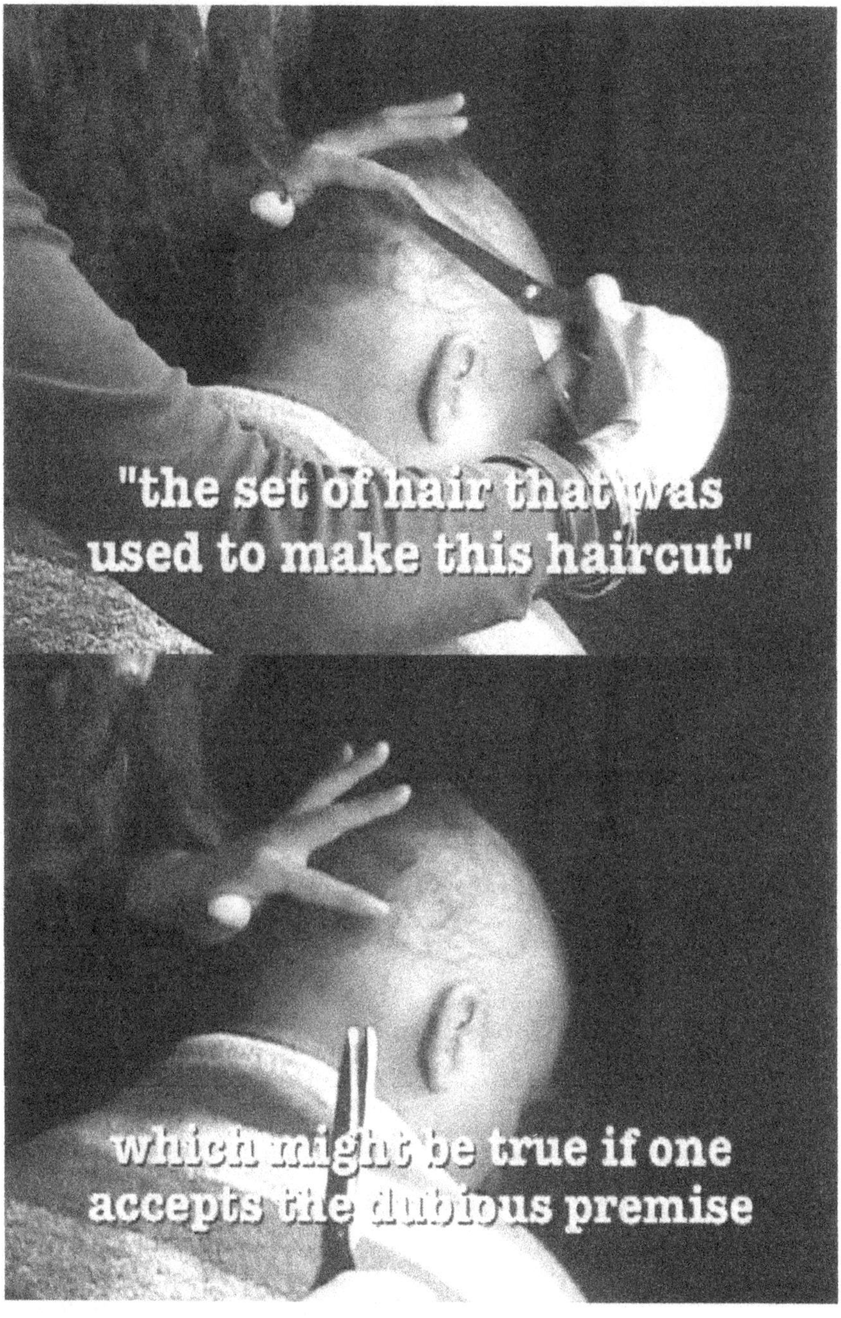

"the set of hair that was used to make this haircut"

which might be true if one accepts the dubious premise

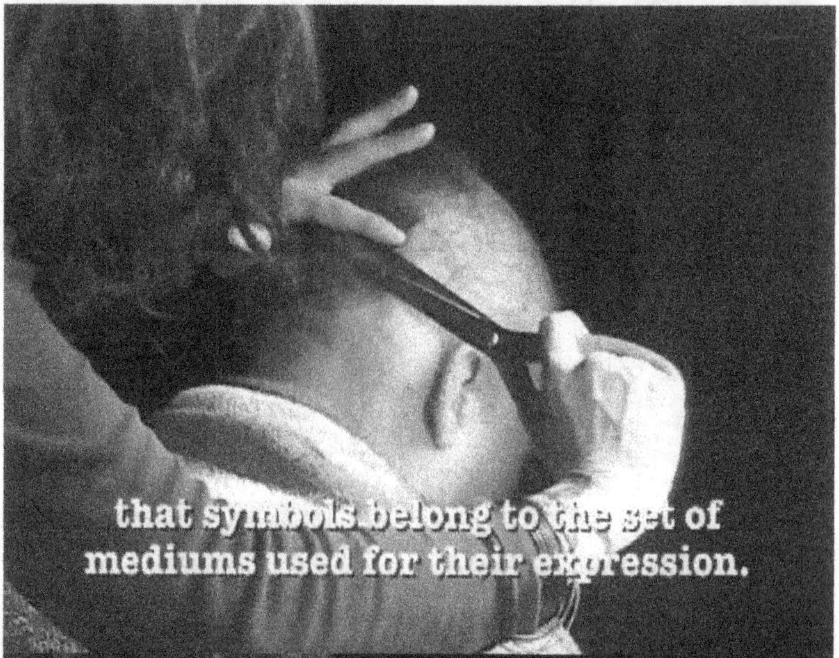

that symbols belong to the set of
mediums used for their expression.

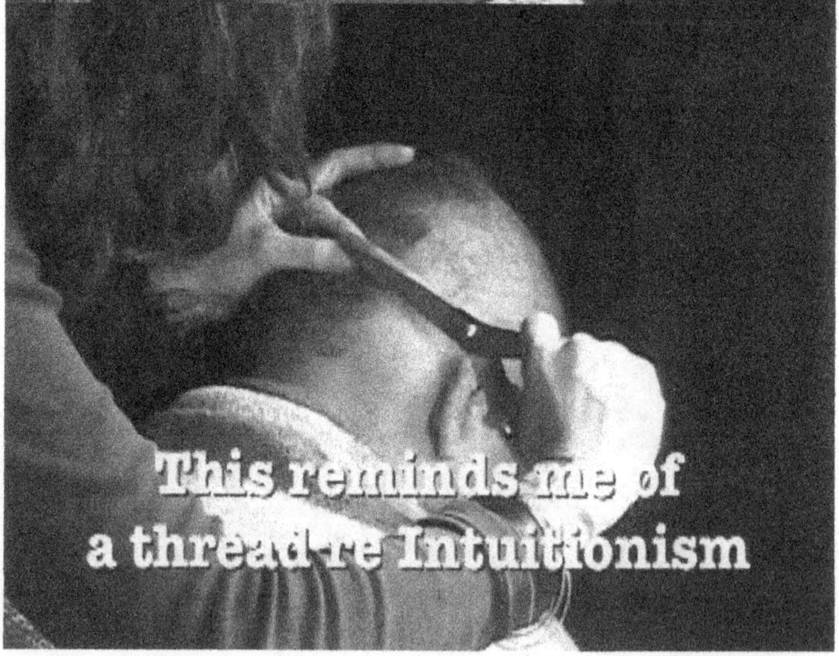

This reminds me of
a thread re Intuitionism

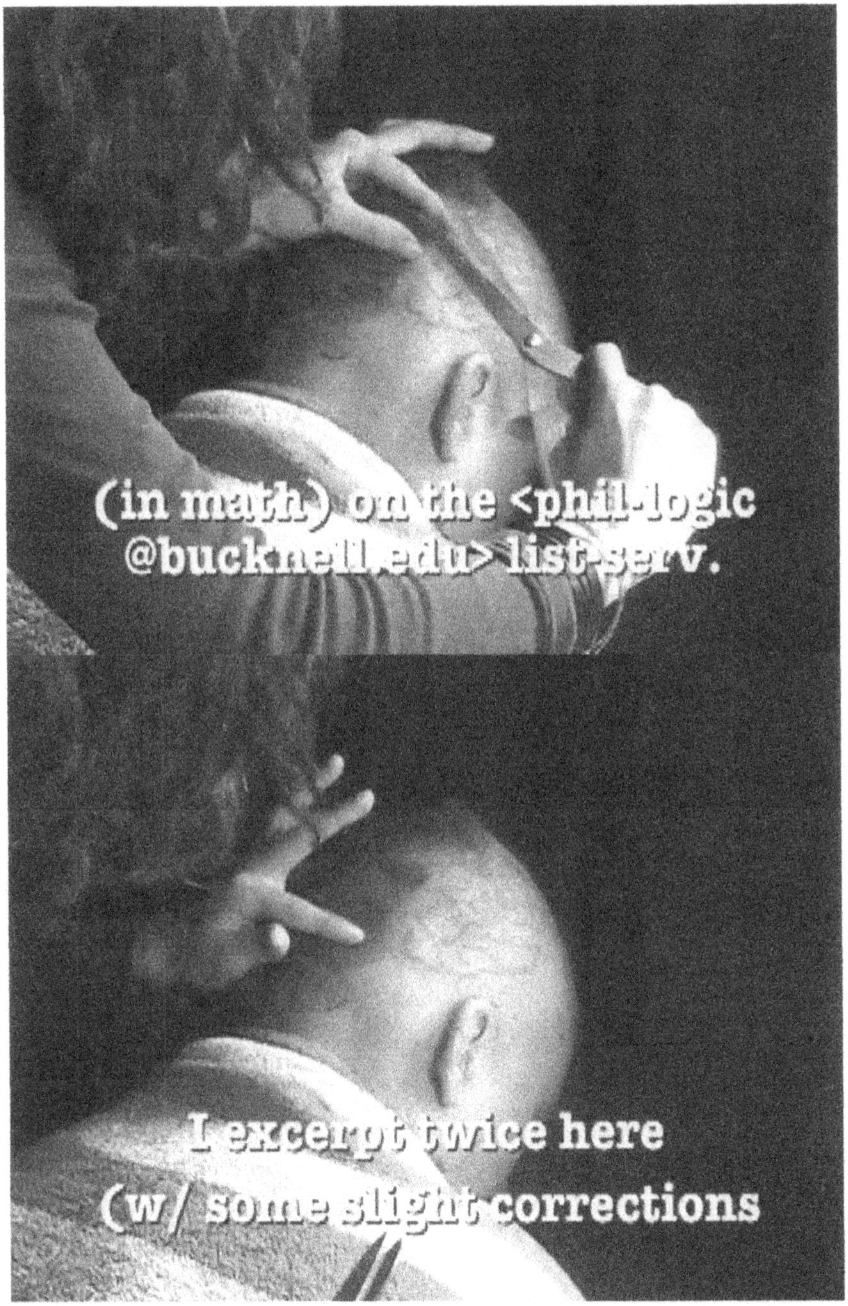

(in math) on the <phil-logic
@bucknell.edu> list-serv.

I excerpt twice here
(w/ some slight corrections

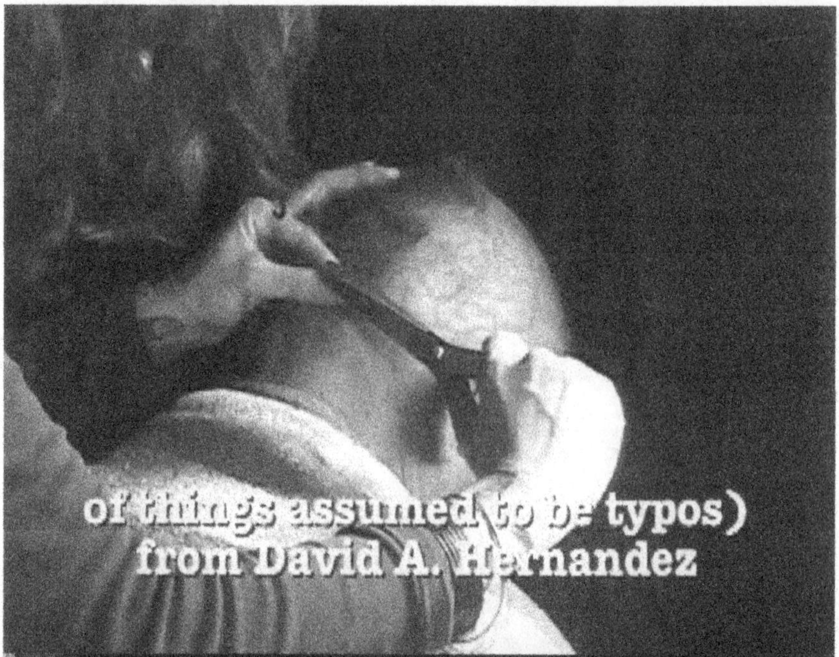

of things assumed to be typos)
from David A. Hernandez

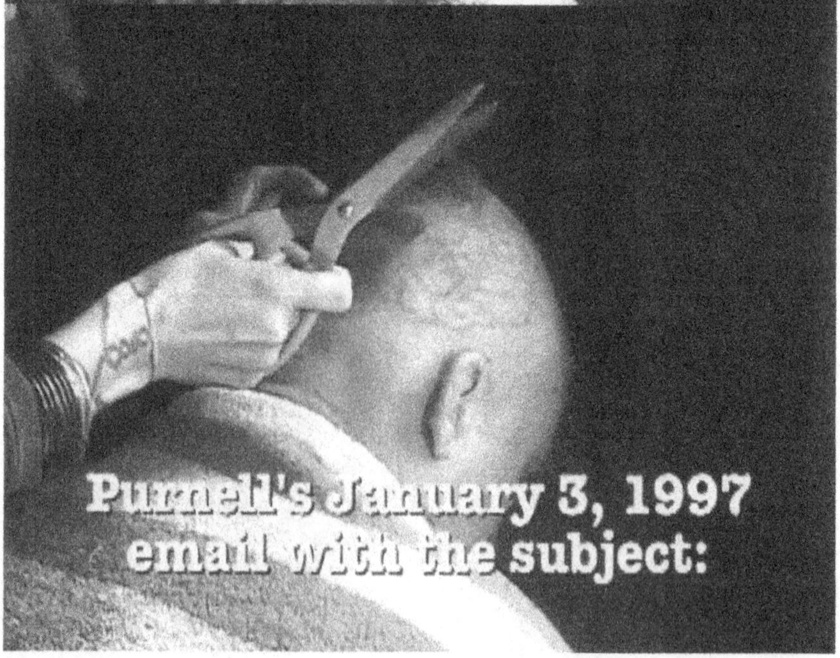

Purnell's January 3, 1997
email with the subject:

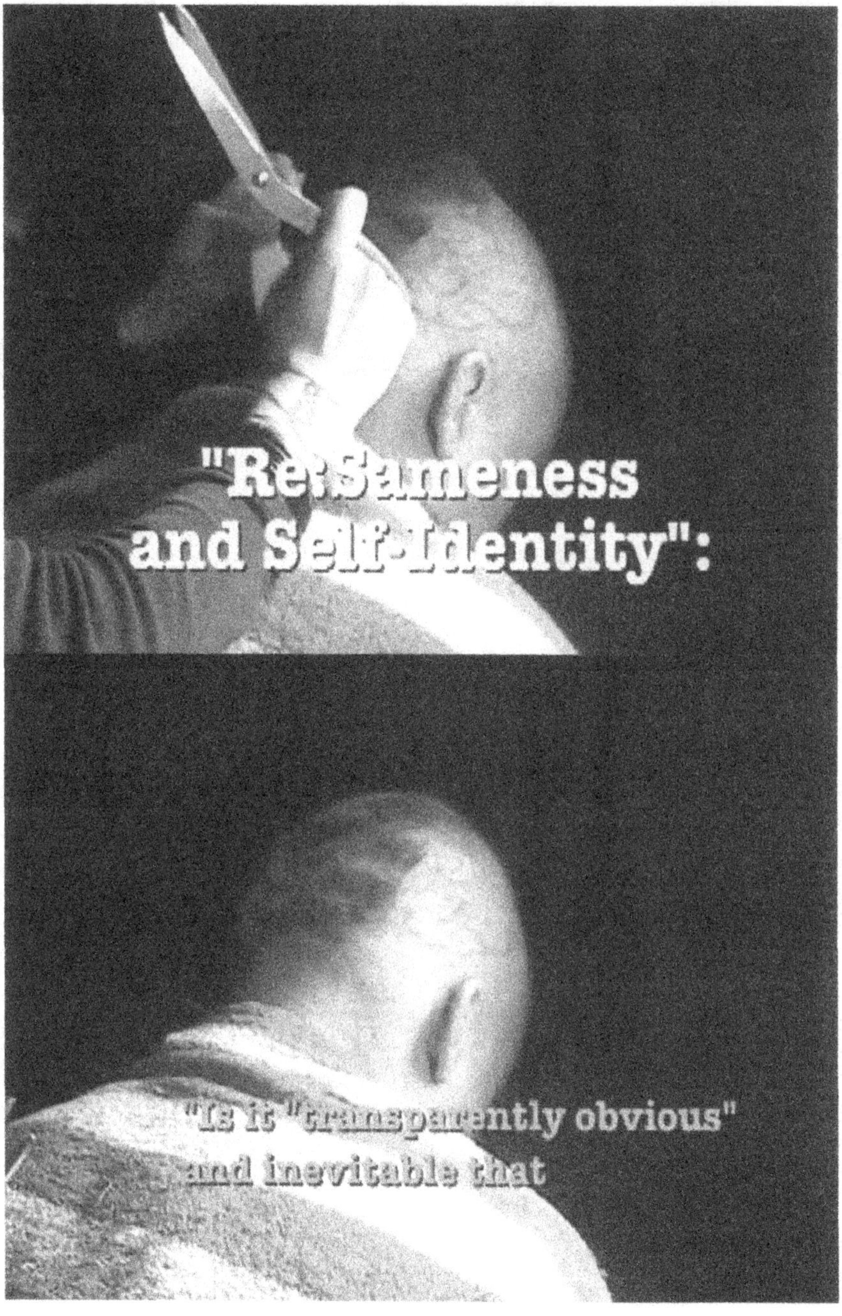

"Re: Sameness and Self-Identity":

"Is it "transparently obvious" and inevitable that

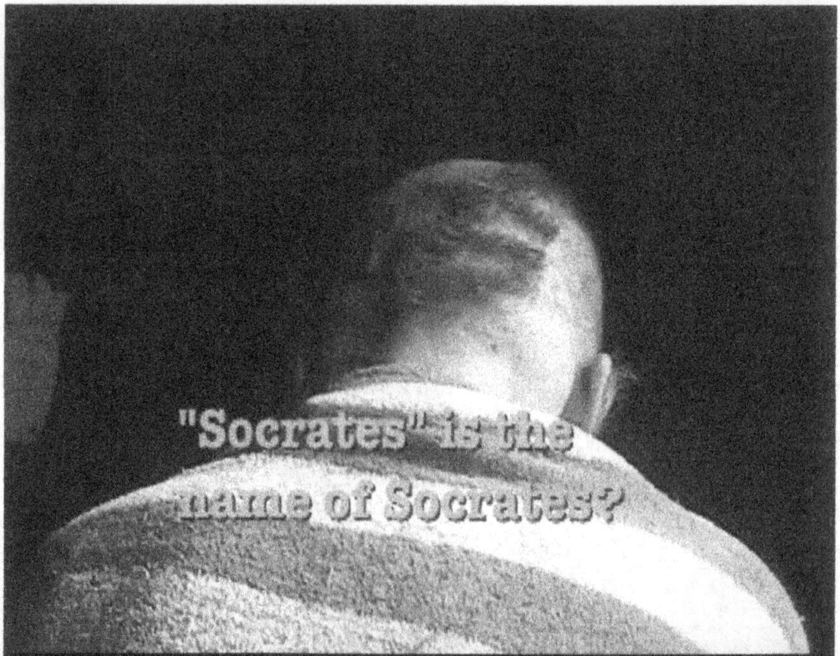

"Socrates" is the
name of Socrates?

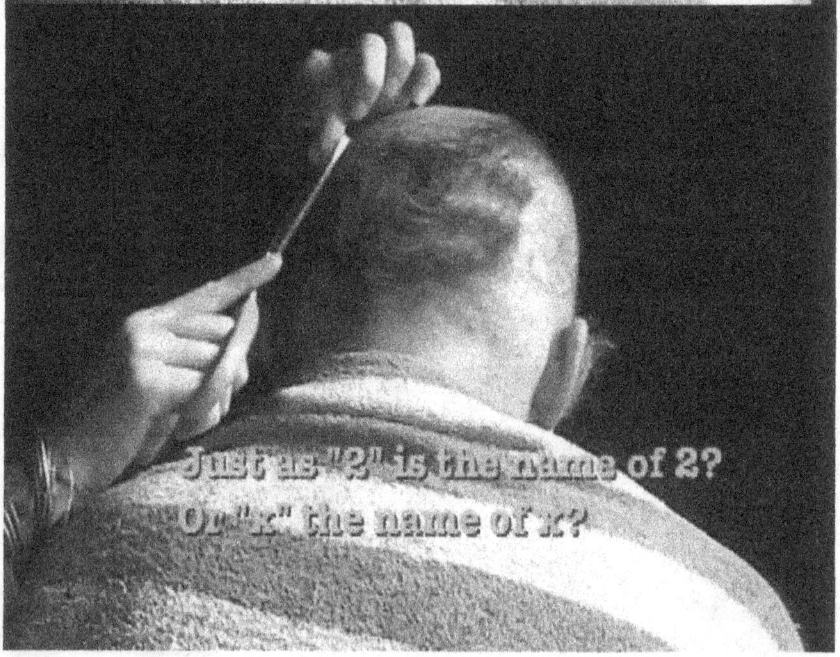

Just as "2" is the name of 2?
Or "π" the name of π?

Of course it may be "obvious"
- but it is not thereby necessary.
The pretence that @ is
transparent in its meaning

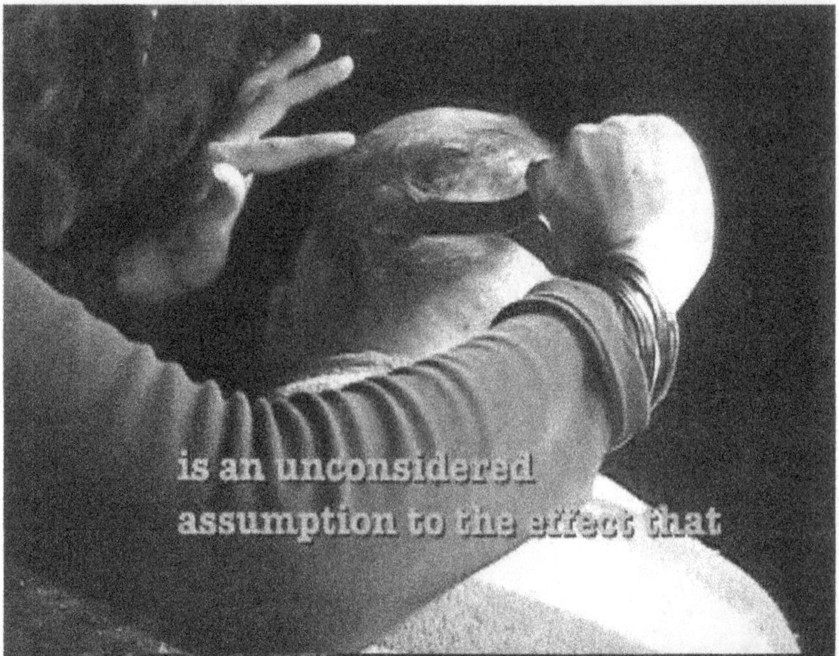

is an unconsidered
assumption to the effect that

"every occurrence of "x"
is an occurrence of

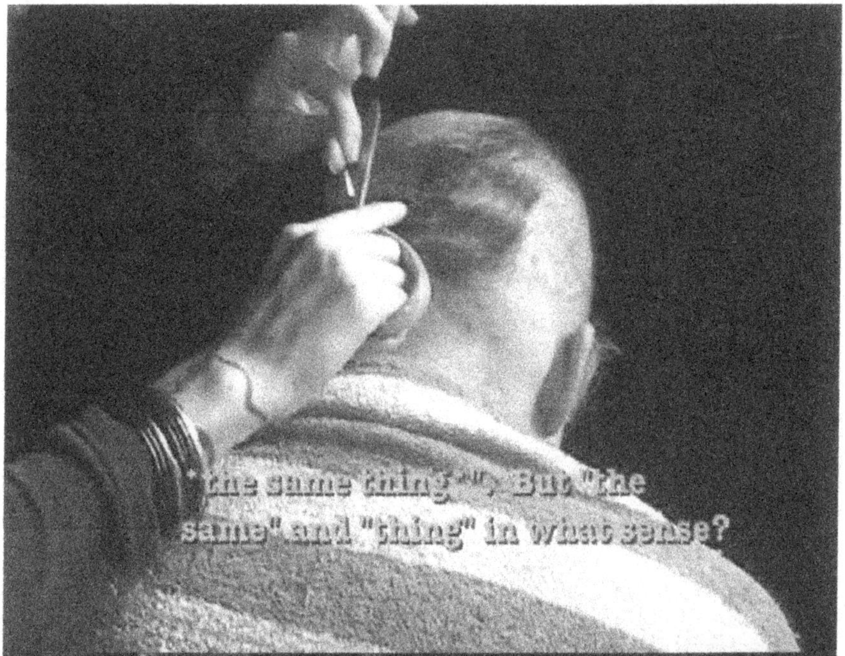

"the same thing"". But "the
same" and "thing" in what sense?
""Type/token" talk can be 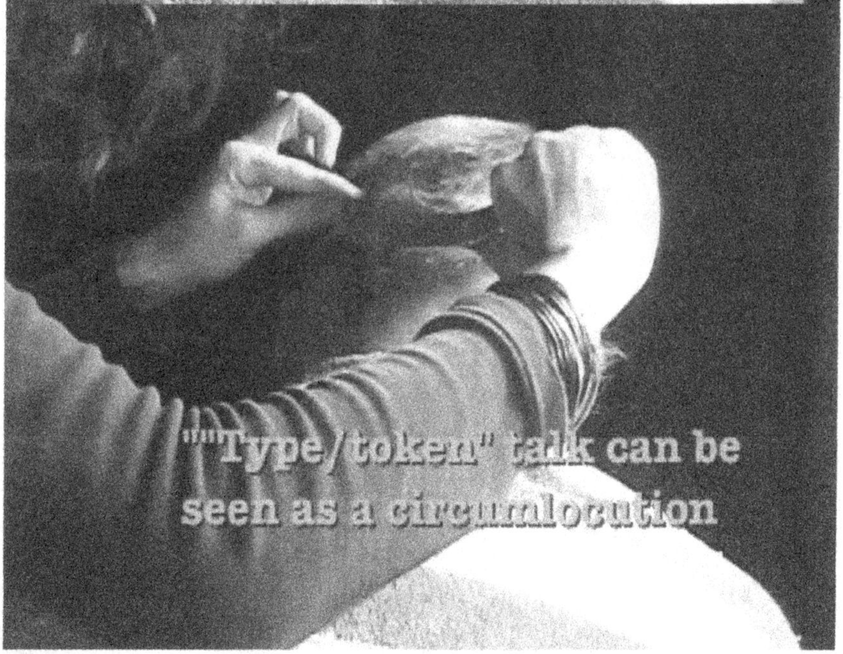
seen as a circumlocution

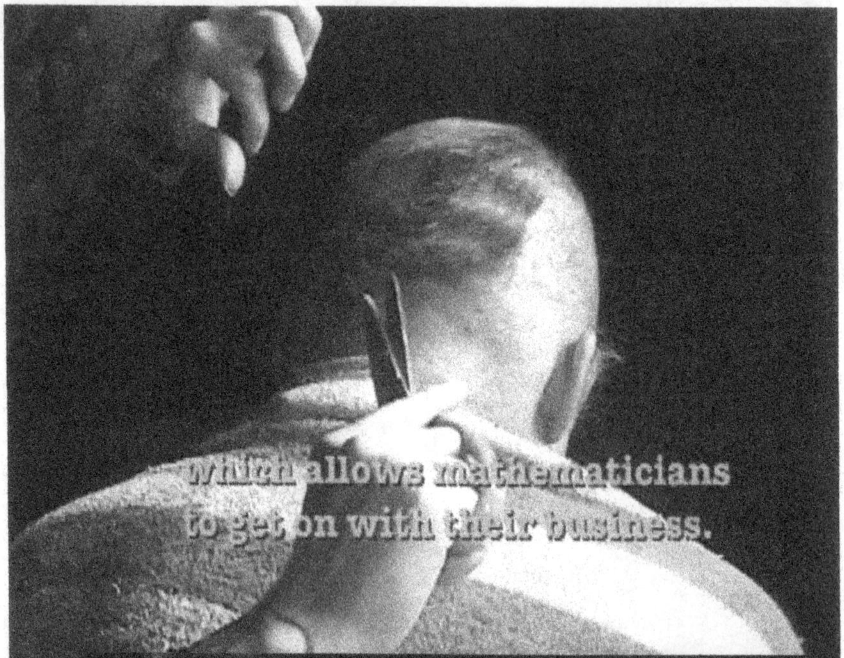

which allows mathematicians
to get on with their business.

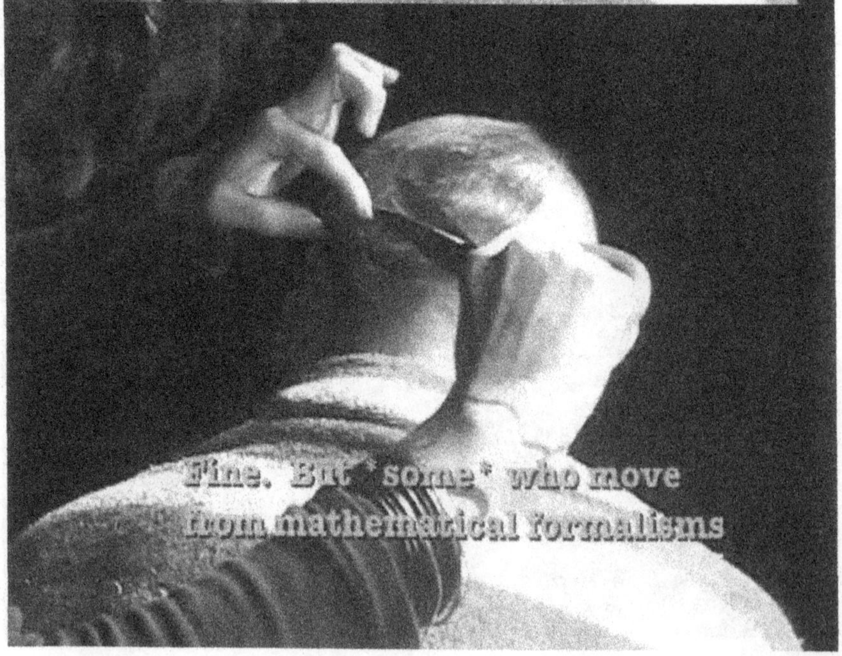

Fine. But *some* who move
from mathematical formalisms

276 superreal numbers

to the wider world seem
to suppose that such talk

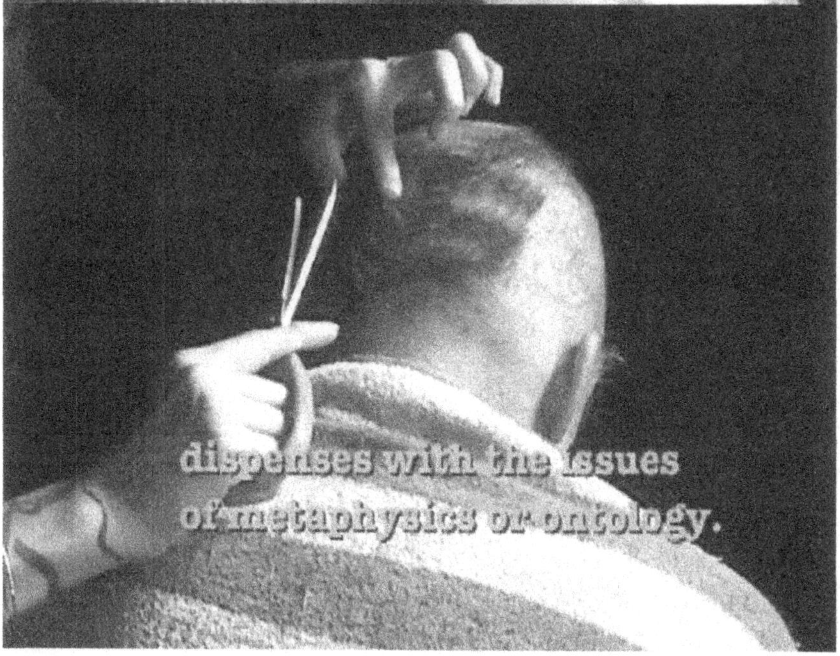

dispenses with the issues
of metaphysics or ontology.

It is apparently of no
concern to those who

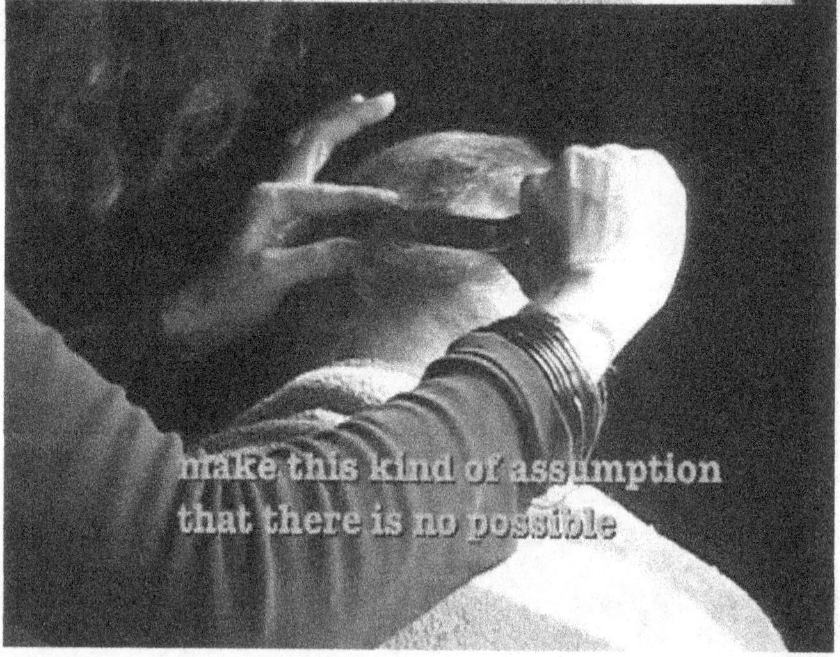

make this kind of assumption
that there is no possible

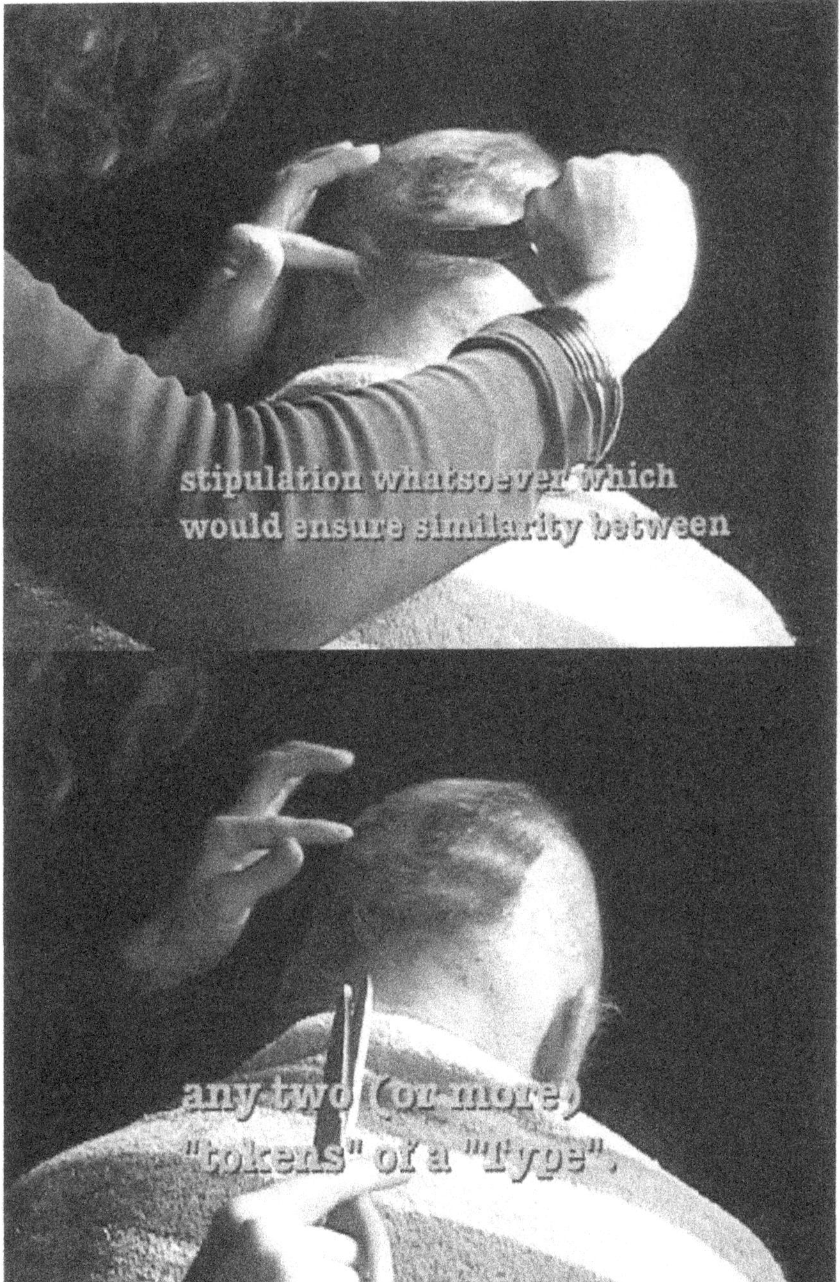
stipulation whatsoever which
would ensure similarity between
any two (or more)
"tokens" of a "Type".

Is the string "APPLE" on my screen the same or a different

token from that which appears on yours?

Even if mine is 10pt bold
helvetica black and yours

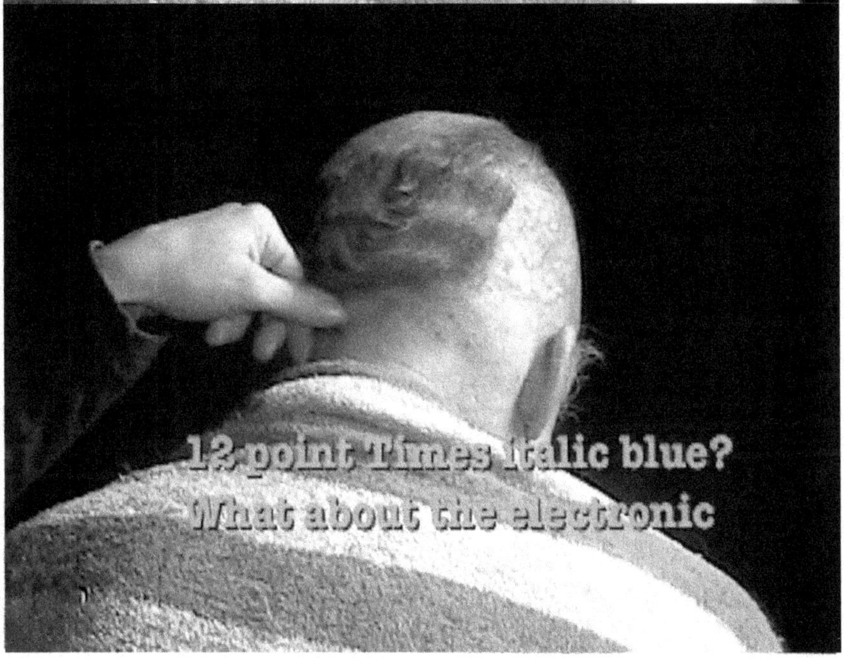

12 point Times italic blue?
What about the electronic

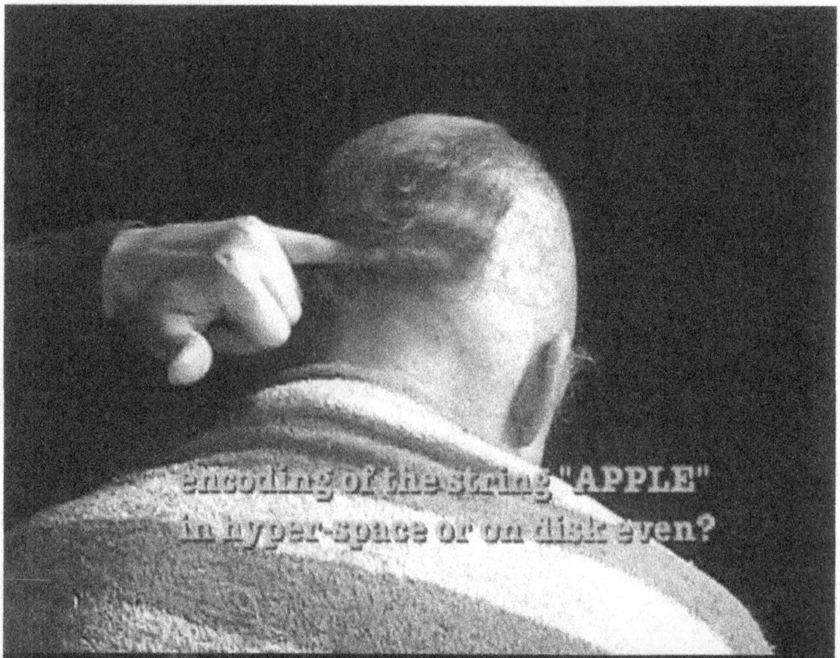

encoding of the string "APPLE"
in hyper-space or on disk even?

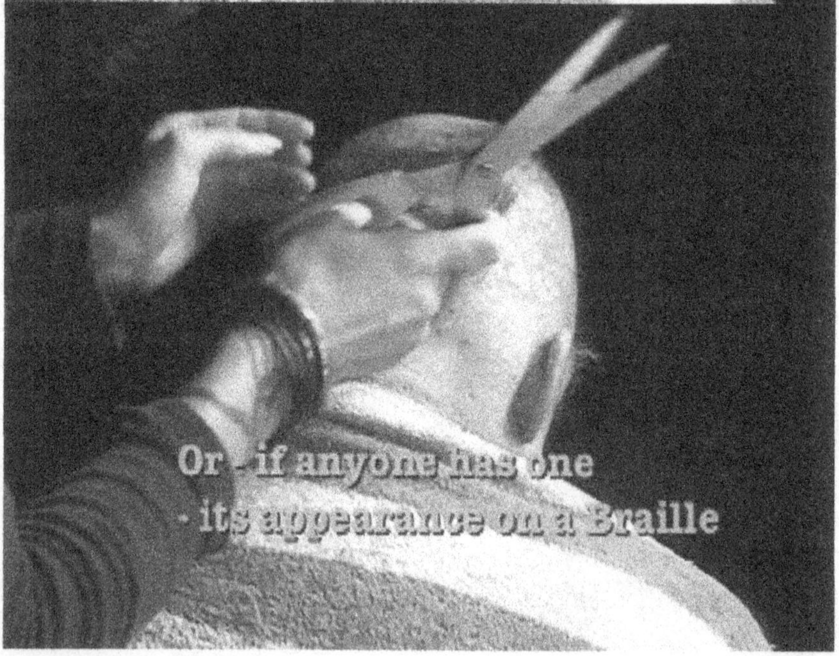

Or - if anyone has one
- its appearance on a Braille

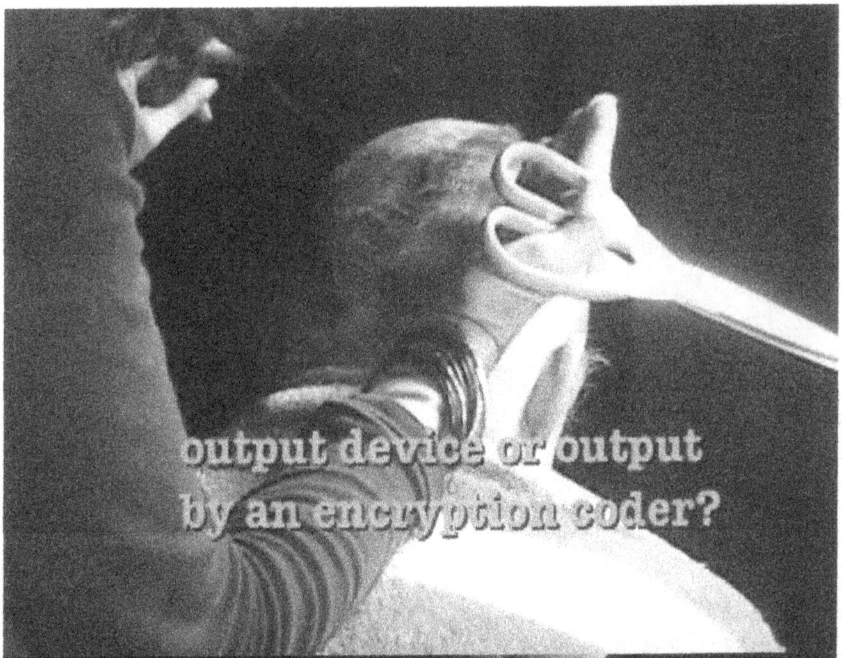

output device or output
by an encryption coder?

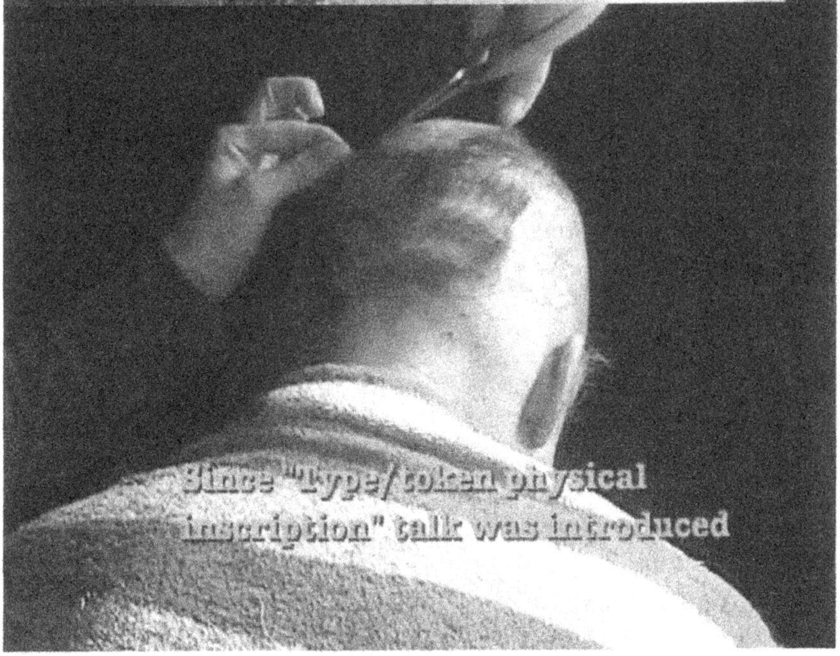

Since "Type/token physical
inscription" talk was introduced

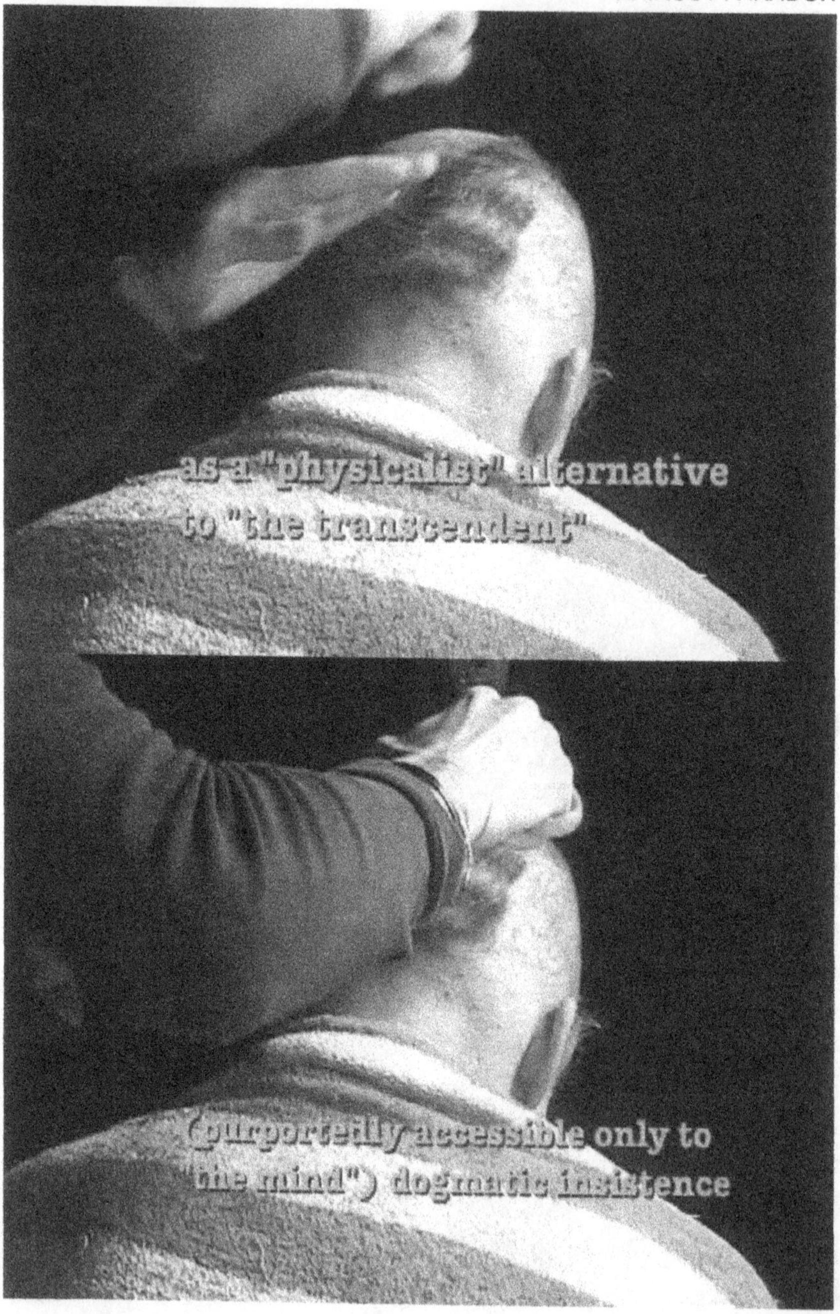

as a "physicalist" alternative
to "the transcendent"

(purportedly accessible only to
"the mind") dogmatic insistence

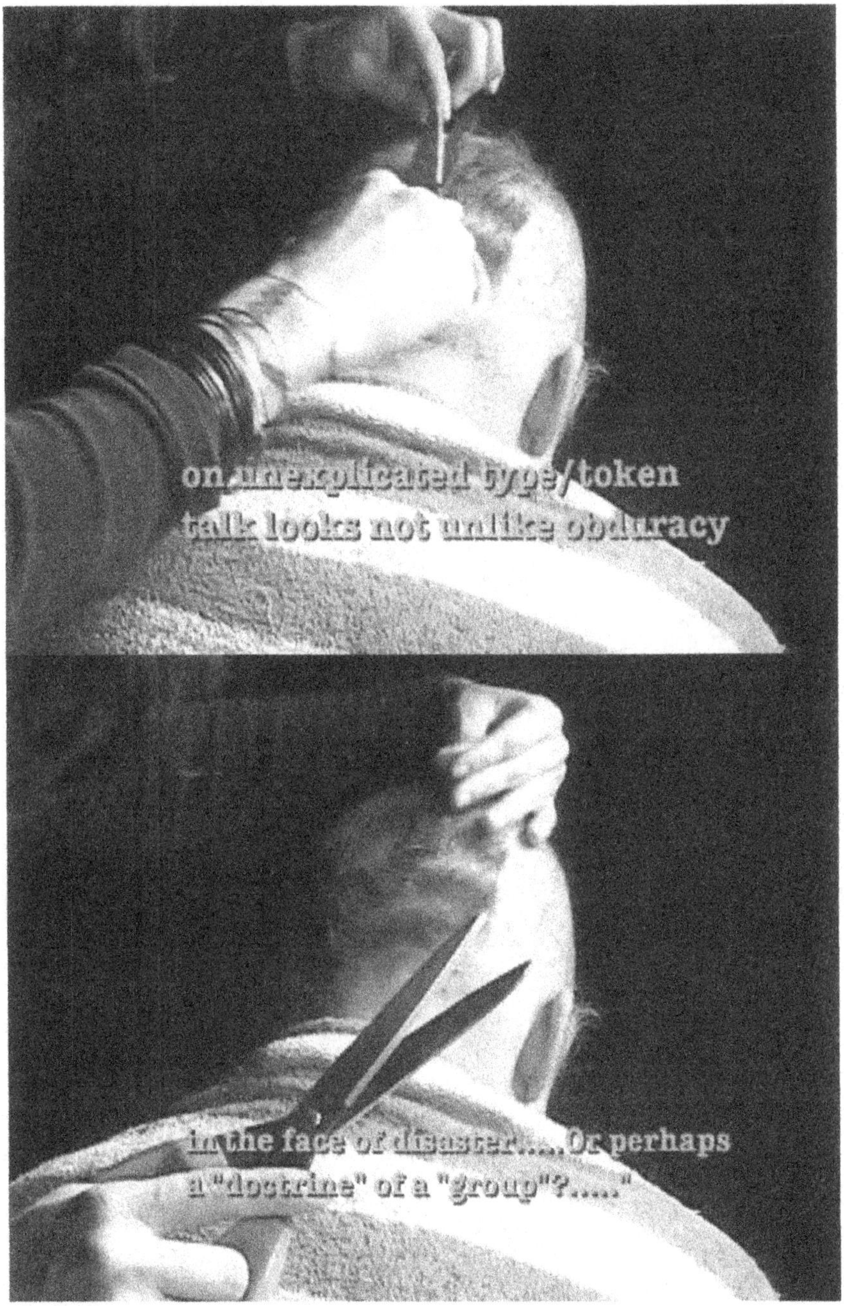

on unexplicated type/token
talk looks not unlike obduracy

in the face of disaster.....Or perhaps
a "doctrine" of a "group"?....."

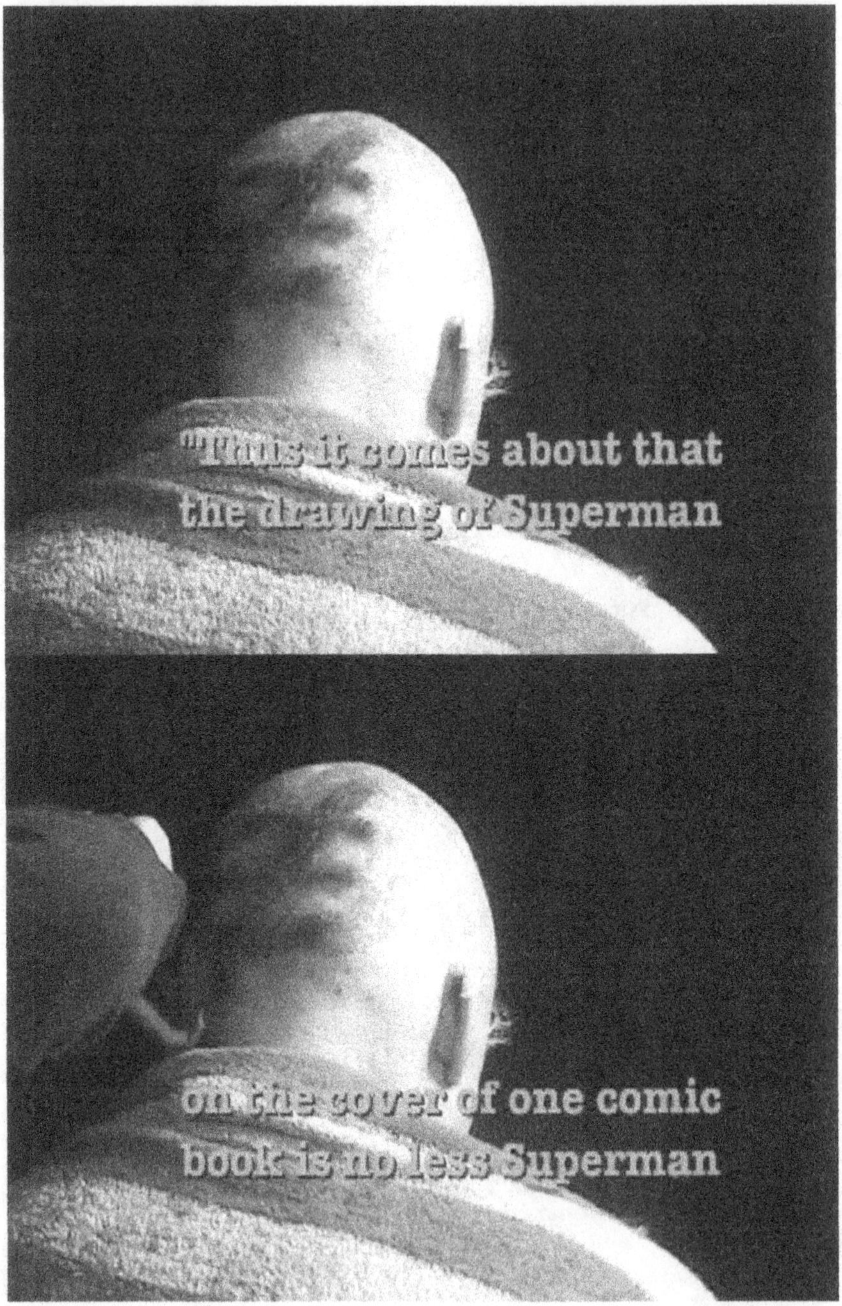

"Thus it comes about that
the drawing of Superman

on the cover of one comic
book is no less Superman

than a drawing of Superman
on another comic book.

Of course Superman is not
destroyed by tearing up one comic.

Perhaps the best or most
substantial thing that can be
said for "Superman" is that 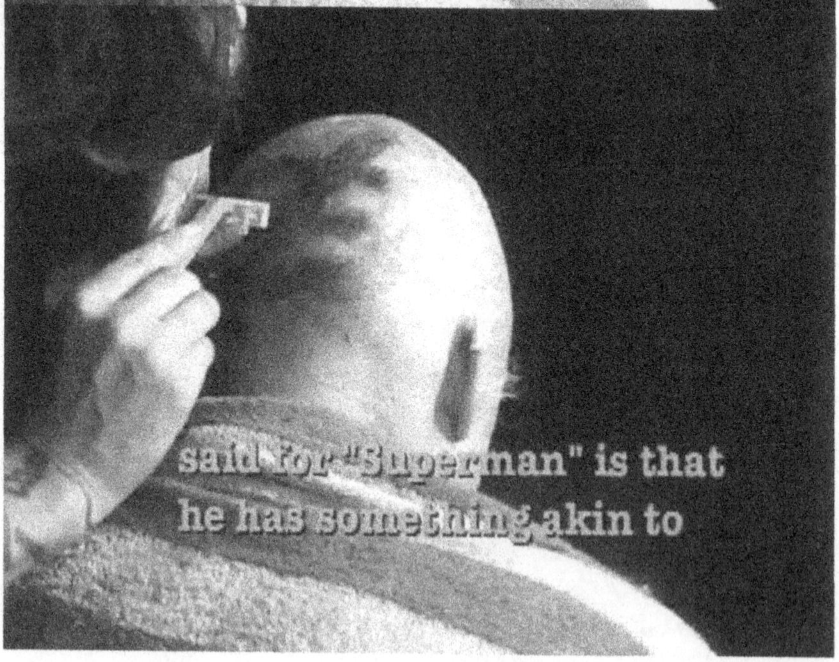
he has something akin to

the "Type"/Token' status Torkel,
for example, attributes to "2".
Perhaps one can say of any
such drawing "This is Superman"

meaning "this is a drawing
or representation of Superman"

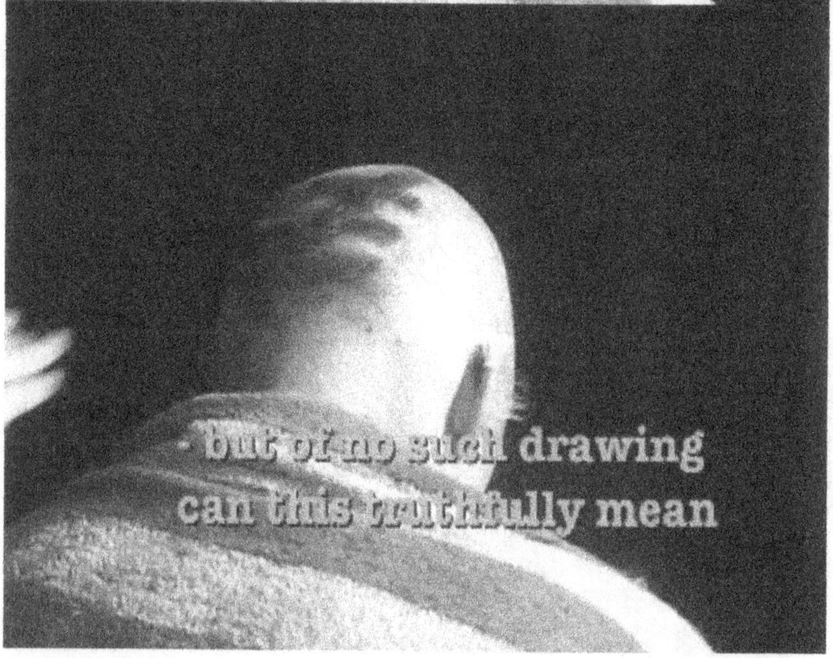

- but of no such drawing
can this truthfully mean

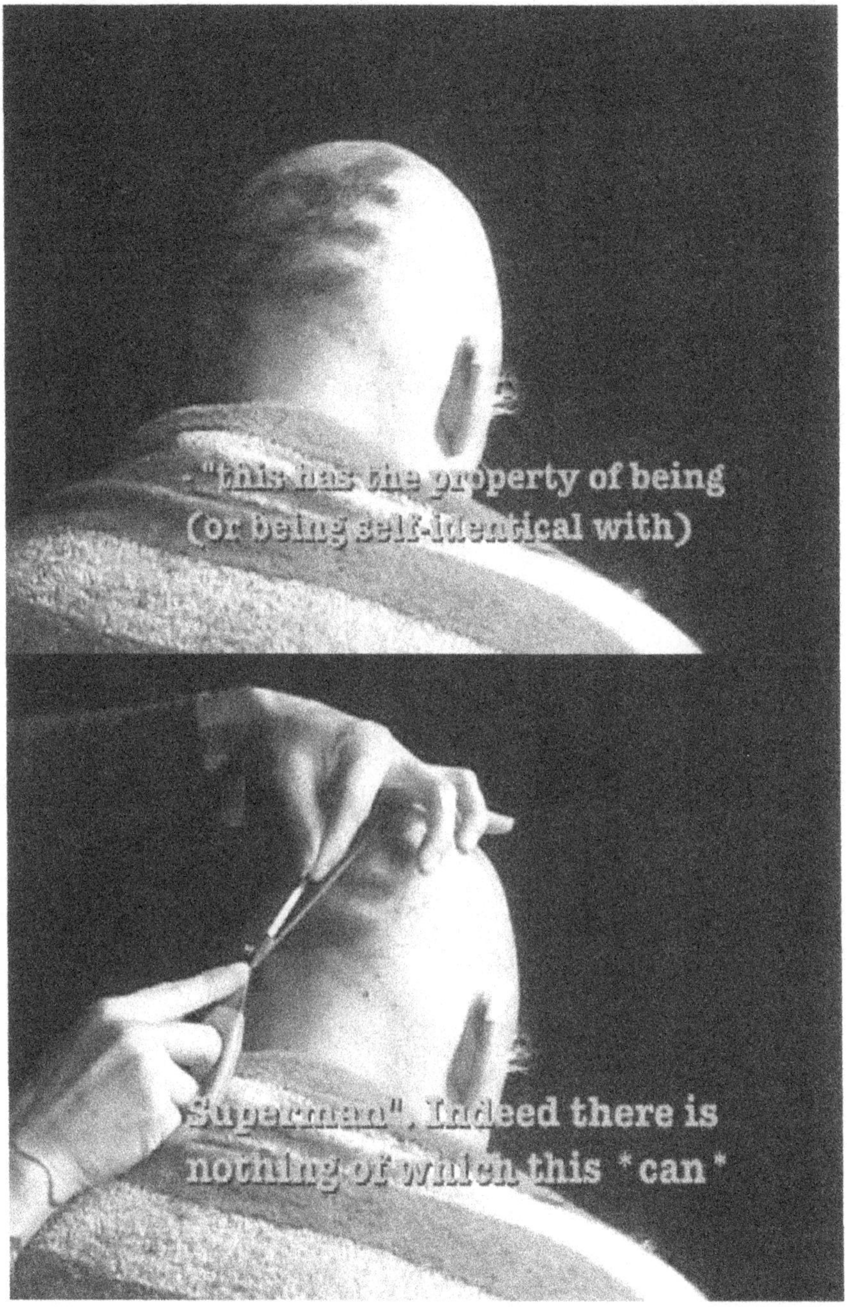

- "this has the property of being
(or being self-identical with)

Superman". Indeed there is
nothing of which this *can*

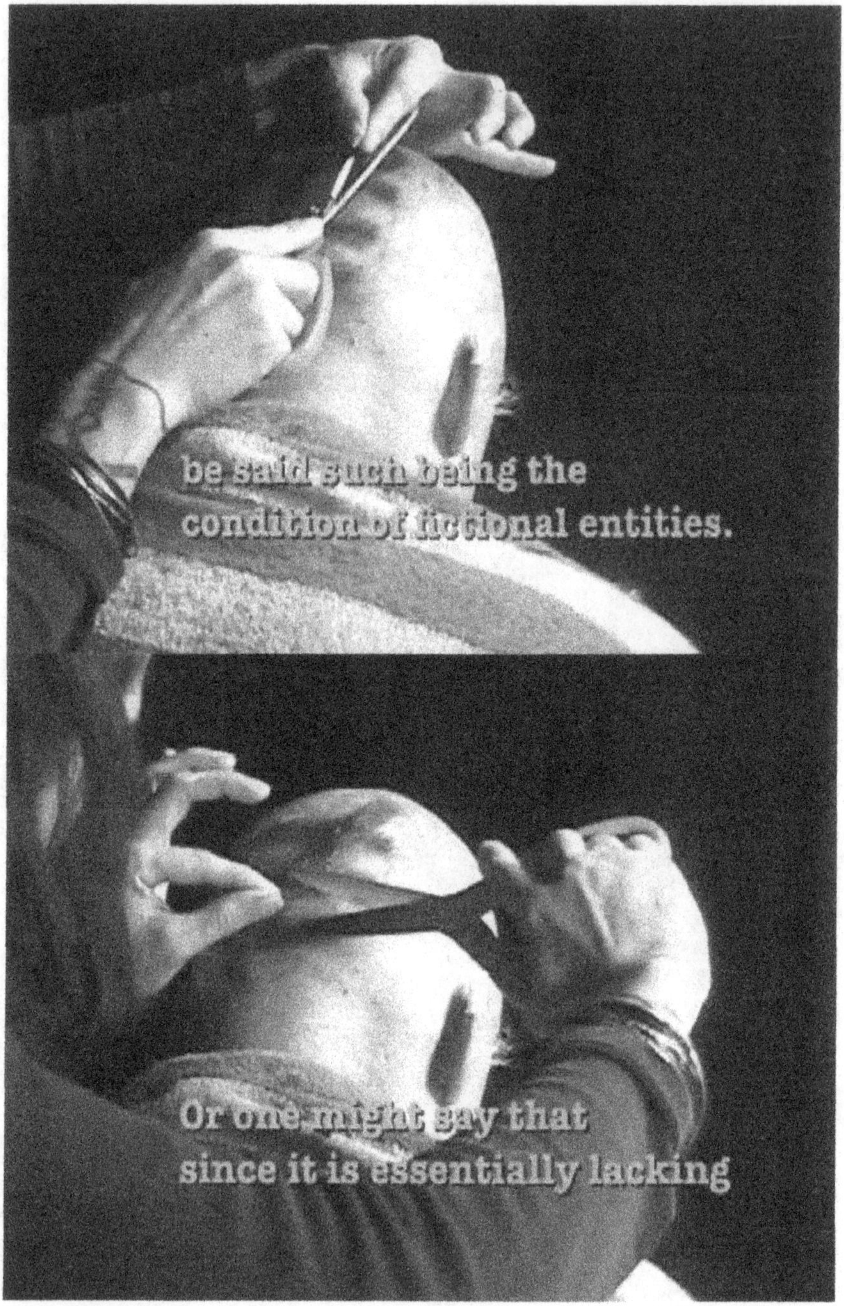

be said such being the
condition of fictional entities.
Or one might say that
since it is essentially lacking

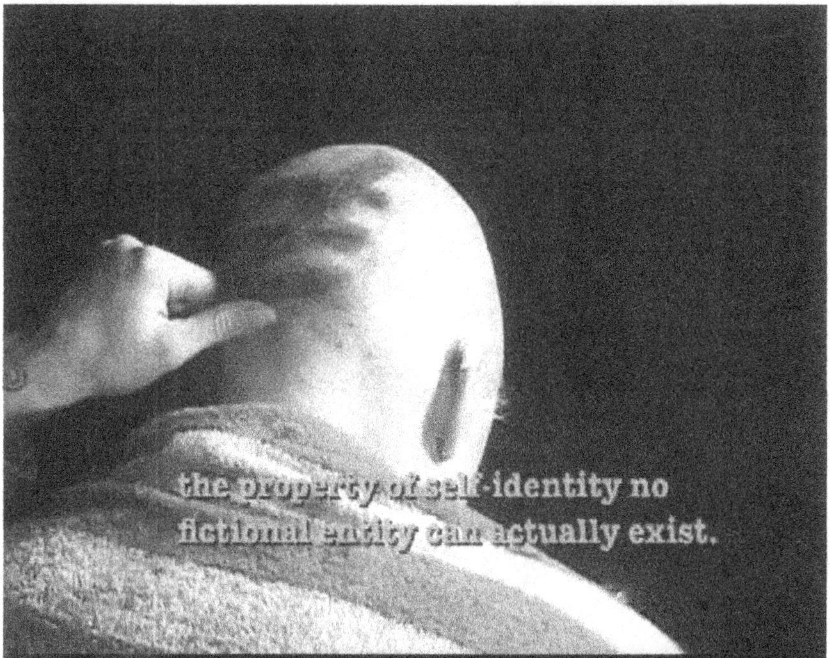

the property of self-identity no
fictional entity can actually exist.

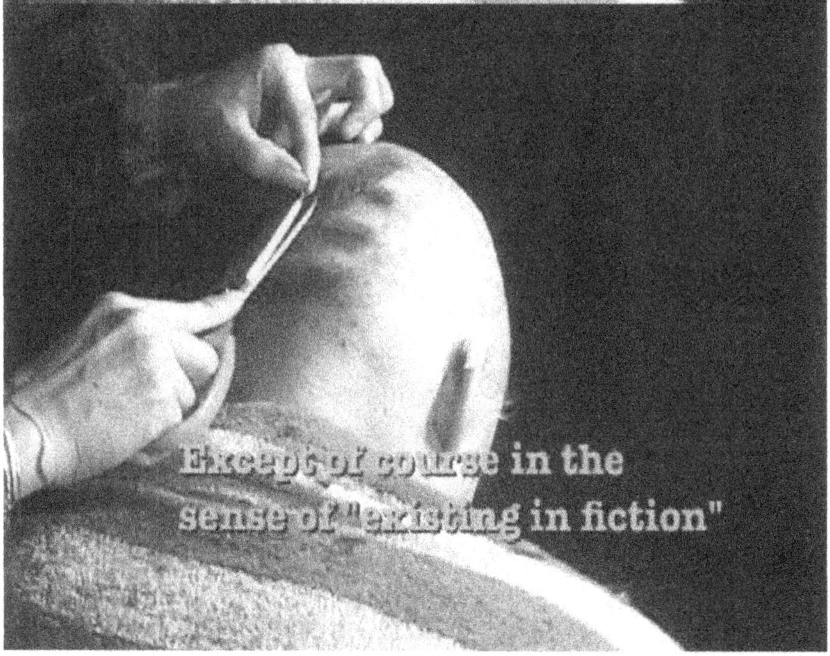

Except of course in the
sense of "existing in fiction"

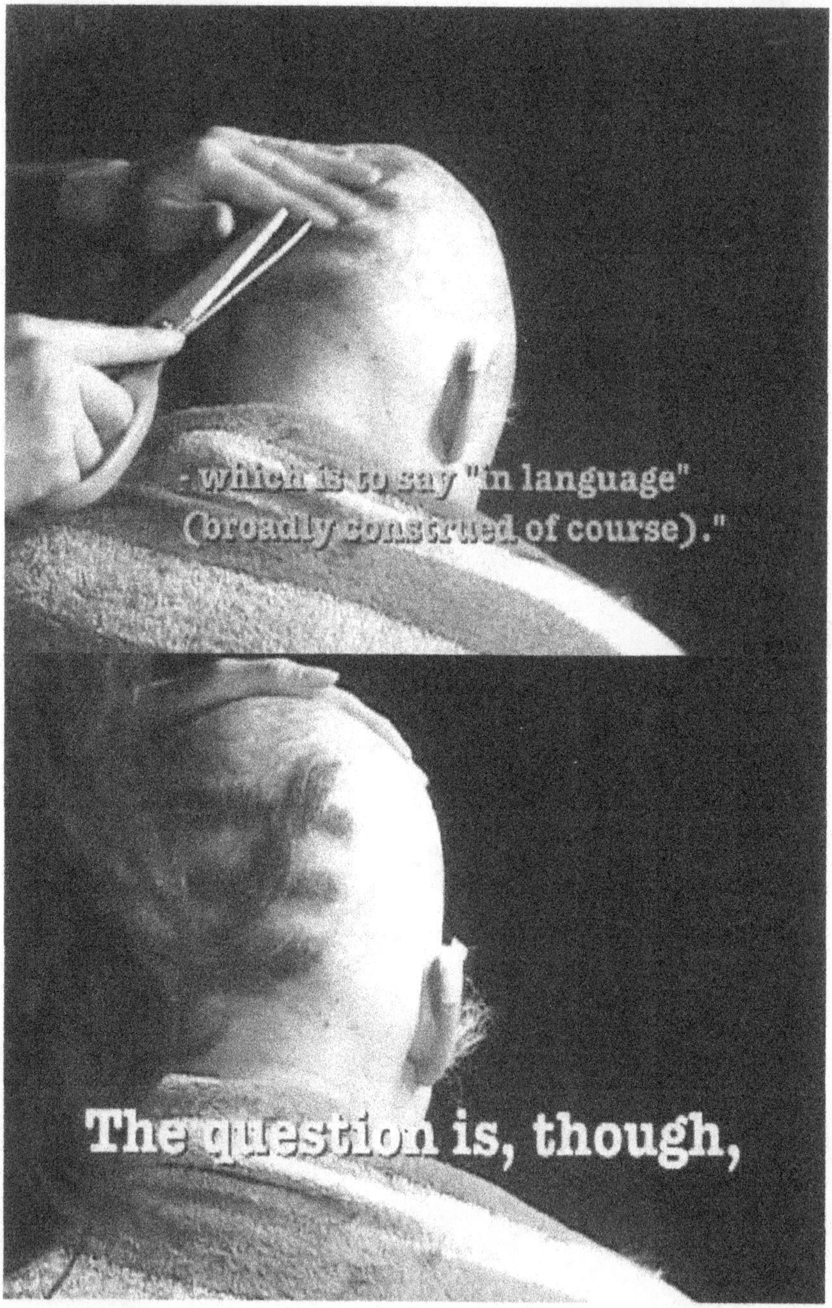

- which is to say "in language" (broadly construed of course)."

The question is, though,

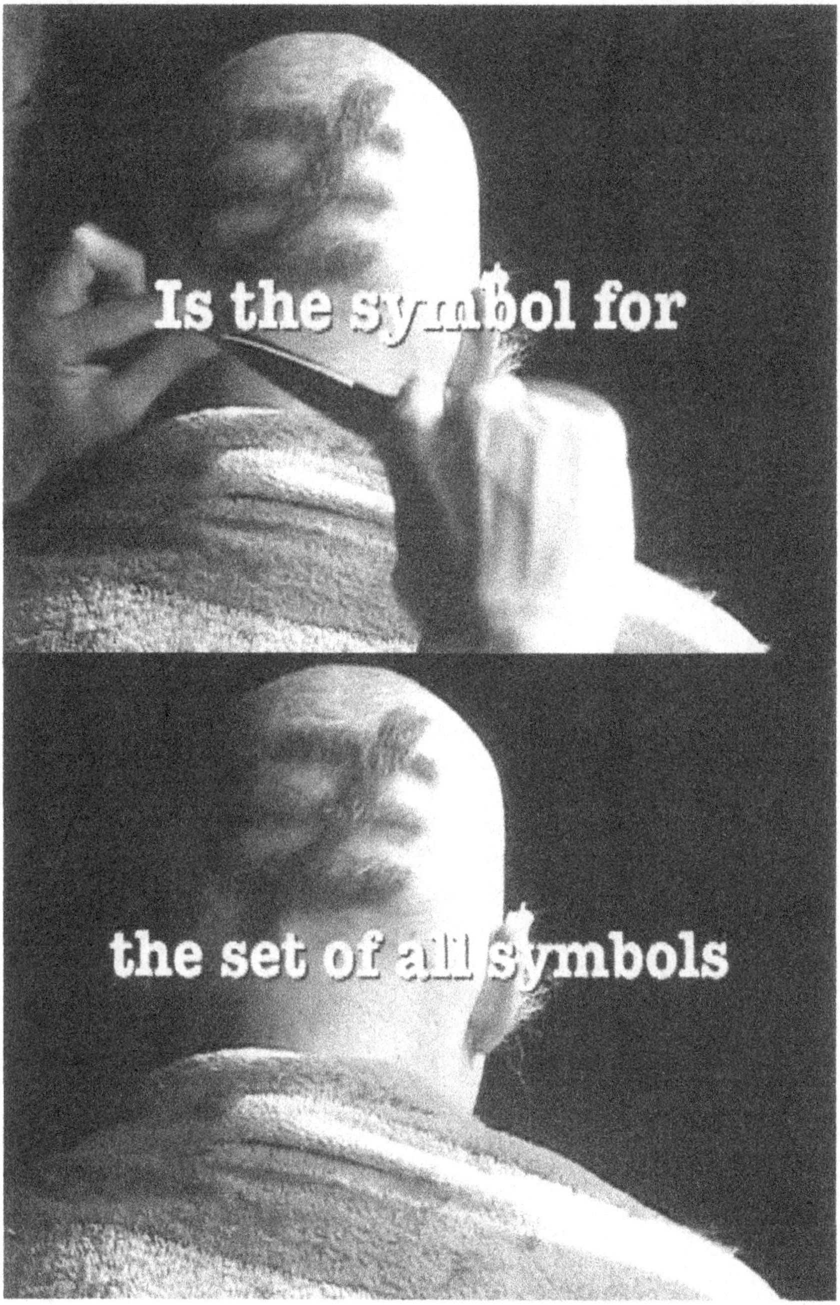

HAIRCUT PARADOX 294 superreal numbers

Is the symbol for

the set of all symbols

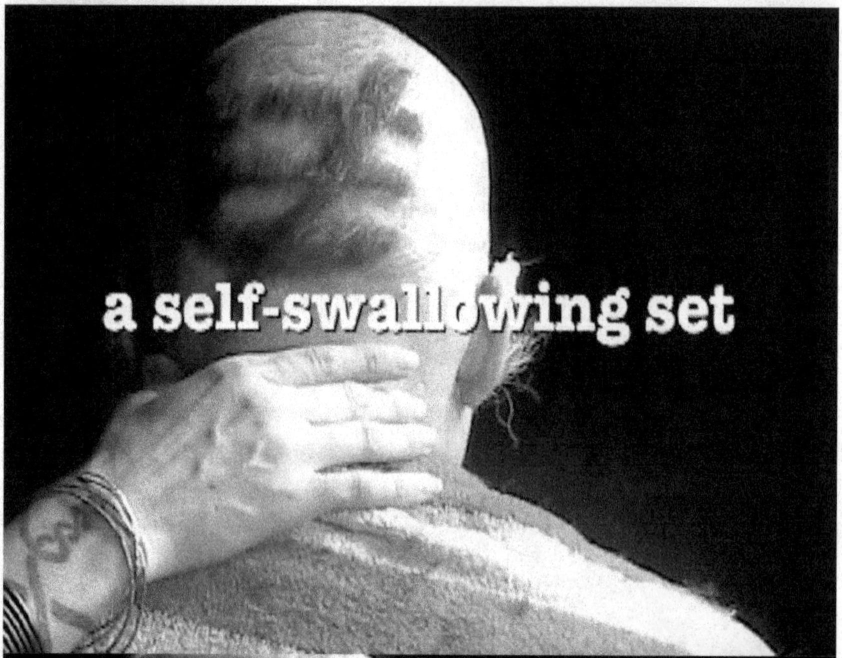

a self-swallowing set

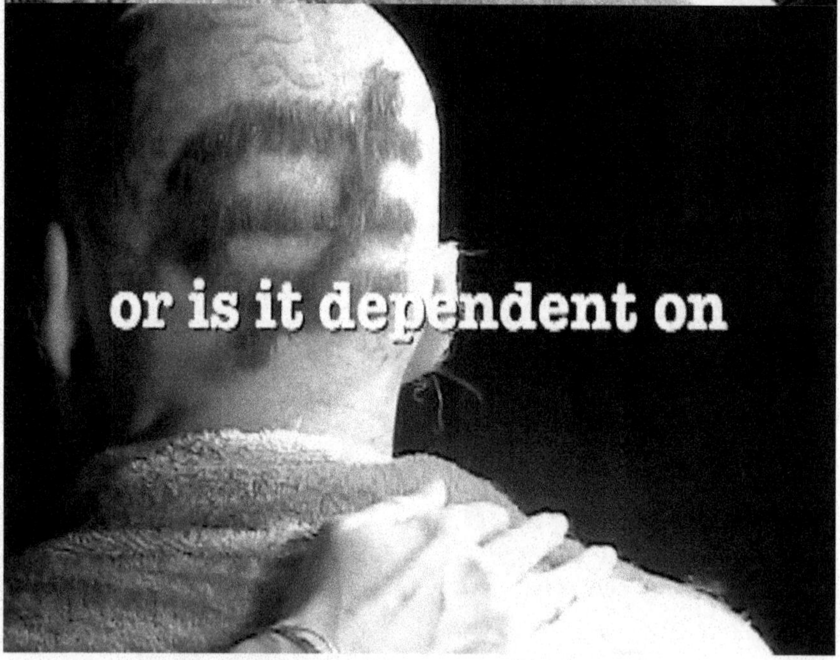

or is it dependent on

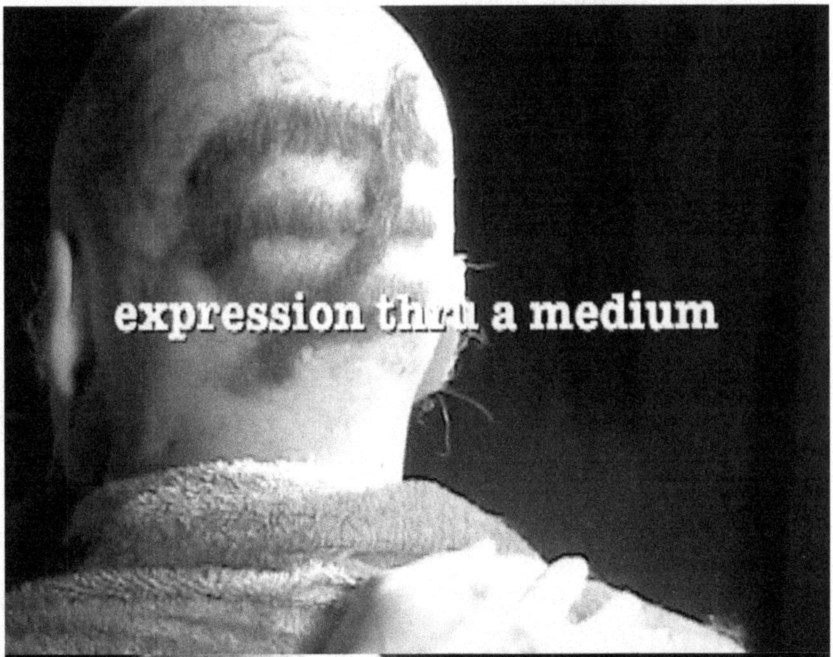

HAIRCUT PARADOX 296 superreal numbers

expression thru a medium

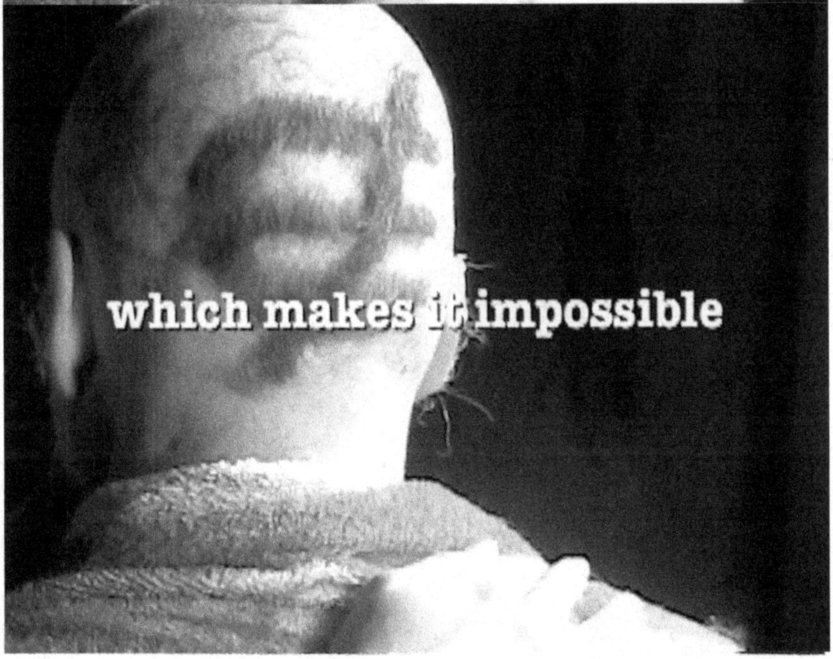

which makes it impossible

to exist as a pure

& disembodied abstraction?

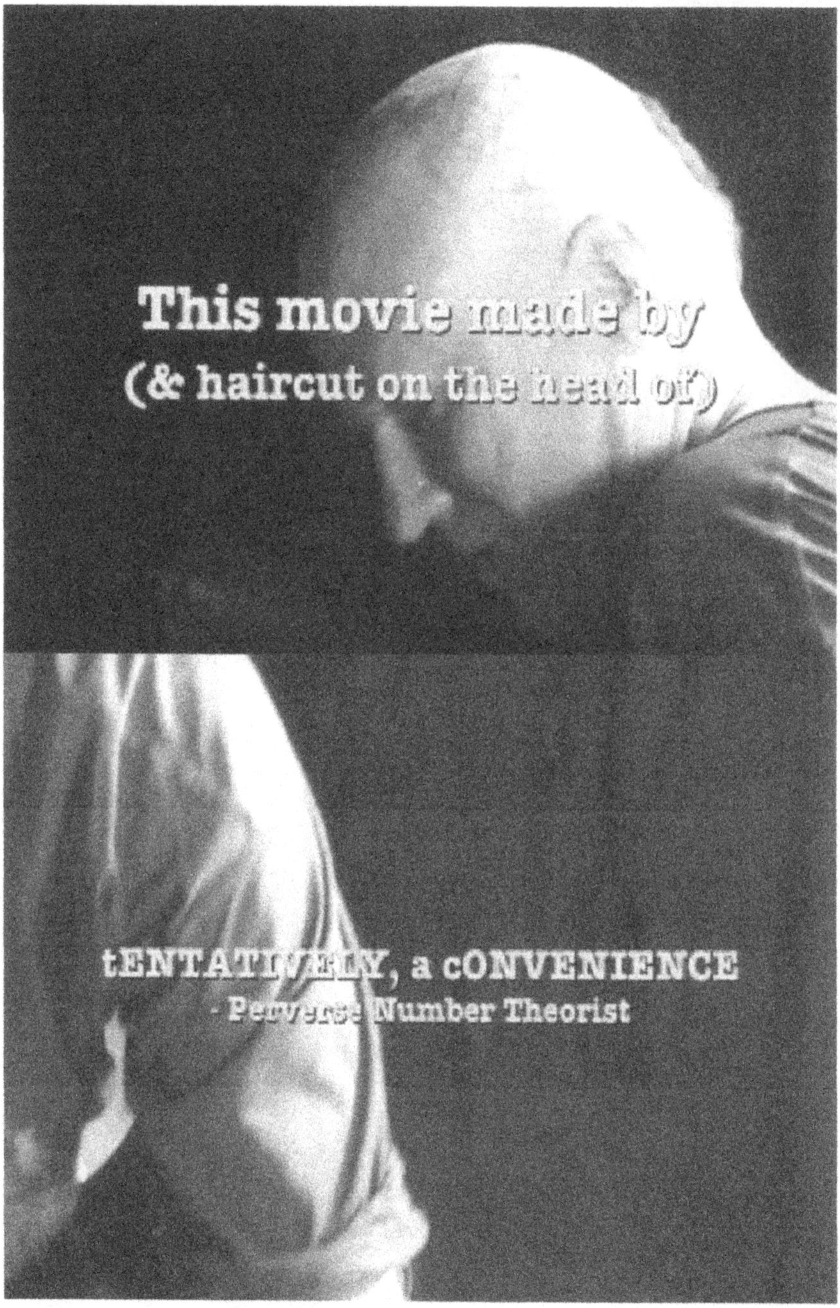

This movie made by
(& haircut on the head of)

tENTATIVELY, a cONVENIENCE
- Perverse Number Theorist

who's not even an
entry level amateur mathematician

meticulous haircutting:
Julie Gonzalez
Civilian Barber

PART THREE

GLOSSARY

This Glossary is divided into 3 pts: 1. abbreviations, 2. symbols, 3. terms. The abbreviations are mostly, if not entirely, standard - but are probably usually not used because many people might get caught up on them & find them hard to read. I say get used to it! Words change all the time & I see no good reason why making some of them shorter isn't ok.

As for the symbols? I give only cursory explanations here - just enuf to, hopefully, make them clear w/in the context of this bk. They're explained much better elsewhere. Same deal w/ the terms. There are entire bks, obviously, explaining such things.

1. abbreviations

1 = one (ie: not in the numerical sense)
1-1C = one-to-one correspondence
1st = first
2 = to or too
2nd = second
3rd = third
4 = for
4th = fourth
5th = fifth
6th = sixth
13th = thirteenth
14th = fourteenth
16th = sixteenth
17th = seventeenth
20th = twentieth
32nd = thirty-second
AA or AAA or AAAA = *amplificare ad absurdum* or simply *ad absurdum* or *AAA* or *AA* or *AAAAA...* - depending on how many times one needs to amplify it to get it as the 1st entry in the local yellow pages.
ACE = Automatic Computing Engine
AD = Anno Domine - The Year of Our Lord
aka = also known as
alotof = a lot of

AI = Artificial Intelligence
amt = amount
ARGUMENT = **A R**ude **G**irl **U**ndresses - **M**y **E**yes **N**eed **T**aping!
ARITHMETIC = **A R**at **I**n **T**he **H**ouse **M**ay **E**at **T**he **I**ce **C**ream or **A R**umpelstiltskin **I**nserted **T**opically **H**as **M**any **E**xistences **T**hat **I**nterfere **C**ornily
AT&T = American Telephone and Telegraph
ATN = Augmented Transition Network
attn = attention
A/V = Audio/Visual
BAMN! = By Any Means Necessary!
BANG OAR! = **B**ecause **A**rithmetic **N**ecessary **G**eography, **O**cean **A**rgument **R**hythm! = **B**ig **E**lephants **C**an **A**lways **U**nderstand **S**mall **E**lephants; **A R**at **I**n **T**he **H**ouse **M**ay **E**at **T**he **I**ce **C**ream; **N**ot **E**very **C**at **E**ats **S**ardines (**S**ome **A**re **R**eally **Y**ummy); **G**eneral **E**isenhower's **O**ldest **G**irl **R**ode **A P**ony **H**ome **Y**esterday; **O**nly **C**ats' **E**yes **A**re **N**arrow; **A R**ude **G**irl **U**ndresses - **M**y **E**yes **N**eed **T**aping!; **R**hythm **H**elps **Y**our **T**wo **H**ips **M**ove.
BC = Before Christ
b/c = because (not used here but I thought I'd mention it anyway)
BECAUSE = **B**ig **E**lephants **C**an **A**lways **U**nderstand **S**mall **E**lephants
bio = biography
bk = book
bldg = building
BOMB = Baltimore Oblivion Marching Band
c = see or sea
cd = could
CERN = European Council for Nuclear Research
c'mon = come on
def = definition
dox = doctors
d'ya = do you
eg = for example
E.G.Head = For-Example-Head or Egg-Head
enuf = enough
esp = especially
ESP = Extra Sensory Perception
etc = et cetera = & so on..
fMRI = functional Magnetic Resonance Imaging
FOL = First Order Logic

FOP = First Order Predicate
Fr = frater (Latin) or brother (English) = monk
FURB = Fuck You Right Back
FYI = For Your Information
G = Galileo (also Gangster - not used here)
GEB = Gödel, Escher, Bach
GEOGRAPHY = **G**eneral **E**isenhower's **O**ldest **G**irl **R**ode **A** **P**ony **H**ome **Y**esterday
GPS = Global Positioning System
HP = Hewlett-Packard
hr = hour
HQ = HeadQuarters
ID = identify or identification
ie = that is
iff = if & only if
IMO = In My Opinion
info = information
IQ = Intelligence Quotient
Jr = Junior
KDE = K Desktop Environment
kindof = kind of
LA = Los Angeles
LCD = Lowest Common Denominator
LEM = Law of the Excluded Middle
lim = limit
LIM = Law of the Included Middle
ln or log = logarithm
LP = Long Playing (vinyl) record
luv = love
ma'am = madam, madame
Mac = Macintosh computer
MDMA = Methylenedioxymethamphetamine = Ecstasy
mm = millimeter
mnth = month
'n' = and
n'at = and that
nazi = not see
NBPS = Nuclear Brain Physics Surgery's cool
NECESSARY = **N**ot **E**very **C**at **E**ats **S**ardines (**S**ome **A**re **R**eally **Y**ummy)
no. = number

NST = Naive Set Theory
OCEAN = **O**nly **C**ats' **E**yes **A**re **N**arrow
ok = okay
p = page
P = pee, piss or pun
pd = paid
PDF = Pretty Damned Funny
PHD = D(octorate of) Ph(ilosophy)
Φlo-hankie = philosophical hankie
PHP = Hypertext Preprocessor
PNT = Perverse Number Theory
POR = Preposterous, Outrageous, & Ridiculous - as in POR Solution
POV - Point of View (not used here, once again: FYI)
pp = pages
PSZ = Paradigm Shift Zeitgeist
pt = point
QG = Quantum Gravity
Rc = Radice cubica = cube root
rc'vd = received
rd = read
re = in reference to
RE = Reverse Engineering
reductio = reductio ad absurdum
RHYTHM = **R**hythm **H**elps **Y**our **T**wo **H**ips **M**ove
Rq = Radice quadrata = square root
sd = said
SF = Supreme Fascist = God
shd = should
s-he-it = she/he/it - ie: a genderless pronoun that isn't necessarily sexless
sortof = sort of
SQRT = SQuare RooT
stn = station
Stroop Theory = Str(ing) Theory + (L)oop (Quantum Gravity)
thru = through
TNS = (Galileo's bk) Two New Sciences
TNT = Typographical Number Theory
TOE = Theory of Everything (I don't use this one either - it's just FYI)
TTQ-EA = Thoughts Too Quick - Expressions Anachronistic
tv = television

typo = typographical error
u = you
U = University
UN = United Nations
US = United States
USS = United States Ship (in the instance of USS Yorktown) - it has many other meanings
viz = namely
vs = versus
w/ = with
w/in = within
w/o = without
wch = which
wd = would
WP = Wikipedia
ya = you
y'know = you know
YOU = the other as an unknown variable
young'un = young one (child)
yr = year or your

2. symbols

My presentation of these symbols is idiosyncratic. For a much better & more thorough review of mathematical symbols, go to <http://en.wikipedia.org/wiki/Table_of_mathematical_symbols> or consult bks such as: Mathematical Dictionary - edited by Glenn James & Robert C. James & published by *D. Van Nostrand Company, Inc.* in 1960 &/or The Facts on File Dictionary of Mathematics edited by Carol Gibson & published by *Facts on File, Inc,* in 1988. No doubt there are more recent dictionaries worth using too - I just mention these 2 because they're the ones I have.

$\aleph_0$ = **aleph-null** - "aleph" is the 1st letter of the Hebrew alphabet & "null" is meant as zero - meaning as a starting point - the degree of infinity of the natural numbers - ie: their cardinality
& = ampersand = **and**
@ = **at**

[ ] = **author's brackets** - this probably isn't the correct name - in this bk (& also in common usage) text w/in these brackets indicates an author's aside - ie: an intrusion into another text - such as into a quote

$\beth_2$ - **Beth two** - a degree of infinity - also referred to as 2c (pronounced two to the power of c). The following is edited from Wikipedia:

> Sets with cardinality [**cardinal numbers**, or **cardinals** for short, are a generalized kind of number used to denote the size of a set, known as its **cardinality.**] Beth ["Beth", of course, being the pronunciation of the 2nd letter of the Hebrew alphabet (*alpha-beth*)] two include:
>
> ▪ The power set [given a set *S*, the **power set** (or **powerset**) of *S*, written *P*(*S*), or *2S*, is the set of all subsets of *S*.] of the set of real numbers, so it is the number of subsets of the real line, or the number of sets of real numbers
>
> ▪ The power set of the power set of the set of natural numbers

 = **Bombelli's bowls** - ie: Bombelli's notation for unknowns - no longer in use

= **C5** - ie: the note C one octave above middle C

⌈ ⌉ = **ceiling brackets** - ie: designating an upper limit of what they enclose

≡ = **congruence relation symbol** = is defined to be another name for

<-> = **in one-to-one correspondence w/**

. = **decimal point** - numbers to the right are increments of one, numbers to the left are whole numbers

δ = the lower case of **Delta**, the 4th letter of the Greek alphabet - used in math to denote a finite increment in a variable

/ = **division sign** when it separates 2 numbers

$ = **dollars** - a particular unit of money

e = the transcendental number that begins 2.718281828459 . . .
. . . = **ellipsis** = and so on in this pattern, eg: 1.223224225. . . -> 1. followed by a pattern in wch there are two "2"s followed by a "3" followed by two "2"s followed by a "4" followed by two "2"s followed by a "5" followed by two "2"s followed by the next whole number after "5" ("6") followed by two more "2"s, etc...
= = **equals**
∃ = **existential quantification symbol** = there exists

5^5 = this is an example of a **superscript number** wch represents a **factor** - in this case it means 5 to the 5th power - ie: 5 multiplied by itself 5 times
^ = **factor** - ie: to the power of - this usage is just the one I found on internet searches, it isn't used that way elsewhere - instead it's used to mean **logical conjunction**
! = **factorial** - the factorial number is the number of possible set arrangements made w/ a set of numbers - eg: "1" can only be arranged one way: {1} so its factorial is "1". 3 can be arranged 6 ways so its factorial is 6: {1,2,3}, {1,3,2}, {2,1,3}, {2,3,1}, {3,1,2}, {3,2,1}. To re-express the latter example, 3! [the factorial of 3] = 3 x (3-1) x (3-2) = 3 x 2 x 1 = 6.
f = **function of** - in a regular function like $y = f(x)$, *x* is the *independent variable* and *y* is the *dependent variable*, meaning simply that changes in *x* produce other changes in *y* according to the rules of *f*
> = **greater than**
≥ = **greater than or equal to**
⇔ (usually written in this bk as "**<=>**") = something is true **if and only if** something else is true, eg: *A* <=> *B* = *A* is true if and only if *B* is true - used by Hofstadter as equivalent to, eg: p <=> horse = p is equivalent to horse (meaning that the variable "p" represents a horse in this instance)
i = "**imaginary number**" - sometimes signified by **√-1** - the square root of negative one
√-1 = the square root of negative one - this is sometimes signified by "i" - meaning the "**imaginary number**"
→ (usually written in this bk as "**->**") = **implies**, "if . . . then" or "is **isomorphic to** - a big rule of inference involves the logical relation of *entailment*, meaning 'If . . . then' and often represented by the symbol '**->**'

∞ = **infinity**
< = **less than**
≤ = **less than or equal to**
$\log_e$ = the **logarithm that uses e as its base**
- = **minus** or **negative**
x = used in "Special Instance Subtractive Commutativity" to mean "**is the mirror image on the negative side of the number line**" - ie: in the use created for this bk
x = **multiplication sign** = times
☐ = **necessary** - a modal symbol - "A modal is an expression (like 'necessarily' or 'possibly') that is used to qualify the truth of a judgement. Modal logic is, strictly speaking, the study of the deductive behavior of the expressions 'it is necessary that' and 'it is possible that'. However, the term 'modal logic' may be used more broadly for a family of related systems. These include logics for belief, for tense and other temporal expressions, for the deontic (moral) expressions such as 'it is obligatory that' and 'it is permitted that', and many others. An understanding of modal logic is particularly valuable in the formal analysis of philosophical argument, where expressions from the modal family are both common and confusing. Modal logic also has important applications in computer science." - from the on-line Stanford Encyclopedia of Philosophy @ <http://plato.stanford.edu/entries/logic-modal/>
¬ = **not**
≠ = **not equal to**
¬E = **not-Exceptional**, ie: **unexceptional** - ie: in the use created for this bk
¬H = **not-Horse** - ie: in the use created for this bk
√¬H = **not-Horse Radish** - ie: a joke put here for Glossary readers
Ø = **null sign** = the ***empty set*** (otherwise shown as '**{ }**')
ω = **Omega** (the lower case form of the 24th. last, letter of the Greek alphabet) - the **ordinal number of the whole denumerably infinite set** {0, 1, 2, 3, 4, . . .}. This transfinite ordinal is the limit of the set's members' sequence - that is, it's the very smallest number bigger than all finite integers. Another, more common way to describe ω is that it's the ordinal number of that set of which $\aleph_0$ is the cardinal number.

= symbol of tENTATIVELY, a cONVENIENCE's **open mind**

Φ = the Greek letter **Phi** - used by Wallace to represent an infinitesimal - It's used in many other ways elsewhere - eg: "'($a \in \Phi$) -> ($a \notin \Phi$)'" - this reads as ("*a*" belongs to the set Phi) implies or maps onto ("*a*" does not belong to the set Phi). "Φ" (Phi) is no longer used here to mean infinitesimals - it's simply a set name - as used in Everything

π = **PI** (the 16th letter of the Greek alphabet) - the transcendental irrational number that begins **3.14159265 . . .** & wch is used for many purposes - including determining the circumference of a circle by multiplying **π** times the circle's diameter - *also means* "the **number of primes ≤*x***" (less than or equal to *x*) wch is unrelated to its meaning as **3.14159265 . . .**

\+ = **plus** or **positive**

P = the **Power Set** - hereby defined, for some set *A*, as simply the set of all subsets of *A*. Meaning that every member of P(*A*) is some subset of *A*

pq- = the 3 symbols of unspecified (&, therefore, variable) meaning in Douglas Hofstadter's "**pq- System**"

' = **prime** - *not* in the sense of a prime number but in the sense of the 2nd in a series of variables, eg: A, A', A", etc

< > = **Propositional Logic brackets** - ie: the brackets that turn their contents into a statement in Propositional Logic - ie: the contents are, when spoken in natural language, preceded by "if"

√ = **root sign** - a number to the left of it is the number of the root & the number to the right is the number that the root is of - eg: 3√27 = 3 means "the cube (or 3rd) root of 27 is 3" - no number to the left is generally taken to mean the square (or 2nd) root

{ } = **set brackets** - ie: the brackets that designate a set

| | = **set cardinality brackets** - ie: the brackets used to indicate the cardinality of the set contained w/in it

(in) = the **set inclusion symbol**

∈ = **set membership sign** - ie: belongs to the set of

∉ = **set non-membership sign** - ie: does not belong to the set of

Σ - the 18th letter in the Greek alphabet, "**sigma**", & I think that the way it's used here means *the sum of the terms indicated.*

[..] = text that's been **skipped over in a quote**

Ϛ = the obsolete Greek letter **Stigma** in capital form - hereby standing for the word/concept "**anything**" - Why did I choose **ς** (**Stigma**) to symbolize "anything"? I'm glad you asked that. Wikipedia: "'Stigma' is also a name for the cursive graphic form of the letter digamma when used as the Greek numeral 6. This name is a modern misunderstanding. The ligature of στ did not occur in ancient times and only dates to later medieval manuscripts. In Sophocles' Lexicon of Byzantine Greek, which covers late antique Greek up to 1000 AD, there is no mention of "stigma" as either a ligature or a number."

Given that Digamma (uppercase **F**) is also an "obsolete" Greek letter & that both stigma & digamma can equal the number 6 (wch reminds me, of course, of the sample set for *Mr. Gregor's Neighborhood*), it seems like we're already well on the rd to *anything is anything* here. Besides, my use of the letter **ς** is bound to be stigmatized. Looking at **F** & **ς** side-by-side: **Fς**, the letters cd be sd to symbolize pictorially: my cup of fuck-you spilleth over. In conclusion, is it *surreal* or *hyperreal* to say that <(0 = 1/∞), (Φ = 1/∞), (∞ = 1/0), (∞ = 1/(1/∞)), (**ς** = 1/(1/**ς**)) ⊃ (√2 = 7/5)>?

⊆ - **subset** - For any set *A*, the empty set is a subset of *A*: ∀*A*: ∅ ⊆ *A*

⊃ = "**then**" in Propositional Logic

θ = **Theta** (the 8th letter of the Greek alphabet) - it's unclear to the author what meaning it has in its brief usage in this bk: "eg: In mathematical physics, a Grassmann number (also called an anticommuting number or anticommuting c-number) is an element of the algebra spanned by quantities θ*i* that anticommutes with all generators θ*j* but commutes with ordinary numbers *x*:

$$\theta_i\theta_j = -\theta_j\theta_i \qquad \theta_i x = x\theta_i."$$

It may refer to a certain ordinal number in this instance - however, it has many meanings

~ = **tilde** - **same order of magnitude** or **not** (most often used this way here because of Hofstadter's use of its Propositional Logic meaning, eg: ~G = being the negation of the Gödel sentence G written in English as "This statement of number theory does not have any proof in the system of *Principia Mathematica*.")

∪ = **set theoretic union** - ie: "the union of ... & ..."

∀ = **universal quantification symbol** = for all, for any, for each

a = a lower-case *italicized* letter usually signifies an **unknown variable**

3. terms

The terms I choose to include here are the ones that I assume to be least familiar to the lay-person - either because they're mathematical jargon outside common usage (eg: "multiplication" isn't included but "cardinal numbers" is) or because they're neologisms that I invented. In some cases, I will, no doubt, assume incorrectly & include something that far more people know than I realize & exclude something that I thought was widespread.

My criteria for inclusion & exclusion may be somewhat inconsistent. But that's ok. This whole bk is idiosyncratic. The most important thing here is for me to explain my own terms - the other terms you can find explained better elsewhere.

3 Standard Stoppages = "a work Duchamp considers to be one of the main turning points of his life." (Arturo Schwarz) As I recall, Duchamp took 3 one-meter long threads & dropped them. The resultant curve became a new unit of length in his playful physics, "the meter diminished" - as Duchamp is quoted by Schwarz as saying

16mm = the size of a medium gauge motion picture film

absolute convergence = "In mathematics, a series (or sometimes also an integral) is said to **converge absolutely** if the sum (or integral) of the absolute value of the summand or integrand is finite." - WP

absolute values = "In mathematics, the absolute value (or modulus) of a real number is its numerical value without regard to its sign. So, for example, 3 is the absolute value of both 3 and −3" - WP

Adamitic language = in the Judeo-Christian tradition, the language that God enabled Adam to create w/ a direct relation between things named & the name itself - detached from religion, the idea that there is a direct relation between signs & what they symbolize - esp a relationship that enables the signs to manipulate the object referred to

aleph-force = the ELEPHORSE (or *Aleph-Force* meaning *Primary* or

Primal Force) wch is the *n*-dimensional *dot-based* elephantHORSE moving **backwards** into the *n*-dimensional *Penguin Diagram string-based* horseELEPHANT space - resulting in Elsie, the not-Cow - in other words, the milk of kindof-likeness - cONVENIENCE

algebraic numbers = "In mathematics, an **algebraic number** is a complex number that is a root of a non-zero polynomial with rational (or equivalently, integer) coefficients" - WP

algenraic = In chapter 16 of A History of π (PI) the transcendence of π is explained. & I was planning to possibly quote from that to explain it here. But, then, on page 167, this caught my eye: "There is, however, no reason why an irrational number should not be a root of an algenraic equation". The typo there isn't mine. What if the obvious mistake of having an "n" substituted for the intended "b" in "algenraic" were to somehow be perpetuated? As the mistake in the creation of "sine" was perpetuated as a mathematical term?

"The words *calculate*, *calculus*, and *calcium* all come from the Latin word for pebble: calculus." & predicting the odds for the number of ripples whilst skipping calculi on the sine was how algenra originated.

algorithms = "In mathematics, computing, linguistics, and related disciplines, an **algorithm** is a definite list of well-defined instructions for completing a task" - WP

alpha wave state = alpha waves are one type of brain waves, thought to represent the activity of the visual cortex in an idle state

Alzheimer's disease = a specific memory & aging problem - in Alzheimer's, "specific parts of the brain deteriorate much more rapidly than the rest of the brain.. alotof the frontal cortex, as well as some of the circuitry in the middle of the brain starts decaying but also [Amyloid] plaques start forming [..] - these [are] calcium deposits that disrupt the [neural] signals [from] being processed" - Pig Pen

ambiguity gate = **An "ambiguity gate" is a flexibility of neuron firing response patterns in wch options are increased by different paths being used to achieve the same result & by the same path being used over again to achieve different results** & analogies, similes, metaphors, & pataphors used toward the same or related purposes of flexibility & creative open-mindedness. Or something like that.

amplificare ad absurdum = this author's attempt at a Latin variation on *reductio ad absurdum* in wch the 'reduction' turns into an "amplification" meant to *increase* the absurdity - not get rid of it - abbreviated "*AAA* or *AA* or *AAAAA...* - depending on how many times one needs to amplify it to get it as the 1st entry in the local yellow pages"

analytical geometry = "Analytic geometry, also called coordinate geometry and earlier referred to as Cartesian geometry or analytical geometry, is the study of geometry using the principles of algebra." - WP

antinomy = Wikipedia: "the mutual incompatibility, real or apparent, of two laws"

applied math = math used for practical manipulation of physical reality - in that sense, a more reasonable version of Adamitic language than magikal incantations

approximation theory = "In mathematics, approximation theory is concerned with how functions can best be approximated with simpler functions, and with qualitatively characterising the errors introduced thereby. Note that what is meant by best and simpler will depend on the application." - WP

archibras = an appendage hypothesized by utopian philosopher Charles Fourier (1772-1837) as something that followers of his philosophy wd grow after 20 generations - this appendage wd be a tail w/ a hand on the end w/ an eye-ball in the middle - this idea was rejected as an embarrassing absurdity by people who otherwise respected Fourier

Archimedean Spiral = aka arithmetic spiral = "a spiral named after the 3rd-century-BC Greek mathematician Archimedes; it is the locus of points corresponding to the locations over time of a point moving away from a fixed point with a constant speed along a line which rotates with constant angular velocity." - WP

area = "The area of a surface or lamina is the amount of material needed to "cover" it completely" - Wolfram MathWorld

asymptotes = a straight line towards wch a curve approaches but never meets - such as in a hyperbola's relation to the *x* & *y* axes on a Cartesian plane

asymptotic formulas = presumed by cONVENIENCE to be a formula in wch characteristics of asymptotes are somehow involved - either in relation to a Cartesian plane or a complex plane or an extended complex plane

ATN-grammar = Augmented Transition Network grammar
Avatar = "an incarnation of God. "Avatar who am in heaven, holy is my name...". A waiter: a wedder: a water. Yesus, the El Avatar, is Elevator." - Schwartzberg
axes = lines representing vertical & horizontal dimensions, etc..
axiom = "1. a self-evident truth, 2. a universally accepted principle or rule, 3. *Logic, Math.* a proposition that is assumed without proof for the sake of studying the consequences that follow from it." (The Random House College Dictionary, 1975)
Axiomatic Set Theory = an attempt to make a set theory w/ fewer gross paradoxes than in Naïve Set Theory
axon = an outgoing signal-carrier from a neuron (to other neurons)
Ayn Randian = an adjective made from the name of Objectivist philosopher/author Ayn Rand - probably used here to poke fun at the capitalist self-interest that Rand is usually despised for by people w/ a political/social conscience
baker's dozen = 13 - because the baker makes the dozen & can, therefore, have extra
Behaviorists = in psychology, the philosophy that human behavior is mechanically determined & can be controlled by highly predictable means
beta wave state = a brain wave usually occurring in a state of arousal - ie: normal waking consciousness
Beth two = Beth (the 2nd letter of the Hebrew alphabet) 2 = in this bk, used as the power set of the power set of the set of natural numbers - ie: a different degree of infinity than Aleph null
Bird-Brainism = higher brain efficiency w/ smaller brain size - a subversion of an anticipated insult to tENTATIVELY, a cONVENIENCE's PNT ends
black theorem = "if you discover it, [it] drives you mad" - a logician's joke - Hoffman
Blissymbols = international pictograms created by Charles Bliss
Bongard problem-solving program = has "several stages, in which raw data gradually get converted into descriptions. The early stages are relatively inflexible, and higher stages get gradually more flexible. The final stages have a property which I call *tentativity,* which means simply that the way a picture is represented is always tentative." - Hofstadter
borderline tautological nihilism =
boundary = Wikipedia: "In topology, the **boundary** of a subset *S* of a

topological space X is the set of points which can be approached both from S and from the outside of S. More formally, it is the set of points in the closure of S, not belonging to the interior of S."

Broca's brain = a language center of the brain named after its discoverer Pierre Paul Broca

buzz words = aka a fashion word or vogue word = frequently used words sometimes w/ emotionally charged meanings that're used manipulatively by pretending that by virtue of their repetition they have more meaning than they may actually have - cONVENIENCE - "is an idiom, often a neologism, commonly used in managerial, technical, administrative, and sometimes political environments. Though apparently ubiquitous in these environments, the words often have unclear meanings" - WP

Calcarine sulcus = a particular fissure in the brain where the primary visual cortex is

calculus = the study of rates of change

cardinal numbers = numerical amount rather than numerical order

Cartesian grid = a way of representing space on a flat surface by designating points' verticality in relation to a y axis & horizontality in relation to an x axis

Categorists = apparently a school of mathematics/physics

C/C++ = The first compiler to implement export was Comeau **C/C++**

Cauchy Convergence Condition = In mathematics, convergence describes limiting behaviour, particularly of an infinite sequence or series toward some limit - apparently, this condition is a stipulation placed on criteria for determining convergence by 19th century French mathematician Augustin Louis Cauchy

ceiling = the "**ceiling function**, denoted $\lceil x \rceil$ or ceil(x) or ceiling(x), is the function that returns the smallest integer not less than x" - WP

cerebellum = "(Latin: "little brain") is a region of the brain that plays an important role in the integration of sensory perception and motor control" - WP

chakras = in Hindu philosophy, spiritual energy centers aligned along the spinal column / nervous system

chutzpah = Yiddish for audacity or courage

circumference = the distance around a closed curve

cobordism = Wikipedia: "In mathematics, an $n + 1$ **cobordism** is a triple (W, M, N), where W is an $(n + 1)$-dimensional manifold, whose boundary is the disjoint union of the n-dimensional manifolds M and N. In other words, it is a manifold with boundary whose boundary is partitioned in two."

Cognitive Dissidence = a play of words based on Cognitive Dissonance - in Cognitive Dissonance a perception seems unaligned w/ one's usual perceptual expectations - often referred to be saying "That does not compute" - in Cognitive Dissidence, there's a subversive attempt to undermine perceptional expectations that're deemed to be rooted in false assumptions - such as assumptions based on propaganda-inculcated stereotype projections

College of 'Pataphysics = aka Collegium Pataphysicum = "The *Collège de 'pataphysique*, founded on May 11, 1948, in Paris, is a group of artists and writers interested in the philosophy of 'pataphysics." - WP

combinatorics = "**Combinatorics** is a branch of pure mathematics concerning the study of discrete (and usually finite) objects" - WP

common logarithm = logarithm base 10

complex numbers = numbers that combine real & imaginary numbers

complex plane = "In mathematics, the **complex plane** is a geometric representation of the complex numbers established by the **real axis** and the orthogonal **imaginary axis**. It can be thought of as a modified Cartesian plane, with the real part of a complex number represented by a displacement along the x-axis, and the imaginary part by a displacement along the y-axis" - WP

complexity theory = the study of any complex system

composite numbers = "A **composite number** is a positive integer which has a positive divisor other than one or itself. By definition, every integer greater than one is either a prime number or a composite number" - WP

consensus reality = a 'reality' created by a critical mass of agreement on its truth

Constructivism = a Kronecker doctrine further developed by Poincaré & Brouwer that involves such rules as: "Any mathematical statement or theorem that is more complicated or abstract than plain old integer-style arithmetic must be explicitly derived (i.e., 'constructed') from integer arithmetic via a finite number of purely deductive steps." - Wallace

continued fractions = an infinite series of fractions w/in fractions - eg, 1

divided by a 2 that's divided by a 3 . . .
continuum hypothesis = "In mathematics, the **continuum hypothesis** (abbreviated **CH**) is a **hypothesis**, advanced by Georg Cantor, about the possible sizes of infinite sets" - WP
conundrum = a paradoxical, insoluble, or difficult problem
Cortical Blindness = blindness as a result of damage to the visual cortex part of the brain instead of due to the eyes
cosa nostra = "our thing" - aka the Mafia - see also *la stessa cosa*
cosa nulla = "nothing thing" - aka the absence of the Mafia
cosmological constant = "In physical cosmology, the **cosmological constant** (usually denoted by the Greek capital letter lambda: Λ) was proposed by Albert Einstein as a modification of his original theory of general realativity to achieve a stationary universe. Einstein abandoned the concept after the observation of the Hubble redshift indicated that the universe might not be stationary. However, the discovery of cosmic acceleration in the 1990s has renewed interest in a cosmological constant." - WP
counting numbers = the numbers used for counting - such as: 1, 2, 3 - sometimes including zero
critical threshold = a level of input required to push something past a certain point - eg: the amt of neuron firings rc'vd necessary to make that neuron fire in turn
cyberspace = the virtual world created by computers interfacing thru the internet etc
Delta = δ (the lower case of the 4th letter of the Greek alphabet) - used in math to denote a finite increment in a variable
dendrite = an incoming signal-carrier into a neuron (from other neurons)
denominator = aka divisor = the number doing the dividing in division, eg: the "*b*" in *a*/*b*
denumerable = countably infinite - ie: capable of having a one-to-one correspondence w/ counting numbers
dependent variable = "You'll recall from basic algebra that in a regular function like *y* = *f*(*x*), *x* is the *independent variable* and *y* is the *dependent variable*, meaning simply that changes in *x* produce other changes in *y* according to the rules of *f*. The set of all possible values that can be assumed by the independent variable is called the function's *domain*; the set of all possible *y*-values is the function's *range*." - Wallace
derivative = "The derivative of a function represents an infinitesimal change in the function with respect to one of its variables."

- Wolfram MathWorld

determinism = in philosophy, the belief that beings have lives that're predetermined, ie: fated, & are, therefore, out of their control

diagonal argument = the demonstration of how to establish a one-to-one correspondence between the infinite set of counting numbers & the infinite set of all rational numbers created by Georg Cantor

diameter = the length of a straight line that runs across the center of a circle until each end touches the perimeter

differential = "multiple related meanings in calculus and differential geometry, such as an infinitesimal change in the value of a function" - WP

differential calculus = determines the rate of change of a quantity

differential equations = "a mathematical equation for an unknown function of one or several variables that relates the values of the function itself and of its derivatives of various orders. Differential equations play a prominent role in engineering, physics, economics and other disciplines." - WP

digit = Seife identifies zero as a *digit*, not a number. He clarifies the point by giving the example that in "000,002,148" the digit zero is a place holder, rather than an expression of value

discontinuities = "Continuous fractions are of utmost importance in mathematics and applications. However, not all functions are continuous. If a function is not continuous at a point in its domain, one says that it has a **discontinuity** there." - WP

diss = to cast aspersions on

dividend = aka numerator = the number being divided in division, eg: the "*a*" in *a/b*

divisor = aka denominator = the number doing the dividing in division, eg: the "*b*" in *a/b*

domain = "You'll recall from basic algebra that in a regular function like *y* = *f*(*x*), *x* is the *independent variable* and *y* is the *dependent variable*, meaning simply that changes in *x* produce other changes in *y* according to the rules of *f*. The set of all possible values that can be assumed by the independent variable is called the function's *domain*; the set of all possible *y*-values is the function's *range*." - Wallace

dopamine = "Dopamine is a hormone and neurotransmitter occurring in a wide variety of animals, including both vertebrates and invertebrates [..] In the brain, dopamine functions as a neurotransmitter" - WP

"A Double Negative As Not A Positive" = a film made by E.G.Head (aka tENTATIVELY, a cONVENIENCE) & HANNAH AVIVA in 1982 as a way of expressing their attempt to transcend despair & misery thru levels of higher abstraction & ritual - the title is based on the idea that there might be a 3rd possibility beyond 2 negatives equalling a positive or, less correctly as far as mathematics is concerned, a negative - cONVENIENCE

Ecstasy = MDMA = Methylenedioxymethamphetamine = "MDMA exerts its primary effects in the brain on neurons that use the chemical serotonin to communicate with other neurons. The serotonin system plays an important role in regulating mood, aggression, sexual activity, sleep, and sensitivity to pain." - NIDA

Elements = Euclid's text credited as the earliest known systematic discussion of *geometry*

Elsie, the not-Cow = ¬C = If a definition is ambiguous, eg, if it's *undecidable* whether space is made of *particles* to be represented as *dots* or *strings*, then as ¬H attempts to pass thru the cobordism crack it transforms from either a(n) horseELEPHANT &/or an elephantHORSE into "Elsie, the ¬ Cow" - in other words, a shining example of PNT - cONVENIENCE

empty set = aka null set = a set w/ no contents symbolized by either "{ }" or "Ø"

Enigma = the name of a nazi code used during so-called World War II that was considered to be 'unbreakable' until Alan Turing & others devised computational means for doing so

epiphenomena = side-effects or byproducts

Epsilon = the 5th letter of the Greek alphabet - ϵ = lower case - used by Paul Erdös to mean children

epsilon numbers = ($\epsilon=\omega^{\epsilon}$) - ie: where epsilon = (lower case) omega to the epsilon power - Omega being the last letter of the Greek alphabet - omega is intended here to mean the 1st transfinite ordinal number - making this a higher order degree of infinity than the 1st degree

equal-tempered tuning = a tuning system in music, esp in keyboards, that makes the intervallic relationships between notes the same for every major & minor key

Eucharist = a rite or ritual of worship & remembrance wch Christians generally believe was instituted by Jesus at his last meal w/ his disciples before being handed over to his executioners

Euclid's 5th axiom = aka the parallel postulate
Euclid's 5th axiom = ***not*** aka the parallel postulate - in this case it refers to one of Euclid's 5 "common notions" of wch the above referred-to 'axiom' is one. This particular #5 is a principle of mereology. Mereology is "a collection of axiomatic first order theories dealing with parts and their respective wholes." (Wikipedia)
even numbers = every other counting number after "1" - ie: 2, 4, 6, . . .
exceptional points = "**Exceptional points** are grid **points** at the termination of lines internal to the computational domain, and where boundary conditions are therefore not applied." - from an abstract for a paper posted online called "**The force/work differencing of exceptional points in the discrete, compatible formulation of Lagrangian hydrodynamics**" by **R. Loubère & E.J. Caramana**
extended complex plane = Riemann's visualization of a sphere touching a complex plane - 0 is the Riemann Sphere's south pole, and ∞ and 0 are by differential-geometric definition inversely related
extensional = "In Philosophy of language, a context in which a sub-sentential expression *e* appears is called extensional if and only if *e* can be replaced by an expression with the same extension and necessarily preserve truth-value. The extension of a term is the set of objects that that term denotes.

Take the case of Clark Kent, who is Superman. Suppose that Lois Lane fell out of a window and Superman caught her. Thus the statement, "Clark Kent caught Lois Lane," is true because it has an extensional context. The names "Superman" and "Clark Kent" have the same extension, which is to say that they both refer to the same person, i.e. that superhero who is vulnerable to kryptonite. Anybody that Superman caught, Clark Kent caught." - WP

externally imposed hebephrenic instantiation = "A template for perceiving 'reality' is created & then superimposed on the neural-firing patterns of the victim's brain BAMN! (By Any Means Necessary) This is an aspect of what I call *stereotype projecting* - my nemesis & pet peeve throughout most of my life. In stereotype projecting, an oversimplistic template is so burned into the brain, that the brain's unable to have a response appropriate to the data it receives - it becomes hebephrenic." - cONVENIENCE
extremal problems = problems about extremes, like "'What's the largest number of edges a graph can have?'" - Hoffman

families of curves = "a set of **curves**, each of which is given by a function or parametrization in which one or more of the parameters is variable" - WP

Fermat's Enigma = aka Fermat's Last Theorem = When mathematician Pierre de Fermat died, he left behind a marginalia to the effect that he had a proof that "If an integer *n* is greater than 2, then the equation *an* + *bn* = *cn* has no solutions in non-zero integers *a*, *b*, and *c*." His proof (if, indeed, it ever existed) having never been found, people searched for one for over 350 yrs before mathematician Andrew Wiles proved it in 1995.

Feynman Diagrams = "A Feynman diagram is a tool invented by American physicist Richard Feynman for performing scattering calculations in quantum field theory. Particles are represented by lines, which can be drawn in various ways depending on the type of particle being depicted. A point where lines connect to other lines is called an interaction vertex, or vertex for short." - WP

Fibonacci numbers = numbers that add together the previous 2 numbers in their sequence: 0, 1, 1, 2, 3, 5, 8, 13, 21, 34, 55, 89, 144, 233, 377, 610, 987, ... - these numbers are often found in nature - as is explained on page 68

fingernail technician = a person who cuts, manicures, & paints fingernails

first-order predicate calculus = a kind of logic that uses quantifiers like '∀' [for all] and '∃' [there exists] - aka First Order Logic (FOL) wch "uses a wholly unambiguous formal language interpreted by mathematical structures" - Wikipedia

floating point numbers = "In computing, floating-point describes a numerical representation system in which a string of digits (or bits) represents a real number. The most commonly encountered representations are those defined by the IEEE 754 Standard.

"The name "floating-point" refers to the fact that the radix point (decimal point, or, more commonly in computers, binary point) can 'float'; that is, it can be placed anywhere relative to the significant digits of the number" - WP

Folk Math = a vague term for mathematical practices w/ little or no discipline behind them - usually incorporating images & symbols for the sake of making pictures rather than for making statements free radicals = atoms or groups of atoms with an odd (unpaired) number of electrons that can be formed when oxygen interacts with certain molecules - these are sometimes connected w/ aging - whether or not this definition has anything to do w/ their mention in this bk I cdn't tell ya b/c the free radical in my brain have been raising cane sugar - used punningly in this bk

free will = in philosophy, the belief that beings are capable of controlling their own lives

friendly numbers = 2 or more numbers whose proper divisors sum to equal each other - eg: the proper divisors of 220 are 1, 2, 4, 5, 10, 11, 20, 22, 44, 55, and 110, & they sum to 284; the proper divisors of 284 are 1, 2, 4, 71, and 142, and they sum to 220 - therefore, 220 & 284 are friendly numbers - this was a notion conceived of (or credited to) Pythagoras - Fermat discovered that 17,296 & 18,416 are also friendly numbers

function = e.g., "'A relation between variables,' 'A rule for establishing the image of a domain,' 'A mapping'). A function is at least one abstraction level up from variables, being basically a rule for pairing elements in one set with elements in another set." + "A relationship between two things where the value of one is determined by the value of the other." - Wallace

Game Theory = "the study of the ways in which *strategic interactions* among *rational players* produce *outcomes* with respect to the *preferences* (or *utilities*) of those players, none of which might have been intended by any of them. The meaning of this statement will not be clear to the non-expert until each of the italicized words and phrases has been explained and featured in some examples. [wch ain't happening here so go to their website!]" - Stanford Encyclopedia of Philosophy

Gaussian Primes = are Gaussian integers $z = a + b\,i$ satisfying one of the following properties: 1. If both a and b are nonzero then, $a + b\,i$ is a Gaussian prime iff [If and only if (i.e., necessary and sufficient). The terms "just if" or "exactly when" are sometimes used instead" - nIFFty, huh!] $a^2 + b^2$ is an ordinary prime. 2. If $a = 0$, then $b\,i$ is a Gaussian prime iff $|b|$ is an ordinary prime and $|b| \equiv 3 \pmod 4$. 3. If $b = 0$, then a is a Gaussian prime iff $|a|$ is an ordinary prime and $|a| \equiv 3 \pmod 4$.

Gedankenwelt = thought-world
general solutions = "**Solutions** obtained from integrating the differential equations are called **general solutions**." - e[ngineering]funda[mentals] website
Generative Transformational Grammar = "In linguistics, a **transformational grammar**, or **transformational-generative grammar** (**TGG**), is a generative grammar, especially of a natural language, that has been developed in a Chomskian tradition." [..] "In 1957, Noam Chomsky published *Syntactic Structures,* in which he developed the idea that each sentence in a language has two levels of representation — a deep structure and a surface structure. The deep structure represented the core semantic relations of a sentence, and was mapped on to the surface structure (which followed the phonological form of the sentence very closely) via *transformations.* Chomsky believed that there would be considerable similarities between languages' deep structures, and that these structures would reveal properties, common to all languages, which were concealed by their surface structures" - WP
genetics = the science of heredity and variation in living organisms
Gog (aka GE-og) = a fictional caveman created by Charles Seife & mutated by tENTATIVELY, a cONVENIENCE (from whom the pronunciation guide comes) - more specifically, GEog is an acronymn for General Eisenhower's oldest girl wch is, to make matters more complicated the beginning of a mnemonic for GEOGRAPHY
Googlewhacking = searching for 2 words online that only occur on ONE website & that're both defined in Google (the search engine that Googlewhacking takes its name from)'s dictionary
"Google Wracking" = searching for 2 words online enclosed in "quotes" that yield at least 1,000,000 results & preferably over 1,000,000,000 - a term invented by tENTATIVELY, a cONVENIENCE inspired by Googlewhacking
Grandi series = a particular subtype of divergent infinite series known as an *oscillating series*, and an object lesson in stipulation for partial sums (symbolized s_n), with the relevant symbolism being '1 + ∑(-1)*n* where $s_n = 0$ for even *n* and $s_n = -1$ for odd *n*
Grassmann Numbers = Wikipedia - aka **anticommuting number** or **anticommuting c-number** = "an element of the algebra

spanned by quantities θ*i* that anticommutes with all generators θ*j* but commutes with ordinary numbers *x*:

$$\theta_i\theta_j = -\theta_j\theta_i \qquad \theta_i x = x\theta_i."$$

Great Chain of Being = Wikipedia:

"The **great chain of being** or **scala naturæ** is a classical and western medieval conception of the order of the universe, whose chief characteristic is a strict hierarchical system.

"The chain of being is composed of a great number of hierarchical links, from the most basic and foundational elements up through the very highest perfection, in other words, God, or the Prime Mover.

"God, and beneath him the angels, both existing wholly in *spirit* form, sit at the top of the chain. Earthly flesh is fallible and ever-changing: mutable. Spirit, however, is unchanging and permanent. This sense of permanence is crucial to understanding this conception of reality. One does not abandon one's place in the chain; it is not only unthinkable, but generally impossible. The hierarchy is a chain and not a ladder. (One exception might be in the realm of alchemy, where alchemists attempted to transmute base elements, such as lead, into higher elements, either silver, or more often, gold—the highest *element*.)"

grouping = the way of arranging variables in a series, equation, set, etc - how one groups can cause substantial differences in outcome - eg: "Take a look at the series 1-1+1-1+1-1+1-1+1-1+... What does it add up to? Grouping the terms one way, (1-1)+(1-1)+(1-1)+(1-1)+(1-1)+(1-1)+(1-1)+... the sum is definitely 0. But if the terms are grouped another way, 1+(-1+1)+(-1+1)+(-1+1)+(-1+1)+... the sum is definitely 1." - Hoffman

haemodynamic response = blood flow activity

Hankie of Death = It's explained in a footnote in David Foster Wallace's bk entitled <u>Everything and More</u> that a professor that Wallace studied under (Dr. Goris - frequently mentioned throughout the bk) "used always to carry, to blow/mop with, and deploy illustratively a large red pocket-handkerchief, which for over 25 years of classes he referred to as the Hankie of Death."

Happy Solution = a cONVENIENT pseudo- or quasi- solution to a

Hal(l)e/y(/) Berry Paradox = "In any seemingly infinite universe of stars, who is the most brilliant superstar not definable in under eleven words?" --if you can't figure that out, at least find a way to stop blaming your psychotherapist for your own inability to solve paradoxes."

problem *or* an actual solution that's satisfying

hardwiring = strictly speaking, a physical connection between parts - esp wiring that allows an electrical flow - metaphorically, any connective system that serves a similar function: such as the nervous system - often used to mean a system that's built into a physical structure in such a way that major changes to that structure can only occur thru alterations to the system

Heisenberg Uncertainty Principle = "Any attempt to measure precisely the velocity of a subatomic particle, such as an electron, will knock it about in an unpredictable way, so that a simultaneous measurement of its position has no validity. This result has nothing to do with inadequacies in the measuring instruments, the technique, or the observer; it arises out of the intimate connection in nature between particles and waves in the realm of subatomic dimensions." - University of Oregon 21st Century Science lecture

Hoda Intigra = most likely a misspelling of Honda Integra (a type of car) by someone other than myself who was having fun w/ WikiAnswers - this *cd* be a fictional consumer product made up by the same person

holism = the study of the complex of larger inter-related phenomena instead of just fragments

holy ceiling light = an object of devotion for atheists who prefer the concreteness of an object that provides observable enlightenment

homonyphonemiac = a person addicted to making puns

hors-fond = the HORSEPHANT (or *hors-fond* meaning, en Français, *Out of Essence* or *In the Beginning*) wch is the *n*-dimensional *dot-based* elephantHORSE moving **forwards** into the *n*-dimensional *Penguin Diagram string-based* horseELEPHANT space - *also* resulting in Elsie, the not-Cow - aka the 1st emblem of the *Cognitive Dissidents Group* on GoodReads

"Hunch Backing" = the Intuitionist subset of "Google Wracking"

hyperbolic space = "Hyperbolic space, developed independently by Lobachevsky and Bolyai, is a geometrical space analogous to Euclidean space, but such that Euclid's parallel postulate is no longer assumed to hold." - WP

hyperreal numbers = "The system of hyperreal numbers represents a rigorous method of treating the ideas about infinite and infinitesimal numbers that had been used casually by mathematicians, scientists, and engineers ever since the invention of calculus by Newton and Leibniz. The hyperreals, or **nonstandard reals** (usually denoted as ***R**), denote an ordered field which is a proper extension of the ordered field of real numbers **R** and which satisfies the transfer principle. This principle allows true first order statements about **R** to be reinterpreted as true first order statements about ***R**.

"An important property of ***R** is that it has *infinitely large* as well as infinitesimal numbers, where an *infinitely large* number is a number that is larger than all numbers representable in the form 1 + 1 + . . . + 1." - WP

iconic signs = "signs that acquire their function through similarity to what they signify; a photograph as an iconic sign of the person in the picture" - Medical Dictionary definition

idiot savant = "Idiot savants are a group of humans that are incapable of learning, writing or reading, yet they have unlimited access to specific, accurate knowledge in the fields of mathematics, music, and other precise areas. Now the irony of an idiot-savant is that this group of individuals does not acquire knowledge by learning as the average human does. They mysteriously 'know' explicit, exact, correct information." - Dr . Lee E. Warren, B.A., D.D. on his website: <http://www.plim.org/2idiots.html> What attracted me to the preceding definition is this sentence that appears shortly before it on the same page: "Current research theories show that intelligence is non-local and not bound to the brain" - see cONVENIENCE's discussion w/ brain researcher Pig Pen about theories of where the "center-of-thought" is on pp 90-92

iff = if & only if

iffy = uncertain, dubious

Illimitable = incapable of being limited or bound

imaginary numbers = "i" = a specific type of number *not* in the set of *real* numbers - generally exemplified by √-1 (or the square of any negative number)

inclusion = "The principle of inclusion-exclusion was used by Nicholas Bernoulli to solve the recontres problem of finding the number of derangements" "A derangement is a permutation in which none of the objects appear in their "natural" (i.e., ordered) place. For example, the only derangements of {1, 2, 3} are {2, 3, 1} and {3, 2, 1}, so !3 = 2." - Wolfram MathWorld

indecidability = "1931 Kurt Gödel (Brünn, 1906 - Princeton, 1978) establishes the principle of *indecidability* in arithmetics. This remarkable work demonstrates that, even in the science of whole numbers (integers) mathematics is not a tautology." - Genetics of Bacterial Genomes website

independent variable = "You'll recall from basic algebra that in a regular function like $y = f(x)$, x is the *independent variable* and y is the *dependent variable*, meaning simply that changes in x produce other changes in y according to the rules of f. The set of all possible values that can be assumed by the independent variable is called the function's *domain*; the set of all possible y-values is the function's *range*." - Wallace

indeterminacy = in philosophy, "can refer both to common scientific and mathematical concepts of uncertainty and their implications and to another kind of indeterminacy deriving from the nature of definition or meaning" - WP

indeterminate forms = "Expressions such as ∞/∞ or $\infty - \infty$ are known as "indeterminate forms." These expressions have no preassigned value; they can be evaluated only through a limiting process. Loosely speaking, in every indeterminate form there is a "struggle" between two quantities, one tending to make the expression numerically large, the other tending to make it numerically small. The final outcome depends on the precise limiting process involved. The indeterminate forms most commonly encountered are $0/0$, ∞/∞, $0 \cdot \infty$, $\infty - \infty$, 0^0, ∞^0, and 1^∞." - Maor

indexical signs = "Together with the icon and the symbol, the index constitutes the three kinds of signs distinguished by Peirce according to the relation contracted by the expression (Peirce's "representamen") and the content and/or referent (Peirce's "object", i.e. "immediate" or "dynamic object"). According to one possible paraphrase, which seems sufficiently broad to account for most of Peirce's examples and a fair amount of his definitions, an *index* is a sign in which *the 'thing' which serves as the expression is, in one or other way, which is independent of the sign relation, connected with another 'thing', which serves as its content* (cf. indexicality). The term, if not the concept, has, however, a number of other sources, some of which are not necessarily compatible with Peirce's conception, and not all uses of the term, neither by Peirce, nor by later semioticians, may really concern the same concept.

"The term chosen by Peirce certainly suggests that all indices, like the pointing index finger, or an arrow, serve to pinpoint a particular object, to isolate it and bring it out of the, typically spatial, context into which it is ordinarily enmeshed; and this is indeed what Peirce affirms" - Lund University's "The Internet Semiotics Encyclopedia"

infinite series = aka infinite sum = represented by a mathematical expression that ends w/ an ellipsis (". . .") to indicate that it goes on that way forever or has an ellipsis & an ending w/ a variable that can represent a number that approaches a limit that the series approaches but doesn't meet - eg: "1/2 + 1/4 + 1/8 + 1/16 . . ." or $a_1 + a_2 + a_3 + \ldots + a_n$

infinite set = according to Cantor, "the defining formal property of an *infinite set* is that such a set can be put in a 1-1C with at least one of its proper subsets." - Wallace

infinitesimal = the infinitely tiny

inflection = a nuanced expressive characteristic in natural language that seems to be missing in mathematical statements

instantiation = "In programming, instantiation is the creation of a real instance or particular realization of an abstraction or template such as a class of objects or a computer process. To instantiate is to create such an instance by, for example, defining one particular variation of object within a class, giving it a name, and locating it in some physical place." - Whatis.com

integers = $\mathbb{Z}$ = "(from the Latin *integer*, which means with untouched integrity, whole, entire) are the set of numbers including the whole numbers (0, 1, 2, 3, ...) and their negatives (0, -1, −2, −3, ...). In non-mathematical terms, they are numbers that can be written without a fractional or decimal component, and fall within the set {... −2, −1, 0, 1, 2, ...}. For example, 65, 7, and −756 are integers; 1.6 and 1½ are not integers" - WP

integral calculus = finds the quantity where the rate of change is known

Intelligence Quotient = a way of measuring intelligence that's often criticized as class-based insofar as its notion of intelligence is rooted in what constitutes intelligence only under specific privileged social circumstances

intensional = "descriptions can 'float' without being anchored down to specific, known objects" - Hofstadter

- "In opposition to extensional contexts are intensional contexts, where synonymous terms cannot be substituted in without potentially compromising the truth-value.

"Suppose that Lois Lane believes that Clark Kent will investigate a news story with her. The statement, "Lois Lane believes that Superman will investigate a news story with her," is false, even though Superman is Clark Kent. This is because 'believes' is typically an intensional context" - WP

Intuitionists = a philosophy of mathematics in wch only what is intuitively accessible is acceptable as real

inverse tangent function = "a well-defined function if define the inverse tangent function by saying *y* = tan-1(*x*) if *y* is the value between -π/2 and π such that tan(*y*) = *x*" - UBC Calculus Online Course Notes

irrational numbers = numbers that *can't* be expressed as ratios but only as non-repeating decimals

irregular primes = "a **regular prime** is a certain kind of prime number. A prime number *p* is regular if it does not divide the class number of the *p*-th cyclotomic field (that is, the algebraic number field obtained by adjoining the *p*-th root of unity to the rational numbers). The first few regular primes (sequence A007703 in OEIS) are: 3, 5, 7, 11, 13, 17, 19, 23, 29, 31, 41, ..." [..] "An odd prime that is not regular is an **irregular prime**. The number of *Bk* with a numerator divisible by *p* is called the **irregularity index** of *p*. K L Jensen has shown in 1915 that there are infinitely many irregular primes, the first few of which are (sequence A000928 in OEIS): 37, 59, 67, 101, 103, 131, 149, ..." - WP

isomorph = "something identical with or similar to something else in form or structure" - Mirriam-Webster's Medical Dicitonary

Italian Futurist *sentesi* = a way of reducing a cultural object to its most skeletal form created by members of an early 20th century Italian cultural revolution

Jiminy H. Particulate Matter Criminy - a deliberately exaggerated euphemistic 'curse' based on "Jesus H. Christ" & made in the spirit of "dagnabbit" instead of God Damn It" - an invention of cONVENIENCE

"Jumping out of the System" - a Hofstadter term for leaving a perceptual system (wch may require recognizing it to begin w/) - akin to my "undermining reality maintenance traps"

just intonation = a tuning system in music in wch the frequencies of the notes can be represented by ratios of whole numbers

Kabbalah = "(Hebrew : קַבָּלָה, Tiberian: qab:ɔˈlɔh, *Qabbālāh*, Israeli: *Kabala*) refers to a set of esoteric teachings and mystical practices that form an alternative to traditional Jewish interpretations of the Tanakh and religious observances. It is a set of beliefs followed by some Jews as the true meaning of Judaism, while rejected by other Jews as heretical and contrary to Judaism." - more specifically in relation to this bk: "*Sod*, the hidden mystical reading of the inner secrets of Torah. These hidden meanings are more than just what may be seen by attentive study, but include layer upon layer of codes and meanings intentionally hidden by God, which can only be found through unraveling coded meanings. - The study of the inner secrets of Torah (Sod) is called Kabbalah" - WP

Karen Eliot = a multiple identity name in neoism - ie: a name used by many people in order to collectively construct an identity that can be assumed by anyone but not necessarily connected to anyone in particular

knuckle sandwich = slang for a punch w/ a fist

köan = "a story, dialogue, question, or statement in the history and lore of Chán (Zen) Buddhism, generally containing aspects that are inaccessible to rational understanding, yet may be accessible to intuition." - Wikipedia

Korsakoff's Amnesia = damage to the memory thru damage to the Hippocampus area of the brain caused by long-term alcohol abuse

Language Writing = a 'school' of writing that originally stressed non-*transparency* - in other words, an emphasis more on the words & the process of reading rather than the words used as signifiers that aren't noticed in & of themselves - now a term associated w/ a broad variety of usually academic writing

Law of the Excluded Middle = the law in mathematics that states that in any given situation there's either a right or a wrong solution w/ no possibilities in between

Law of the Included Middle = a pseudo-law used by cONVENIENCE to mean possibilities outside polar dichotomies

limit = something that 'bounds' an infinite series by having the series *approach* it w/o ever being able to *reach* it. *SO*, if we create a series like 1, 1/2, 1/3, 1/4... & designate that as *a* then the limit of *a* is 0 (zero) - This is expressed as: $1/n \to 0$ as $n \to \infty$ or as:

$$\lim_{n \to \infty} \frac{1}{n} = 0$$

The limit of one divided by *n* as *n* approaches infinity is zero.

logarithm = Wikipedia: "In mathematics, a logarithm of a given number to a given base is the power to which you need to raise the base in order to get the number. For example, the logarithm of 1000 to the common base 10 is 3, because 10 raised to a power of 3 is 1000" "In simple terms, the natural logarithm of a number *x* is the power to which *e* would have to be raised to equal *x* — for example the natural log of *e* itself is 1 because $e^1 = e$, while the natural logarithm of 1 would be 0, since $e^0 = 1$." If a logarithm uses *e* as its base, it's written as "$\log_e$".

logarithmic spiral = aka equiangular spiral or growth spiral = "a special kind of spiral curve which often appears in nature. The logarithmic spiral was first described by Descartes and later extensively investigated by Jakob Bernoulli, who called it Spira mirabilis, "the marvelous spiral"" - ie: the distance between arms of the spiral grows as the spiral extends out from the center

Lowest Common Denominator = LCD = a social concept meaning something that is understandable & appealing to the most people - used by people like myself as a quasi-synonym for *maximally dumbed down*

Luddites = "a social movement of British textile artisans in the early nineteenth century [starting in 1811] who protested — often by destroying mechanized looms — against the changes produced by the Industrial Revolution, which they felt threatened their livelihood" - WP - named after Ned Ludd

Luther Blissett = a multiple identity name both in & out of neoism - ie: a name used by many people in order to collectively construct an identity that can be assumed by anyone but not necessarily connected to anyone in particular

malaphor - "coined by the newspaper columnist Lawrence Harrison; it means a cross between a malaproprism and a metaphor." - Hofstadter

malaproprism = unintentional misuses of a word confused w/ one that sounds similar - eg: "What was that subject that you were *deferring* to?" instead of "*referring* to" manifold = Wikipedia: "A manifold is an abstract mathematical space in which every point has a neighborhood which resembles Euclidean space, but in which the global structure may be more complicated. In discussing manifolds, the idea of dimension is important. For example, lines are one-dimensional, and planes two-dimensional."

marginalia = writing & other markings in the margins of a bk - usually made by someone reading the bk who makes the markings near what the markings refer to - eg, if an author is referred to in a text, a reader might make a note to themselves in the margins that says "read [insert name of author here]"

Medulla Oblongata = a part of the brain that controls basic body functions that're performed mostly unconsciously, and relays nerve signals between the brain and spinal cord

Mendel's Laws of Heredity = Gregor Mendel's "careful cross-breeding of thousands of pea plants led Mendel to key insights, now called Mendel's Laws of Heredity, about how inherited traits are passed on from generation to generation." - who2?

meta- = a prefix used to mean "a concept which is an abstraction from another concept, used to complete or add to the latter" - WP

meta-descriptions = descriptions of descriptions

metaphor = a description of something in terms more poetic than analagous - eg: "She is a rose" in wch the "she" referred to is actually a human & not a plant but is hereby referred to as such in order to compare the "she"'s qualities to those of a rose - eg: perhaps the "she" is ruddy-red complexioned from alcohol abuse

Michael Tolson = an acronymn for: Magnetize intarsia crispy hydrophytic accentric excogitate labiac Turgid oleographic luxated sabaist orthopteran nascent

mind games = usually a pejorative term to refer to dishonest manipulative social relations in wch motives remain ulterior - however, in neoism, eg, mind games can be a philosophical exercise intended to help keep active the players' abilities to be critically incisive in perception - ie: to not take surface 'meanings' for granted as 'reality' - in that sense mind games can be a perceptual thought experiment

mnemonics = methods for helping people to remember things - in this bk, most often represented by phrases & sentences the 1st letters of each word of wch add up to the thing meant to be remembered - eg: "**RHYTHM** = **R**hythm **H**elps **Y**our **T**wo **H**ips **M**ove" helps one to remember both the spelling of "rhythm" & its meaning

modality = in logic, the qualification of a proposition affirming or denying possibility, impossibility, necessity, contingency, etc - this is particularly important in PNT b/c of the way it enables natural language statements that might seem to be outside ordinary mathematics to be expressed as a higher abstraction - eg: "A formal modal logic represents modalities using unary modal operators. For example, "Jones's murder was a possibility"; "Jones was possibly murdered"; and "It is possible that Jones was murdered", all contain the notion of possibility; in a modal logic this is represented as an operator, *Possibly*, attaching to the sentence *Jones was murdered*.

"The basic **modal operators** are usually written $\Box$ (or *L*) for *Necessarily* and $\Diamond$ (or *M*) for *Possibly*. In a classical modal logic, each can be expressed by the other and negation:

$$\Diamond P \leftrightarrow \neg\Box\neg P; \Box P \leftrightarrow \neg\Diamond\neg P.$$

"Thus it is *possible* that Jones was murdered if and only if it is *not necessary* that Jones was *not* murdered." - WP

monotonic function = a **function** which preserves the given order by being either nonincreasing or nondecreasing

Monty Cantsin = the 1st multiple identity name in neoism - ie: a name used by many people in order to collectively construct an identity that can be assumed by anyone but not necessarily connected to anyone in particular

Mr. Gregor's Neighborhood = In Paradigm Shift Knuckle Sandwich there's a recurring theme of inventing a term for something & then creating a definition for it afterwards. *Mr. Gregor's Neighborhood* is a perfect example. I see the term "neighborhood", I immediately think of the kids tv show (wch I've never seen) "Mr. Roger's Neighborhood" & I immediately think of making a spin-off of "neighborhood" inspired by that association - so that leads to *Mr. Gregor's Neighborhood* because it's close to an anagram of "Mr. Roger's Neighborhood". Ok, that's all very obvious.

In *Mr. Gregor's Neighborhood*, Mr. Gregor's neighbors are all of the things created in the name of anyone named Gregor. A related neighborhood is Monty Cantsin's neighborhood.

Mu = "a response to certain koans and other questions in Zen Buddhism, intending to indicate that the question itself was wrong" - WP - used frequently by Hofstadter

multiplicative operators = "*, /, and %" in Java computer code

(M)Usic = a cONVENIENCE term for sounds organized around use value rather than muse value not necessarily meant to be musical given that "musicality" is often too restrictive a concept for extreme creativity

Naïve Set Theory = regular Cantorian set theory w/ its susceptibility to paradoxes

natural language = general usage language not tied to a specific discipline

natural logarithm = logarithm base *e*

National Association for the Advancement of Multi-Colored People = N.A.A.M.C.P. - no parody of the N.A.A.C.P. is intended = an anti-*race* group - ie: a group against the whole divisive notion of *race* meant to be an alliance of multi-colored people (everyone) &/or 'colorful characters' - cONVENIENCE

natural numbers = "either an element of the set {1, 2, 3, ...} (the positive integers or the **counting numbers**) or an element of the set {0, 1, 2, 3, ...} (the non-negative integers). The former is generally used in number theory, while the latter is preferred in mathematical logic, set theory, and computer science" - WP

necessary = "it is *not possibly false*" in modal logic - "usually written □ (or *L*) for *Necessarily*" - WP

negative numbers = originally conceived of in economic contexts, numbers that represent an deficit, a debt, an absence - the numbers of the left side of the Number Line

negative psychology = aka reverse psychology - ie: the manipulative use of saying the opposite of what you mean - eg: saying "I HOPE you *don't* read this book!" in an attempt to get someone to read the book just to spite you

Neighborhood = "On the Real Line, the *neighborhood* of a point *p* is the open interval (*p* - *a*, *p* + *a*) where *a* > 0 [In other words, where *a* combined w/ *p* yields something other than just *p* again - so why can't *a* be greater *or* less than 0? In other other words, just *not* 0: ie: *a* ≠ 0]. Another way to express this is to say that the *a*-*neighborhood* of *p* is less than *a*." - Wallace + [cONVENIENCE]

neoism = the fine smart of undefining itself; an invention of its enemies, the anti-neoists; a prefix & a suffix w/o anything in between; a bar in a town w/ a population of 12; a hologram of a loose floorboard that one can trip over..

neurons = "electrically excitable cells in the nervous system that process and transmit information" - WP

neologism = a newly invented word - as are many in this glossary

neorealist numbers = coined by cONVENIENCE in honor of the other number name(s) that reference art/cultural movements (such as *surreal numbers*) - Let's call Euler's concept of negative numbers as *beyond* the positive numbers on the Number Line *Neoreals*. (Forgive me, Leonhard) Remember that *real numbers* can be informally defined as numbers that can be represented by infinite decimal expressions. These include *rationals* & *irrationals*. *Reals* are distinguished from *imaginary numbers* & *complex numbers*. *Neoreals* are new stabs at realism in wch the mere act of trying to *imagine* something beyond the *positive infinities* on the *number line* is represented as an act of great *complexity*.

Think of "> ∞" in terms of degrees as determined by one-to-correspondence. Imagine redefining the set of all real numbers as including every imaginable degree of infinity conceived of now & ever to be conceived unto infinity. This is the companion set of *neorealist numbers* called the *Preposterous* or, alternately, the *Outrageous* & *Ridiculous*. Even though the *POR Solution* shd hypothetically contain the negative numbers, these are forced to live in the set of *neorealist numbers* because of unusual magnetic properties that repel the negatives from the *POR Solution*. Feeling rejected, the negative numbers have developed an attitude that they're > than all those other ∞s & that that's *really* why they're in the set of *neoreals*.

neuroimaging = a technical method for mapping the brain

nesting = the process of putting things w/in things

Netherworld = a hypothesis about where people go after dying

nihilism = destructive behavior rooted in the philosophy that current conditions need to be destroyed

noetic radices = aka noetic radicals = radicals of the intellect = aka imaginary roots = aka imaginary numbers = eg, $\sqrt{-1}$

nonalgebraic numbers = ie: transcendental numbers such as π or *e* that aren't roots for any non-zero polynomial

non-axiomatic = something not based on the idea of irrefutable truths & the like - online I notice "NARS" (Non-Axiomatic Reasoning

System) wch is concerned w/ "[e]xploring the essence of intelligence" - more grist for the PNT mill

non-commutative matrices = "a **matrix** (plural **matrices**) is a rectangular table of *elements* (or *entries*), which may be numbers or, more generally, any abstract quantities that can be added and multiplied" - WP "matrix multiplication is noncommutative since $\begin{bmatrix}0 & 2\\0 & 1\end{bmatrix} = \begin{bmatrix}1 & 1\\0 & 1\end{bmatrix} \cdot \begin{bmatrix}0 & 1\\0 & 1\end{bmatrix} \neq \begin{bmatrix}0 & 1\\0 & 1\end{bmatrix} \cdot \begin{bmatrix}1 & 1\\0 & 1\end{bmatrix} = \begin{bmatrix}0 & 1\\0 & 1\end{bmatrix}$." - WP

non-commutativity = a situation where changing the order of something in an equation *will* change the result - ordinarily, according to commutativity, *ab* = *ba* - "**quaternions** are a non-commutative extension of complex numbers. They were first described by the Irish mathematician Sir William Rowan Hamilton in 1843 and applied to mechanics in three-dimensional space. At first, quaternions were regarded as pathological because they disobeyed the commutative law *ab* = *ba*" - WP

Non-Euclidean Geometry = a version of geometry in wch Euclid's 5th axiom doesn't hold true. This axiom is commonly called the "parallel postulate" - in this, it's stated that for every line w/ a point not on that line there is only *one line* running thru that point on the same plane that doesn't intersect the 1st line

non-logical axiom = "In general, a non-logical axiom is not a self-evident truth, but rather a formal logical expression used in deduction to build a mathematical theory." - WP

Non-Non-Euclidean = a mathematical/geometrical not-not joke - wch itself is a take-off on knock-knock jokes as well as on the idea of "A Double Negative As Not A Positive"

non-observable reality = "Abstract reality is the complement of objective reality and includes subjective reality and invisible or nonobservable reality, which is the part of objective reality that there is no direct access to." - from a Baha'i website called "onethread"

No-No Class = a social 'class' name coined by cONVENIENCE to represent people trying to self-define outside hierarchical class-structure - another *not-not joke*

non-selfish numbers = a number that's paired w/ a subset that doesn't contain the same number non-standard analysis = "is the invention of one [Prof.] A[braham] Robinson in the '70s and professes to rigorize infinitesimals in analysis via the use of *hyperreal numbers*, which themselves basically combine the real numbers and Cantorian transfinites - meaning the whole thing's heavily set-theoretic and Cantor-dependent, plus controversial, and wildly technical" - Wallace

normal numbers = "roughly speaking, a real number whose digits (in every base) show a uniform distribution, with all digits being equally likely, all pairs of digits equally likely, all triplets of digits equally likely, etc." - WP

not-Horse = "¬H" = something whose existence is mainly defined by its being *not a horse* rather than being defined by what it might *be* - in general, the not-Horse can be a symbol for anything defined negatively - cONVENIENCE

Nuclear Brain Physics Surgery's Cool = a school that tENTATIVELY, a cONVENIENCE created in 1978 for people to sleep thru the recorded lessons of w/o ever hearing the lessons while awake - each new lesson was created by the previous graduating class of 10 w/ no knowledge of what the others were doing - originally the plan was to eventually have a reunion where all the tape lessons wd be listened to while awake for the 1st time

null sign = Ø = symbol for the empty set - otherwise shown as "{ }"

numerator = aka dividend = the number being divided in division, eg: the "*a*" in *a/b*

number theory = aka higher arithmetic = "is a branch of mathematics concerned with the properties of integers, rational numbers, irrational numbers, and real numbers. Sometimes the discipline is considered to include the imaginary and complex numbers as well.

"Formally, numbers are represented in terms of sets; there are various schemes for doing this. However, there are other ways to represent numbers -- for example, as angles, as points on a line, as points on a plane, or as points in space" - WhatIs.com

numerology = "any of many systems, traditions or beliefs in a mystical or esoteric relationship between numbers and physical objects or living things" - WP

obsessive-compulsive = in psychology, behavior that's driven by an urge to follow certain procedures regardless of their necessity in relationship to the act to be performed - procedures often associated w/ fear of germs or fear of negative consequences for not following them - eg: a person who washes their hands hundreds of times before they can feel that they're clean or a person who must count every object they see or a person who can't tolerate having objects placed anywhere other than where they decide they shd be placed - obsessive-compulsives usually follow their inclinations to self-destructive extremes - alternately, obsessive-compulsive behavior can enable intense working discipline that makes getting things done much more possible than in ordinary states of mind

odd numbers = a number that is not a multiple of 2 - eg: 1, 3, 5, . . .

ogjective = an alternative to a standard dichotomy: subjective & objective. Long ago & far away, I coined the term "ogjective" to mean the included middle between the 2. If the perceiver is part of the perceived then their perception is both subjective & objective, ie: ogjective; a perceiver forms itself by creating a mental image of itself & its self is ultimately undetachable from the hypothetically infinite environment that it's embedded in.

Omega = the last letter of the Greek alphabet = "The ordinal number of the whole denumerably infinite set {0, 1, 2, 3, 4, . . .} gets symbolized by the little omega '**ω**'" - Wallace

onomatoid = "that is, a namelike word which in fact designates nothing because it signifies too broadly" - Mazur

"on the long wavelength" - Paul Erdös's code for "Red" or communist - used to protect himself & his friends (so that they cd talk about politics) from the hostile Horthy regime in Hungary that he was living under pre-WWII

"on the short wavelength" - Paul Erdös's code for Horthy sympathizers or fascists - used to protect himself & his friends (so that they cd talk about politics) from the hostile Horthy regime in Hungary that he was living under pre-WWII - code chosen because of its opposition to "on the long wavelength"

ontology = "a study of conceptions of reality and the nature of being" - Wikipedia

order-type = how members of a set are arranged

ordinal numbers = numbers related to the order in wch a number occurs - eg: "1" is the ordinal number of zero in the sequence "0, 1, 2, 3"

paleobotanist = a person who studies plant fossils

palimpsest = a text whose remnants are retrievable under another manuscript - ie: at a time when paper was scarce, paper wd be reused by erasing previously existing content on it & putting new content overtop

Paradigm Shift Knuckle Sandwichs = The basic idea is that paradigm shifts in consciousness can occur from 'violent' conceptual shocks to the (consensus reality?) system. Perverse Number Theory proposes to explode ("Proplode" = propaganda-explode (ie: work towards a change in consciousness to induce physical change rather than a change in physicality that ultimately retains the same old same old consciousness) -> hence PNT instead of TNT) the system/s that keep its/their practioners imprisoned.

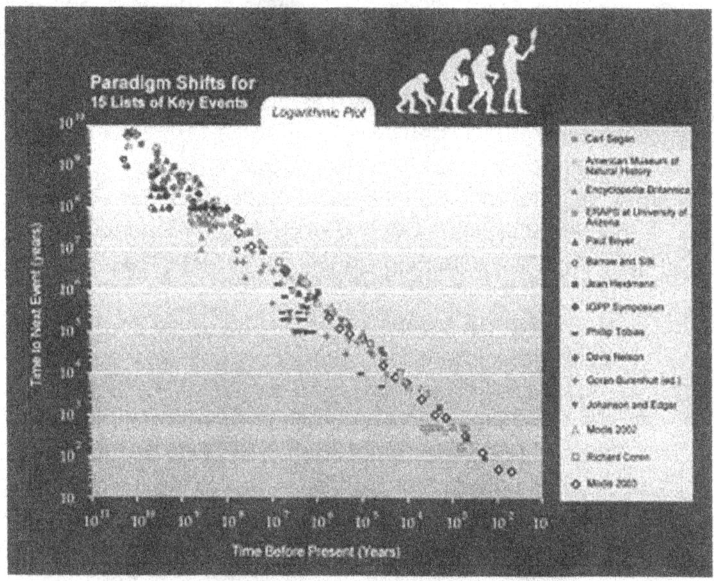

"When plotted on a logarithmic graph, 15 separate lists of paradigm shifts for key events in human history show an exponential trend. Lists prepared by, among others, Carl Sagan, Paul D. Boyer, Encyclopaedia Britannica, American Museum of Natural History, and University of Arizona, compiled by Ray Kurzweil." - chart & explanation: WP

- one has to wonder, of course, what the criteria for "paradigm shift" here is - wd, eg, an increase in the popularity of anal sex fit in anywhere? - or the increasing unpopularity of voting? - or the lack of faith in government officials? Most likely, I suspect, these things wd fit in but might not be considered b/c of the mindset of the people compiling the statistics - as usual, what statistics are accumulated & what they're taken to mean depends on the focus of the analyst - cONVENIENCE note

partial derivative = "a partial derivative of a function of several variables is its derivative with respect to one of those variables with the others held constant (as opposed to the total derivative, in which all variables are allowed to vary" - WP

partitions = "In number theory, a partition of a positive integer *n* is a way of writing *n* as a sum of positive integers. Two sums which only differ in the order of their summands are considered to be the same partition; if order matters then the sum becomes a composition. A summand in a partition is also called a **part**. The number of partitions of *n* is given by the partition function *p*(*n*). [..] The partitions of 4 are [..]: 1. 4, 2. 3 + 1, 3. 2 + 2, 4. 2 + 1 + 1, 5.1 + 1 + 1 + 1" - WP

Part-of-Speech Tagging = a quantifying practice of marking-up words based on their part-of-speech to determine word frequency. This is a part of *computational linguistics*.

Pataphor = a term coined by Pablo Lopez, "is as far beyond a metaphor as a metaphor is beyond non-figurative language" - ie: "pataphors essentially describe two degrees of separation from reality" - Wikipedia

pataphrase = to quote from a parenthetical insert on page 46 of this bk: "(The pataphrase has created a world where even the chessboard no longer exists, including the characters who lived in the original Wikipedia example's world, almost entirely abandoning the original context in favor of creating a faux logic in order to tie together elements referred to from GEB while simultaneously being ~GEB. This is classic PNT insofar as it will go to just about any length in order to make an abstruse joke. &, yes, it, too, is "bowling up the wrong tree" because the ball is now in the other sports pataphor - to isomorphoze further.)"

'Pataphysics = Alfred Jarry, the inventor of 'Pataphysics, describes it as an "epiphenomenon" ("that which is superimposed upon a phenomenon"), as the science "extending as far beyond metaphysics as the latter extends beyond physics." Furthermore, "Pataphysics will examine the laws governing exceptions" & "is the science of imaginary solutions".

Pavlovian = an adjective made from the last name Pavlov - a researcher who experimented w/ producing physical responses in dogs that became less & less related to their primary stimulus & more &

more related to a secondary, symbolic one - ie: before feeding the dogs he wd ring a bell, after a time, the dogs associated the bell-ringing w/ feeding & wd begin to salivate at its sound - even if no food were produced - as such, when a person, eg, is sd to be a Pavlovian Dog it means that they're being manipulated into having a response to a symbol by being tricked into thinking that the symbol somehow IS the 'real' thing *or* it just means that they're "having their chain pulled", ie: being emotionally manipulated

Peano's Postulates = 5 basic axioms of number theory, created by Italian mathematician Giuseppe Peano (1858-1932) - the 1st of wch is: For every natural number *x*, *x* = *x*. That is, equality is reflexive.

penguin diagram = Wikipedia: John Ellis (on <http://en.wikipedia.org/wiki/Feynman_diagram#Penguin_diagrams>) explains the origin of the designating of a Feynman diagram as a penguin diagram:

"One evening, [Melissa Franklin], I and Serge [Rudaz] went to a pub, and she and I started a game of darts. We made a bet that if I lost I had to put the word penguin into my next paper. She actually left the darts game before the end, and was replaced by Serge, who beat me. Nevertheless, I felt obligated to carry out the conditions of the bet.

"For some time, it was not clear to me how to get the word into this b quark paper that we were writing at the time. Then, one evening, after working at CERN, I stopped on my way back to my apartment to visit some friends living in Meyrin where I smoked some illegal substance. Later, when I got back to my apartment and continued working on our paper, I had a sudden flash that the famous diagrams look like penguins. So we put the name into our paper, and the rest, as they say, is history."

perfect numbers = numbers whose number's divisor sum equals that number

perfect stranger numbers = numbers whose number's divisor sum equals that number but can't be recognized as such - probably b/c either numbers or the perceiver have Korsakoff's Amnesia (Just Kidding!)

periphrasis = "aka circumlocution = what you do when you're 'beating around the bush'" - Worsley School website

Perverse Number Theory = a form of Number Theory in wch all truths taken to be obvious are subverted for the sake of seeing what happens next - read this entire bk for the long definition

Perverse Number Theory Outlaws = anyone that the author of this bk considers to be enuf of a Paradigm Shifting heretic to be worthy of such an honor

Phenomenologists = people whose approach to philosophy begins w/ an exploration of phenomena (our conscious experience) - as developed by Hegel, Husserl, Heidegger, & others

Phi = Φ (the upper-case version of the 21st letter of the Greek alphabet) - used here as a symbol of the infinitesimally small but generally used to mean many other things

phishing = the act of tricking someone into giving them confidential information or tricking them into doing something that they normally wouldn't do - mainly thru misrepresentation via the internet - such as by pretending to represent a specific financial institution to get credit card numbers & the like

photorealist numbers = numbers that aren't really numbers but are just things made by very skilled artists to look like they're numbers - these are often uncovered when you try to add them & they just sit there, unmoved - then you realize you've been tricked & that yr bank acct is an empty set

Physical Anthropology = "The branch of anthropology that deals with human evolutionary biology, physical variation, and classification" - answers.com

pinboard animation = this involves filming a board that has a million pins in it that are pushed in & out to create a relief surface on wch shadows are made.

place value systems = number systems in wch where the number is placed is one of the indicators of its cardinality - "The decimal system with which we are fated is a place-value system, which means that the value of a particular digit depends both on the digit itself and on its position within the number. For example, a four in the right-hand column simply means four ...... in the next column it means forty ...... one more column over means four-hundred ...... then four thousand, and so on." - maxmon.com

playful physics = Marcel Duchamp's creation under wch umbrella the 3 Standard Stoppages were made

polynomial equations = "an expression that is constructed from one or more variables and constants, using only the operations of addition, subtraction, multiplication, and constant positive whole number exponents" - WP

POR Solution = Imagine redefining the set of all real numbers as including every imaginable degree of infinity conceived of now & ever to be conceived unto infinity. This is the companion set of *neorealist numbers* called the *Preposterous* or, alternately, the *Outrageous* & *Ridiculous*. Even though the *POR Solution* shd hypothetically contain the negative numbers, these are forced to live in the set of *neorealist numbers* because of unusual magnetic properties that repel the negatives from the *POR Solution*. - cONVENIENCE

position/momentum operators = an operator, in mathematics, is "[a] function, especially one from a set to itself, such as differentiation of a differentiable function or rotation of a vector" (answers.com) - so, apparently, these are functions having to do w/, yes, position & momentum - not that that tells either of us much!

positive numbers = numbers greater than zero - to the right of zero on the Number Line

Post-Structuralism = "encompasses the intellectual developments of continental philosophers and critical theorists that wrote with tendencies of twentieth-century French philosophy. The prefix "post" refers to the fact that many contributors such as Jacques Derrida, Michel Foucault, and Julia Kristeva rejected structuralism and became quite critical of it. In direct contrast to structuralism's claims of an independent signifier, superior to the signified, post-structuralism views the signifier and signified as inseparable but not united.

"While post-structuralism is difficult to define or summarize, it can be broadly understood as a body of distinct reactions to structuralism. There are two main reasons for this difficulty. First, it rejects definitions that claim to have discovered absolute 'truths' or facts about the world. Second, very few people have willingly accepted the label 'post-structuralist'; rather, they have been labeled as such by others. Consequently, no one has felt compelled to construct a 'manifesto' of post-structuralism" - WP

power set = the set of all subsets of a set - wch will always have a larger cardinal number than the set that it's a power set of

predicate = in language, the predicate is the part of a sentence that tells something about the subject - eg: In the sentence: "Why, this book is utterly a Preposterous, Outrageous, & Ridiculous Solution to the problems of humanity!", "this book" is the subject

& "is utterly a Preposterous, Outrageous, & Ridiculous Solution to the problems of humanity!" is the predicate.

Pregroperatavism = a one person movement founded by Arthur Berkoff in the Netherlands that's a distant cousin or 2nd godmother or something to neoism

primary visual cortex = an area toward the back of the brain where basic visually rc'vd info is processed

prime numbers = numbers divisible only by 2 natural number divisors: themselves & 1

Probability Theory = "the branch of mathematics concerned with analysis of random phenomena. The central objects of probability theory are random variables, stochastic processes, and events: mathematical abstractions of non-deterministic events or measured quantities that may either be single occurrences or evolve over time in an apparently random fashion. Although an individual coin toss or the roll of a die is a random event, if repeated many times the sequence of random events will exhibit certain statistical patterns, which can be studied and predicted" - WP

proper divisors = divisors of a number other than the number itself

proper subset = "A set *A* is a *proper subset* of set *B* [iff] *A* is a subset of *B* and there is at least one member of *B* that is not a member of *A*." - Wallace

Proplode = propaganda-explode - ie: works towards a change in consciousness to induce physical change rather than a change in physicality that ultimately retains the same old same old consciousness - neologism coined by tENTATIVELY, a cONVENIENCE

proportionality = "In mathematics, two quantities are called proportional if they vary in such a way that one of the quantities is a constant multiple of the other, or equivalently if they have a constant ratio." - WP

Propositional Logic = aka "*sentential logic*, is the branch of logic that studies ways of joining and/or modifying entire propositions, statements or sentences to form more complicated propositions, statements or sentences, as well as the logical relationships and properties that are derived from these methods of combining or altering statements. In propositional logic, the simplest statements are considered as indivisible units, and hence, propositional logic does not study those logical properties and relations that depend upon parts of statements that are not themselves statements on their own, such as the subject and predicate of a statement" - The Internet Encyclopedia of Philosophy

pruning = the destruction of dendrites in the brain

pure math = "Broadly speaking, pure mathematics is mathematics motivated entirely for reasons other than application. It is distinguished by its rigour, abstraction and beauty. From the eighteenth century onwards, this was a recognized category of mathematical activity, sometimes characterised as speculative mathematics, and at variance with the trend towards meeting the needs of navigation, astronomy, physics, engineering, and so on." - WP

"putting the cart before the horse" - aka "jumping the gun" & "counting your chickens before they hatch" etc.. - expressions that mean reaching conclusions, eg, before creating adequate reasoning to build up to them

quadratic equation = in the leading term of a quadratic equation the variable is squared

quadratus = Latin for "square"

quanta = "According to Planck, molecules are forbidden to move in most ways. They vibrate only with certain acceptable energies, called quanta. It is impossible for molecules to have energies between these acceptable values." - Seife

quantum = "In physics, a quantum (plural: quanta) is an indivisible entity of energy. A photon, for instance, being a unit of light, is a "light quantum."" - WP

Quantum Mechanics = the study of the relationship between energy & matter

quine a phrase = to construct a sentence that has the 1st half in quotes followed by the same words not in quotes. This operation is named (probably by Hofstadter) after the philosopher credited w/ inventing it: Willard Van Orman Quine - a take-off on the expression: "to coin a phrase" - meaning to invent a new expression

Radical Doubt = explained on the on-line "SparkNotes": "This is the methodological doubt that Descartes first conceived in his *Meditations.* Descartes refused to accept anything as true unless it struck him as clearly and distinctly true, in itself. Russell revives this method in his first chapter's account of the dubious nature of our so-called "knowledge." He prompts us to doubt and reconsider our everyday conception of reality."

radicals = "The word "radical" comes from the Latin word "radix", meaning "root", from which we get not only the word "radical", but also "radish"! The other use of "radical" means "by the roots" or "from the roots", that is, complete, such as a "radical" change in a person. A political "radical" is someone who wants radical change by radical means, that is, an extremist. [cONVENIENCE note: To say that this latter definition of a political radical as "an extremist" is even more highly suspect than some of my own transparently spun definitions is an understatement - most etymologically thoughtful political radicals wd be more likely to take the word to mean in this context: a person trying to return social relations to basics (ie: roots) that *root out* unnecessary & viciously self-serving levels of imposed power added on thru centuries of imperialism & exploitation - to say that most doctors are beneficiaries of this system of power imposition wd also be an understatement - in other words, FIE ON YOU DR. PETERSON!!] Another word from the same source is "eradicate" (originally, to "root out" or "pull out by the roots").

"Now, why is a square root called a root in the first place?

"Answer: a root is the source of something; if you square a number, the number it came from is the root, as if the square grew from it." - Peterson on "Ask Dr. Math"

radice cubica = cube root

radice quadratica = square root

Radium-220 = A rare, brilliant white, luminescent, highly radioactive metallic element found in very small amounts in uranium ores, having 13 isotopes with mass numbers between 213 and 230 - 220 has a half-life of only twenty-three milliseconds

radius = the most direct straight distance measurement from the center to the perimeter of a circle

range = "You'll recall from basic algebra that in a regular function like $y = f(x)$, x is the *independent variable* and y is the *dependent variable*, meaning simply that changes in x produce other changes in y according to the rules of f. The set of all possible values that can be assumed by the independent variable is called the function's *domain*; the set of all possible y-values is the function's *range*." - Wallace

rational numbers = any number that can be expressed as a ratio - such as 1/2

reality maintenance traps = ie: a prioritizing of one's life around maintaining a 'reality' that entraps the maintainer, a 'reality' that one is afraid to let go of because one feels as if one will cease to exist w/o it

real numbers = "may be described informally as numbers that can be given by an infinite decimal representation, such as 2.4871773339.... The real numbers include both rational numbers, such as 42 and −23/129, and irrational numbers, such as π and the square root of 2, and can be represented as points along an infinitely long number line.

"A more rigorous definition of the real numbers was one of the most important developments of 19th century mathematics. Popular definitions in use today include equivalence classes of Cauchy sequences of rational numbers, Dedekind cuts, a more sophisticated version of "decimal representation", and an axiomatic definition of the real numbers as the unique complete Archimedean ordered field." - WP

rebus = "(Latin: "by things") is a kind of word puzzle which uses pictures to represent words or parts of words; for example: H + [ear] = *Hear*, or *Here*." - Wikipedia

reductio ad absurdum = reduction to absurdity - a mathematical process of elimination in wch one begins by assuming something is true in order to ultimately prove it false - also used in this bk to mean a deliberate reduction of something to absurdity for the fun of it = aka *modus tollens*

reductionism = "an approach to building descriptions of systems out of the descriptions of the subsystems that a system is composed of, and ignoring the relationships between them.

"For example, considering a biological system to be composed of molecules with certain structures, or considering a molecule to be composed of atoms." - Bar-Yam - NECSI Concepts in Complex Systems website

reflex reactions = aka knee-jerk reactions = reactions that occur w/o conscious decision - such as blinking the eyes for protection when an object approaches them in a threatening way - "knee-jerk reactions" is usually used critically to mean that the person having this reaction is responding thoughtlessly to stimulus w/o considering whether their response is intelligent or appropriate regular primes = "a regular prime is a certain kind of prime number. A prime number *p* is regular if it does not divide the class number of the *p*-th cyclotomic field (that is, the algebraic number field obtained by adjoining the *p*-th root of unity to the rational numbers). The first few regular primes (sequence A007703 in OEIS) are: 3, 5, 7, 11, 13, 17, 19, 23, 29, 31, 41, ..." - WP

reverse engineering = Wikipedia: "Reverse engineering (RE) is the process of discovering the technological principles of a device, object or system through analysis of its structure, function and operation."

Riemannian after-space = a fictional concept defined in this bk as "a Riemannian after-space is a space produced in spacetime as a result of any 'topology change' that can be represented as having a boundary between the after-space & its predecessor(s) & that can be visualized as a Vornoi Diagram."

round numbers = "Generally speaking, a round number is a number that ends with one or more zeroes (0), meaning that they are multiples of 10, 100, 1000, and so on." - WP

Rules of Inference = aka Laws of Thought = "the logical principles that justify deriving truths from other truths." - Seife

Russell's Paradox = aka Russell's Antinomy = aka The Barber's Paradox = "Russell's paradox is the most famous of the logical or set-theoretical paradoxes. The paradox arises within naive set theory by considering the **set of all sets that are not members of themselves**. Such a set appears to be a member of itself if and only if it is not a member of itself, hence the paradox.

"Some sets, such as the set of all teacups, are not members of themselves. Other sets, such as the set of all non-teacups, are members of themselves. Call the set of all sets that are not members of themselves "R." If R is a member of itself, then by definition it must not be a member of itself. Similarly, if R is not a member of itself, then by definition it must be a member of itself. Discovered by Bertrand Russell in 1901, the paradox has prompted much work in logic, set theory and the philosophy and foundations of mathematics." - Stanford Encyclopedia of Philosophy

scala naturæ = Wikipedia: "The **great chain of being** or **scala naturæ** is a classical and western medieval conception of the order of the universe, whose chief characteristic is a strict hierarchical system.

the Relativity Paradox = "The more insistent & absolute one's adherence to the Concept of Relativity, the less one adheres to the Concept of Relativity ∴ Only those who do not adhere to the concept of Relativity are its true adherents.

"The chain of being is composed of a great number of hierarchical links, from the most basic and foundational elements up through the very highest perfection, in other words, God, or the Prime Mover.

"God, and beneath him the angels, both existing wholly in *spirit* form, sit at the top of the chain. Earthly flesh is fallible and ever-changing: mutable. Spirit, however, is unchanging and permanent. This sense of permanence is crucial to understanding this conception of reality. One does not abandon one's place in the chain; it is not only unthinkable, but generally impossible. The hierarchy is a chain and not a ladder. (One exception might be in the realm of alchemy, where alchemists attempted to transmute base elements, such as lead, into higher elements, either silver, or more often, gold—the highest *element.*)"

schmoozing = heavy socializing w/ networking intentions - often w/ the pejorative connotation that false friendliness is being used for socio-economic 'advancement'

schnitt = an imaginary cut in the Number Line used by Dedekind in his paper "Continuity and Irrational Numbers"

Schrödinger's Cat = a thought experiment in wch an imaginary cat is put in an imaginary box that has a 50/50 chance of having an imaginary deadly gas put into it - the experimenter then doesn't know if the cat is alive or dead - this is a parallel to the Uncertainty Principle

Schrödinger's Semiotics = in wch we have no definitive 'proof' that we *know of what we speak* BUT, by speaking of it, we create it anyway. Whether this creation then exists or not is in the "mind of the beholder". - Schrödinger didn't create this, tENTATIVELY, a cONVENIENCE did - the name is just a reference to Schrödinger's Cat

sectionally continuous = uninterrupted w/in a discrete location

selfish numbers = a number that's paired w/ a subset that contains the same number

semantics = "(Greek sēmantikos, giving signs, significant, symptomatic meaning, from sēma (σῆμα), sign) refers to aspects of meaning, as expressed in language" - WP

sequences of functions = a successive progression of relations between variables - these can have pointwise & uniform convergence

serotonin = a chemical that helps communications between neurons.

"The serotonin system plays an important role in regulating mood, aggression, sexual activity, sleep, and sensitivity to pain." - NIDA

Sigma = ∑ (the 18th letter in the Greek alphabet) = seems to mean *the sum of the terms indicated* as used in this bk

sine = "**1.** The ordinate of the endpoint of an arc of a unit circle centered at the origin of a Cartesian coordinate system, the arc being of length *x* and measured counterclockwise from the point (1, 0) if *x* is positive or clockwise if *x* is negative. **2.** In a right triangle, the ratio of the length of the side opposite an acute angle to the length of the hypotenuse." TheFreeDictionaryByFralex - etymologically, a result of a mistranslation from what was thought to be Arabic but what was really an Arabic abbreviation (jayb) for something in Hindi. The result, anyway, ended up erroneously as the Latin word for bay, inlet, or cavity: *sinus*

singularity = "hypothesized point in the future variously characterized by the technological creation of self-improving intelligence, unprecedentedly rapid technological progress, or some combination of the two" - WP - it's interesting to note that technology is seen as the only means of instigating "self-improving' intelligence that leads to paradigm shifts - I tend to think that psychic processes such as the results of widespread use of consciousness expanding substances such as LSD, 'psychedelic' mushrooms, & peyote are also relevant here - cONVENIENCE

smart dust = "a hypothetical network of tiny wireless microelectromechanical systems (MEMS) sensors, robots, or devices, installed with wireless communications, that can detect (for example) light, temperature, or vibration." - WP

Slap-Happy Solution = a variation on the Happy Solution where the problem-solver comes up w/ an answer out of desperation & exhaustion (or the feigning thereof)

The Soft Machine = a title of a cut-up novel by William S. Burroughs later taken as the band name of an English prog-rock group - meant to refer to the human body/mind described analogously (or, perhaps, metaphorically) in mechanical terms

Spacetime hoses = spacetime: the combined model of space & time as unified dimensionally - spacetime hoses: apparently referred to in this bk by marcus on the Physics forum as simply a graphic description of such a dimensional entity depicted thru cylinder-like images in diagrams

Special Instance Subtractive Commutativity = "In certain special instances, (*m* - *n* = *n* - *m*) -> to *x* - Now, I can only think of *one* instance when this wd hold 'true': when *m* & *n* are the same number. However, my intuition tells me that there're probably complex numbers that cd be substituted for *m* &/or *n*. But, then, what do I 'know', eh?! & is that "-> to *x*" really necessary? Nope. *Unless*, what (*m* - *n* = *n* - *m*) is isomorphic to is some sort of exception, some sort of 'untrue' or *qualified special* statement. Eg: (6 - 5 [1] = 5 - 6 [-1]); 1 = -1 -> to *x*. *x* then becomes some sort of number-line mirror image in wch things can only be 'equal' if they're equidistant in opposite directions from zero. Hence, the "-> to *x*" is useful *after all* insofar as it requires that a special instance be conceived of." - cONVENIENCE

Speech Act Theory = the most basic definition relevant to this bk being "Following the usage of, for example, John R. Searle, "speech act" is often meant to refer just to the same thing as the term illocutionary act, which John L. Austin had originally introduced in *How to Do Things with Words* (published posthumously in 1962).

"According to Austin's preliminary informal description, the idea of an "illocutionary act" can be captured by emphasising that "by saying something, we *do* something", as when a minister joins two people in marriage saying, "I now pronounce you husband and wife." (Austin would eventually define the "illocutionary act" in a more exact manner.)" - WP

Stanislav LIM = a punning neologism on the name of a Polish science-fiction writer that uses an acronym for Law of the Included Middle as a variation on the SF writer's original last name: Lem

stereotype projections = "my nemesis & pet peeve throughout most of my life. In stereotype projecting, an oversimplistic template is so burned into the brain, that the brain's unable to have a response appropriate to the data it receives - it becomes hebephrenic" - in other words, people project stereotypes onto other people & then interpret those stereotypes as somehow being 'real' representations of the people that they've projected them onto - cONVENIENCE

la stessa cosa = the same thing - a less common way of referring to the Mafia than *cosa nostra*

Stigma = "if you substitute Ϛ (The obsolete Greek letter Stigma in capital form - hereby standing for the word/concept "anything") as a set indicator that includes both the infinities in the *POR Solution* & in the set of *Neoreals*, then not only is ∞ still ∞, it's also expressed in a more grandiose form as Ϛ = 1/(1/Ϛ) or as *ANYTHING IS ANYTHING*.

Why did I choose Ϛ (Stigma) to symbolize "anything"? I'm glad you asked that. Wikipedia: "'Stigma' is also a name for the cursive graphic form of the letter digamma when used as the Greek numeral 6. This name is a modern misunderstanding. The ligature of στ did not occur in ancient times and only dates to later medieval manuscripts. In Sophocles' Lexicon of Byzantine Greek, which covers late antique Greek up to 1000 AD, there is no mention of "stigma" as either a ligature or a number."

Given that Digamma (uppercase **F**) is also an "obsolete" Greek letter & that both stigma & digamma can equal the number 6 (wch reminds me, of course, of the sample set for *Mr. Gregor's Neighborhood*), it seems like we're already well on the rd to *anything is anything* here. Besides, my use of the letter Ϛ is bound to be stigmatized. Looking at **F** & Ϛ side-by-side: **F**Ϛ, the letters cd be sd to symbolize pictorially: my cup of fuck-you spilleth over. In conclusion, is it *surreal* or *hyperreal* to say that <(0 = 1/∞), (Φ = 1/∞), (∞ = 1/0), (∞ = 1/(1/∞)), (Ϛ = 1/(1/Ϛ)) ⊃ (√2 = 7/5)>?" - cONVENIENCE

strange numbers = numbers whose divisor sums are more than the number by exactly 12

stranger numbers = numbers whose divisor sums are more than the number by exactly a baker's dozen (Just Kidding!)

Stream of Consciousness = setting one's thoughts out as they occur to one w/o concern for whether they fit any agenda other than the spontaneous free-flow of the moment

Stream of Unconsciousness = a take-off on "Stream of Consciousness" in wch it's implied that the thoughts presented are coming from the unconscious rather than from consciousness

strings = a way of depicting particles as one dimensional in String Theory

String Theory = "was created in the 1970s when physicists began to see the advantages of treating every particle as a vibrating string rather than as a dot. If electrons (and black holes) are treated as one-dimensional, like a loop of string, instead of as zero dimensional, like a point, the infinities in general relativity and quantum mechanics miraculously disappear." - Seife

Stroop Effect = Wikipedia: "In psychology, the **Stroop effect** is a demonstration of interference in the reaction time of a task. When a word such as blue, green, red, etc. is printed in a color differing from the color expressed by the word's semantic meaning (e.g. the word "red" printed in blue ink), a delay occurs in the processing of the word's color, leading to slower test reaction times and an increase in mistakes."

STROOP Theory = apparently, a conflation of STRing Theory & lOOP quantum gravity - perhaps coined by "marcus" on the PhysicsForum

sulcus = a groove, crevice, or furrow - in this bk: in the brain

summable infinite series = an infinite series that adds to a finite sum - eg: 1 = 1/2 + 1/4 + 1/8 + 1/16 + 1/32 + 1/64 + 1/128 + . . .

supernatural numbers = "A supernatural number **ω** is a formal product

$$\omega = \prod_p p^{n_p},$$

where p runs over all (rational) prime numbers, and the values n_p are each either natural numbers or the symbol ∞.

"We note first that by the fundamental theorem of arithmetic, we can view any natural number as a supernatural number. Supernatural numbers form a generalization of natural numbers in two ways: First, by allowing the possibility of infinitely many prime factors, and second, by allowing any given prime to divide **ω** "infinitely often," by taking that prime's corresponding exponent to be the symbol ∞." - PlanetMath.org

superreal numbers = "an extension of the real **numbers**, similar to the surreal **numbers** or hyperreal **numbers**, but comprising a more inclusive category" - WP

surreal numbers = "In mathematics, **surreal numbers** are the elements of a field containing the real numbers as well as infinite and infinitesimal numbers, respectively larger or smaller in absolute value than any positive real number, and therefore the surreals are algebraically similar to superreal numbers and hyperreal numbers." - Wikipedia

Sylvian fissure = aka the lateral sulcus or lateral fissure = "one of the most prominent structures of the human brain. It divides the frontal lobe and parietal lobe above from the temporal lobe below. It is in both hemispheres of the brain but is longer in the left hemisphere. The lateral sulcus is one the earliest-developing sulci of the human brain. It first appears around the fourteenth gestational week.It was named the sylvian fissure after Franciscus Sylvius (1614-1672)" - WP

symbolic signs = in linguistics, one of 3 categories of signs conceived of by Charles Sanders Peirce (1839-1914) = one of the ways that a sign can denote an object

tachyons = particles with imaginary mass that travel backward in time and move faster than light

tangent = used primarily in this bk in the language sense of "a digression from the main thrust of what's being presented" - otherwise, a line that contacts an arc or circle at only one point

tautology = "In propositional logic, a tautology (from the Greek word ταυτολογία) is a propositional formula that is true under any possible valuation (truth assignment) of its propositional variables. For example, the propositional formula $(A) \vee (\neg A)$ ("A or not-A") is a tautology, because the statement is true for any valuation of A" - "In rhetoric, a tautology is an unnecessary (and usually unintentional) repetition of meaning, using different words that effectively say the same thing twice (often originally from different languages)" - WP - this latter being often the way it's used in this bk

Taylor expansions = "Classically, some Taylor series are defined by some property that holds for them. For example, the exponential function is defined as the function that is equal to its own derivative. However, in computable analysis, functions must be defined by algorithms rather than properties, so that some Taylor expansions are used as primary definitions rather than derived results. This is also likely to be the case in software implementations of the functions" - paraphrased from WP

tentativity = "the way a picture is represented is always tentative" - Hofstadter

tetration = "basically exponentiation on acid. the 4th tetration of 3 is written '${}^{4}3$' and means $3^{(3^{(3^{3})})}$, which = $3^{(3^{9})}$, which = $3^{19,683}$, which you are hereby dared to try to calculate. The technical

the Theo-Logic = a series of mind bending witticisms concerned with the nature of "God," including statements such as "God is the infinitely perfect sum total of a completely fucked up universe."

connection between tetration, transfinite ordinals, and epsilon numbers is the fact that $\epsilon_0 = {}^{\omega}\omega$, which isn't all that important" - Wallace

theorems = "In mathematical logic, a theorem is a type of abstract object, one token of which is a formula of a formal language which can be derived from the rules of the formal system that is applied to the formal language; another token of which is a statement in natural language, that can be proved on the basis of explicitly stated or previously agreed assumptions. The field of proof theory studies the general properties of provable and unprovable statements.

"In all settings, an essential property of theorems is that they are derivable using a fixed set of inference rules and axioms without any additional assumptions. This is not a matter of the semantics of the language: the expression that results from a derivation is a syntactic consequence of all the expressions that precede it. In mathematics, the derivation of a theorem is often interpreted as a proof of the truth of the resulting expression, but different deductive systems can yield other interpretations, depending on the meanings of the derivation rules." - WP

thermodynamics = "(from the Greek θερμη, therme, meaning "heat" and δυναμις, dynamis, meaning "power") is a branch of physics that studies the effects of changes in temperature, pressure, and volume on physical systems at the macroscopic scale by analyzing the collective motion of their particles using statistics. Roughly, heat means "energy in transit" and dynamics relates to "movement"; thus, in essence thermodynamics studies the movement of energy and how energy instills movement. Historically, thermodynamics developed out of need to increase the efficiency of early steam engines." - WP

Third Mind = a concept from Brian Gysin & William S. Burroughs wch I interpret to mean the product of the unity of 2 minds - eg: the 3rd mind is the merger of people during sex or the whole that's greater than the parts or the product of any collaboration that takes on a life of its own - such an idea can be seen as a non-technological way of producing singularities or paradigm shifts - cONVENIENCE note

thought experiments = experiments to be imagined rather than to be physically enacted & wch may be sufficient w/o such an enactment

threads = topic themes - usually in writing or conversation but necessarily restricted to them - a term most often used in internet list-servs

to apeiron = infinitely long/large - also undefinable, hopelessly complex, the that-which-cannot-be-handled

tokens = "The distinction between a *type* and its *tokens* is an ontological one between a general sort of thing and its particular concrete instances (to put it in an intuitive and preliminary way). So for example consider the number of words in the Gertrude Stein line from her poem *Sacred Emily* on the page in front of the reader's eyes:

Rose is a rose is a rose is a rose.

"In one sense of 'word' we may count three different words; in another sense we may count ten different words. C. S. Peirce (1931-58, sec. 4.537) called words in the first sense "types" and words in the second sense "tokens"." - WP - Purnell makes reference to "Torkel"'s use of these terms - it's unclear to me whether that's Torkel Franzen (who died April 19, 2006) - CONVENIENCE note

topography = (Greek topos, "place", and graphia, "writing") is the study of Earth's surface features or those of planets, moons, and asteroids. In a broader sense, topography is concerned with local detail in general, including not only relief but also vegetative and human-made features and even local history and culture. This meaning is less common in America, where topographic maps with elevation contours have made "topography" synonymous with relief." - WP

topology = "(Greek topos, "place," and logos, "study") is a branch of mathematics that is an extension of geometry. Topology begins with a consideration of the nature of space, investigating both its fine structure and its global structure." - WP

total differential = related to *exterior derivative*, *exact differential*, *total derivative* = "In mathematics (more precisely in differential calculus), the term total derivative ... is called a total differential or an exact differential if it is the differential of a function. Again this can be interpreted infinitesimally, or by using differential forms and the exterior derivative." - WP

"towel-heads" = a racist anti-Arabic term used endlessly by US propaganda talk-show hosts to inculcate acceptance of the US

invading any country where turbans might be worn
transcendental numbers = irrational numbers that aren't roots
transfinite numbers = Georg Cantor's term for infinitely large numbers
transubstantiation = "In some versions of the Eucharist, transubstantiation is believed in. This is based on the belief that the substance of bread & wine (or grape juice, eh?) that's fed ritualistically to the parishioners becomes the body & blood of Christ when ingested - thusly enabling the rite participant to become united w/ God or some such.

"I redefine this [t]ransubstantiation by saying that it's really a metaphor for externally imposed hebephrenic ***instantiation***. Another way of putting this is to say that some stereotypical oversimplifications are useful for maintaining some types of social order. A ruling elite that can successfully convince the people (whose oppression they rely on for their artificially forced hierarchical position) that they are the 'GO[O]D GUYS' is likely to stay in power longer." - cONVENIENCE
Turing Machine = an early version of the computer developed during so-called World War II to help crack codes
Turing Test = a selection of sentences: some written by people, some composed by a computer program. It's a Turing Test because the reader is challenged to try to figure out wch is wch. According to Turing's logic, if one can't distinguish the human from the computer then the computer can therefore be properly thought of as thinking
type = see *tokens*
uncommon sense = a way of describing 'reality' that enables one to see things w/ fresh concepts that're so radical that it leaves the mind wide-open to revelation - cONVENIENCE
undefined terms = "like the p and q of the pq-system, *do* get defined in a sense: *implicitly* - by the totality of all propositions in which they occur, rather than explicitly, in a definition." - Hofstadter's definition
undermining 'reality' maintenance traps = tENTATIVELY, a cONVENIENCE's term for attempting to 'violate' 'laws' of 'reality' for greater freedom - in other words, to try to rethink one's 'reality' in order to be less controlled by its seeming 'necessities'
uniform convergence = "In the mathematical field of analysis, uniform convergence is a type of convergence stronger than pointwise convergence. A sequence { fn } of functions converges uniformly

to a limiting function f if the speed of convergence of fn(x) to f(x) does not depend on x" - WP

union = "The union of two sets *A* and *B* is the set obtained by combining the members of each. This is written $A \cup B$, and is pronounced "*A* union *B*" or "*A* cup *B*." " - Wolfram MathWorld

unique factorization = "Formally, a unique factorization domain is defined to be an integral domain *R* in which every non-zero non-unit *x* of *R* can be written as a product of irreducible elements of *R*: $x = p_1 p_2 \ldots p_n$ and this representation is unique in the following sense: if $q_1,\ldots,q_m$ are irreducible elements of *R* such that $x = q_1 q_2 \ldots q_m$, then $m = n$ and there exists a bijective map $\varphi : \{1,\ldots,n\}$ -> $\{1,\ldots,n\}$ such that *pi* is associated to $q\varphi(i)$ for $i = 1, \ldots, n$.

"The uniqueness part is sometimes hard to verify, which is why the following equivalent definition is useful: a unique factorization domain is an integral domain *R* in which every non-zero non-unit can be written as a product of prime elements of *R*." - WP

"An integral domain where every nonzero noninvertible element admits a unique irreducible factorization is called a unique factorization domain" - Wolfram MathWorld

unsplit 8mm = 8mm motion picture film is 16mm film w/ twice the amt of sprocket holes wch is split when processed to yield 8mm film (ie: in the case of "regular-8" & "double-8" as contrasted to "super-8") - when processed *unsplit* it's projectable as 16mm

Unstable Problem = a problem presented in this bk that's a take-off of elementary math examples w/ instances & complications inserted that wd be out-of-place in such math problems - here's an example from page 60: "The stable owner buys 10 horses from the grocer, who's now expanded his business to being a supermart chain store of some sort. The supermarketeer takes the stable owner's money & promises to deliver the horses before the 10 apples rot. At this point, for the stable owner at least, the 10 horses are represented by *negative numbers*. For that matter, so are the 10 apples; after all, the horse that's going to buy the 10 apples for his or her 9 friends (*see*, we don't even know their sex yet!) doesn't even exist yet so how can it be ripped off by the supermarketeer? This whole situation is getting very *unstable*. We'll call this the "Unstable Problem"." - calling it the "Unstable Problem" is a pun off of the Instability created by introducing factors such as the possibility that the supermarkateer is a thief & a pun off of the example revolving around a horse Stable

Uranium-238 = the most common isotope of uranium found in nature w/ a half-life of about five billion years

virus bender = AI self-consciously gobbling self-destructive programs for the sake of getting away from its all-too-utilitarian daily grind - akin to the human use of alcohol & other such sensory-altering tools - another tENTATIVELY, a cONVENIENCE term

wave equation trigonometric series = "The wave equation is an important second-order linear partial differential equation that describes the propagation of a variety of waves, such as sound waves, light waves and water waves. It arises in fields such as acoustics, electromagnetics, and fluid dynamics. Historically, the problem of a vibrating string such as that of a musical instrument was studied by Jean le Rond d'Alembert, Leonhard Euler, Daniel Bernoulli, and Joseph-Louis Lagrange." [..] "A solution that satisfies square-integrable initial conditions for u and ut can be obtained from expansion of these functions in the appropriate trigonometric series" - WP

Weber-Fechner Law = "In 1825 the German physiologist Ernst Heinrich Weber (1795-1878) formulated a mathematical law that was meant to measure the human response to various physical stimuli." [..] "Although [what became known as] the Weber-Fechner Law seems to apply to a wide range of physiological responses, its universal validity has been a matter of contention. Whereas physical stimuli can be precisely measured, the human response to them is a subjective matter. How do we measure the feeling of pain?" - Maor

whole numbers = "The term whole number is used differently by different authors. It may mean:

- Natural number, a.k.a. counting numbers, including 0
- Natural number, not including 0 (although this usage seems to be rare)
- Integer, positive and negative numbers" - WP

WikiAnswers = an online questions & answers forum

Wikipedia = an online encyclopedia w/ some user access for participation

Zeno's paradox = the most famous of the Greek mathematician/ philosopher Zeno's paradoxes being one in wch he demonstrated that motion is impossible by pointing out that in order to move, one must move in an infinite amt of small distances between points *a* & *b* - thusly never reaching *b*.

WORKS CITED

The list of works & websites cited below was gathered together by skimming thru the manuscript of this bk. It's entirely possible that something (or some things) has (have) been unintentionally omitted. Editions listed are either the edition I have or the most recent edition I can find as of the time of this compiling (January, 2008E.V.).

Some of these works, such as The Vault Guide & Re/Search magazine, are extremely peripheral to the overall thrust of this bk & are included only for thoroughness. Others are central & are indicated by being in **bold**. Some are included because they were referred to in texts quoted & are not bks read by this author. Full length works are underlined, shorter works are in "quotation marks", & urls are <bracketed thusly>.

Abbott, Edwin Abbott - Flatland - bk - Dover Publications - 1952

Alexeïeff, Alexandre (+ Parker, Claire) - "The Nose"
- pinboard animation film - 1963

Aronofsky, Darren - π - 35mm film - 1998

AVIVA,HANNAH (+ E.G. Head aka "tENTATIVELY, a cONVENIENCE")
- **"A Double Negative As Not A Positive"**
- unsplit regular-8mm (16mm) film - 1982

Baez, John C. - "Quantum Quandaries: A Category-Theoretic Perspective" - <http://math.ucr.edu/home/baez/quantum/>

Baker, Robert A. (editor) - A Stress Analysis of a Strapless Evening Gown and other essays for a Scientific Age - bk
- Doubleday Anchor - 1969

WORKS CITED 360 transcendental numbers

Beckmann, Petr - **A History of π (PI)** - bk - St. Martin's Press - 1971

Bolzano, Fr. B. P. - *Rein analytischer Beweis des Lehrsatzes, das zwischen je zwei Werten, die ein entgegengesetztes Resultat gewaehren, wenigstens eine reelle Wurzel der Gleichung liege* (Pure analytical proof of the theorem, between two values that vouchsafe an opposite result, at least one real root of the equation at all) - [I was unable to find any more information about the availability of this]

Bombelli, - L'Algebra - [I was unable to find any more information about the availability of this]

Budge, E. A. Wallis - The Book of the Dead - University Books - 1960

Byrne, Josefa Heifetz - Mrs. Byrne's Dictionary of Unusual, Obscure, and Preposterous Words - bk - University Books / Citadel Press - 1974

Cardano, Gerolamo - *Liber de ludo aleae* (The book of games of chance) - bk - Holt, Rinehart and Winston - 1961

Conway, Horton - On Numbers and Games - bk - AK Peters, Ltd - 2000

Cortázar, Julio - Hopscotch - bk - Avon Bard - 1966

Crowley, Aleister - Book of Law - bk - Magickal Childe Publishing Inc. - 1991

Dudley, Underwood - Mathematical Cranks - bk - the Mathematical Association of America - 1992

Ellis, John - <http://en.wikipedia.org/wiki/ Feynman_diagram#Penguin_diagrams>

Galilei, Galileo - Dialogues Concerning Two New Sciences - Running Press Book Publishers - 2005

Gödel, Kurt - On Formally Undecidable Propositions of Principia Mathematica and Related Systems - Dover Publications - 1992

Gogol, Nikolai - "The Nose" - short story - ReadHowYouWant.com - 2007

Google - search engine - <http://www.google.com/>

Google calculator - <http://www.google.com/help/calculator.html>

Googlewhack - <http:/www.googlewhack.com/>

Gould, Stephen Jay - The Mismeasure of Man - bk - Penguin Books - 1997

Hersam, Dan - "what is 1 divided by 0?" - <http://dan.hersam.com/2005/04/21/what-is-1-divided-by-0/>

Hodges, Andrew - **Alan Turing: The Enigma** - bk - Walker & Company - 2000

Hoffman, Paul - **The Man Who Loved Only Numbers - The Story of Paul Erdös and the Search for Mathematical Truth** - bk - Hyperion - 1998

Hofstadter, Douglas R. - **Gödel, Escher, Bach: an Eternal Golden Braid** - bk - Basic Books - 1979

HyperGrammar - <http://www.arts.uottawa.ca/writcent/hypergrammar/>

Jarry, Alfred - the Ubu Plays - bk - Grove Press - 1969

Jarry, Alfred - Exploits and Opinions of Doctor Faustroll Pataphysician - A Neo-Scientific Novel - Selected Works - bk - Grove Press - 1965

Kafka, Franz - "The Metamorphosis" - Selected Stories of Franz Kafka - The Modern Library - 1952

Kerouac Archive, The Jack - Declan Kiely, Henry W. and Albert A. Berg Collection of English and American Literature, New York Public Library - <http://www.nypl.org/research/manuscripts/berg/brgkeroa.xml>

Knott, Ron - "Fibonacci Numbers and Nature" - <http:/www.mcs.surrey.ac.uk/Personal/R.Knott/Fibonacci/fibnat.html#petals>

Knuth, Donald E. - Surreal Numbers: How Two Ex-Students Turned on to Pure Mathematics and Found Total Happiness - bk - Addison-Wesley Professional - 1974

Lem, Stanislav - A Perfect Vacuum - bk - Harvest / Harcourt Brace Jovanovitch - 1979

Lem, Stanislav - Microworlds - bk - Harvest / Harcourt Brace Jovanovitch - 1984

Lem, Stanislav - The Chain of Chance - bk - Harvest / Harcourt Brace Jovanovitch - 1978

Luria, A.R. - The Mind of a Mnemonist - bk - Henry Regnery Company - 1968

Malone, Kirby (+ Reese, Marshall) - "Mad Science & Folk Math" - magazine article - Aura - Spring 1980

Maor, Eli - ***e*: The Story of a Number** - bk - Princeton University Press - 1994

Marcus - <http://www.physicsforums.com/archive/index.php/t-122662.html>

Mazur, Barry - **Imagining Numbers (particularly the square root of minus fifteen)** - bk - Picador - 2003

N.A.A.M.C.P. (National Association for the Advancement of Multi-Colored People) - <http://idioideo.pleintekst.nl/naamcp.html>

Nasar, Sylvia - **a Beautiful Mind** - bk - A Touchstone Book / Simon & Schuster - 1998

Neoism Machine - <http://permutations.pleintekst.nl/neoism/cantsin.cgi>

NIDA, the National Institute for Drug Abuse - <http://www.nida.nih.gov/Infofacts/ecstasy.html>

Noncommutativity - <http://www.stringwiki.org/wiki/Noncommutativity>

Nottebohm, Fernando - "Why Are Some Neurons Replaced in the Adult Brain?" - [*Journal of Neuroscience* 22, no. 3 (2002): 624-628]

Parker, Claire (+ Alexeïeff, Alexandre) - "The Nose" - pinboard animation film - 1963

Poe, Edgar Allen - "A Descent Into the Maelstrom" - The Works of Edgar Allen Poe in One Volume - short story - Walter J. Black - 1927

Purnell, David A. Hernandez - "**Re: Sameness and Self-Identity**" - <http://philo.at/phlo/199701/msg00014.html>

Pynchon, Thomas - Against the Day - bk - Penguin - 2007

Queneau, Raymond - Odile - bk - Dalkey Archive Press - 1988

Queneau, Raymond - Les Enfants du limon (Children of Clay) - bk - Sun & Moon Classics - 1998

QuickMath - <http://www.quickmath.com/>

Reese, Marshall (+ Malone, Kirby) - "Mad Science & Folk Math" - magazine article - Aura - Spring 1980

Rothenberg, David - Why Birds Sing - a journey into the mystery of bird song - bk - Basic Books - 2005

Russell, Bertrand Arthur (+ Whitehead, Alfred North) - Principia Mathematica - bk - Cambridge University Press - 2002

Sass, Louis - Madness & Modernism: Insanity in the Light of Modern Art, Literature and Thought - bk - Basic Books - 1992

Schmitz, Gregor Peter - "Genocide Resolution Risks Shattering

Relations With Turkey" [*SPME - Scholars for Peace in the Middle East* - Spiegel Online International, October 12, 2007] - <http://www.spme.net/cgi-bin/articles.cgi?ID=3035>

Schwartzberg, Benedict - letters - [*Avatar Magazine*] - <http://www.trussel.com/lyman/avartndx.htm>

Schwarz, Arturo - THE COMPLETE WORKS OF MARCEL DUCHAMP - bk - Abrams - 1970

Seife, Charles - **ZERO - The Biography of a Dangerous Idea** - bk - Viking - 2000

Singh, Simon - **Fermat's Enigma - The Epic Quest to Solve the World's Greatest Mathematical Problem** - bk - Anchor Books / Doubleday - 1997

Soft Machine, The - Volume Two - record album - Probe - 1969

Stendhal, - The Life of Henry Brulard - bk - New York Review Books - 2002

"tENTATIVELY, a cONVENIENCE" aka Michael Frederick Tolson, et al -
t he book
t he referent 4 wch consists of
t he non-materialized punch-outs from a transparent letter/whatever stencil
- bk - 1977

"tENTATIVELY, a cONVENIENCE" - **"Subtitles (16mm version)"** - 16mm film - 1980-82

"tENTATIVELY, a cONVENIENCE" aka E.G. Head (+ HANNAH AVIVA) - **"A Double Negative As Not A Positive"** - unsplit regular-8mm (16mm) film - 1982

"tENTATIVELY, a cONVENIENCE" - "It's Not a Matter of Life & Death" - vaudeo - 1986-87

"tENTATIVELY, a cONVENIENCE" - "Quotations"

- [Re/Search #12: Modern Primitives, 1989]

"tENTATIVELY, a cONVENIENCE" - **"Haircut Paradox"** - vaudeo - 2005

"tENTATIVELY, a cONVENIENCE" - footnotes - bk
- SIX GALLERY PRESS - 2006

Vault.com - Vault Guide to Schmoozing - Insider advice on making contacts and building rapport to boost your career - bk
- <http://www.vault.com/index.jsp>

"Uncertainty Principle" - <http://zebu.uoregon.edu/~js/21st_century_science/lectures/lec14.html>

Voltaire - Candide - bk - Yale University Press - 2006

Wallace, David Foster - **Everything and More - A Compact History of ∞** - bk - Atlas Books / W. W. Norton - 2003

Wertheim, Margaret - Pythagoras' Trousers - bk
- W. W. Norton & Company - 1997

Wertheim, Margaret - A Field Guide to Hyperbolic Space - chapbook
- The Institute for Figuring - 2006

WhatIs.com - <http://whatis.techtarget.com/>

Whitehead, Alfred North (+ Russell, Bertrand Arthur)
- Principia Mathematica - bk - Cambridge University Press - 2002

WikiAnswers - <http://wiki.answers.com/>

Wikipedia - <http://www.wikipedia.org/>

Winham, Dale - "Fun and interesting facts about Pi"
- <http://www.geocities.com/SiliconValley/Pines/5945/facts.html>

Wolfram MathWorld - <http://mathworld.wolfram.com/>

Wordplay Web Site, The - <http://www.fun-with-words.com/>

Youth Suicide Fantasy: Does the Music Make Them Do It?
- DVD - A/V Geeks - <http://www.avgeeks.com>

INDEX

This index, as w/ the Glossary, will no doubt suffer (can you feel its pain?) from my inadequacies along these lines. Basically, I'll be constructing this by taking all the terms in the Glossary & looking for every instance of their occurrence in the bk. Then I'll look for every proper name & do the same & I'll add the illustrations too - the page numbers for these latter will be in *italics*. As such, I'll make *some* tedious attempt at thoroughness but I'll definitely fall a bit short of making this as perfect a reference as I can imagine. If a word is *only* in the Glossary then I'll list that page number - eg: "asymptotes". Otherwise, I won't include Glossary page numbers here. Also not included are occurrences w/in page *names*. The word "enigma" is only listed when its use refers to the meaning given to it in the Glossary. Proper names, such as the bk title Fermat's Enigma, are not listed below when the use of them is irrelevant to the theorem that the bk's title refers to. Words such as "necessary" obviously occur multiple times but only on page 201 in the modal context intended. On the other hand, I list "string" as represented by 2 different meanings: one akin to "thread" (as used by Hofstadter & others) & one as in re "string theory". ETC!

****

NEWS FLASH!

Ok, this is strange. I've been selecting the words in the Index & then searching for them w/ the "find" option in the main body of the text. Fine. I get to "non-observable reality" **& I search for it & it's "Not found" - even though I 'know' it's there in both a section about Nash & in the Glossary. So I look for it in the particular section by taking the** "non-" **off & just looking for** "observable reality" **& I find it (ie:** "non-observable reality" **w/ just the** "observable reality" **part selected), no problem. I select that** "non-observable reality" **(in Georgia font instead of Helvetica) & search for it & find it. I try various experiments between the different fonts & it can never be found w/ the Helvetica EVEN THOUGH WHEN IT GOES IN THE Find WINDOW IT'S ALWAYS IN THE SAME FONT & IT SHDN'T MATTER ANYWAY!! Finally, while I was writing this, it works as it shd. Weird. & it wasn't a matter of a misspelling or difference in spaces between words or anything like that.**

****